Electrica

Ian McKenzie Smith
B.Sc., Dip.A.Ed., C.Eng., M.I.E.E.,
M.I.E.R.E., F.I.T.E.

Electrical and electronic principles

Level 3

Longman Scientific & Technical
Longman Group UK Limited,
Longman House, Burnt Mill, Harlow,
Essex CM20 2JE, England
and Associated Companies throughout the world

First published 1979
Second impression 1982
Third impression 1988

British Library Cataloguing in Publication Data

McKenzie Smith, Ian
Electrical and electronic principles,
level 3. – (Longman technician series:
electrical and electronic engineering).
1. Electric engineering
I. Title
621.3 TK145 78-41316

ISBN 0-582-41199-8

Produced by Longman Singapore Publishers Pte Ltd
Printed in Singapore

Contents

Symbols and abbreviations

	Symbol	Unit abbreviation
Acceleration, linear	a	m/s^2
Area	A	m^2
Capacitance	C	F
Charge	Q	C
Current	I	A
Efficiency	η (eta)	—
Electric field strength	E	V/m
Electric flux	Q	C
Electric flux density	D	C/m^2
Electric potential	V	V
Electromotive force	E	V
Energy	W	J
Force	F, f	N
Form factor	k_f	—
Frequency	f	Hz
Frequency, angular	ω (omega)	rad/s
Frequency, resonant	f_r	Hz
Gain	G	(dB)
Impedance,	Z	Ω (omega)
Inductance	L	H
Inductance, mutual	M	H
Length	l	m
Magnetic field strength	H	At/m
Magnetic flux	Φ (phi)	Wb
Magnetic flux density	B	T
Magnetic flux linkage	Ψ (psi)	Wb t
Magnetic potential difference	F	At
Magnetomotive force	F	At
Mass	m	kg
Period	T	s
Permeability	μ (mu)	H/m

	Symbol	Unit abbreviation
Permittivity	ε (epsilon)	F/m
Phase angle	ϕ (phi)	rad
Pole pair	p	—
Power, active	P	W
Power, apparent	S	VA
Power, reactive	Q	var
Reactance, capacitive	X_C	Ω (omega)
Reactance, inductive	X_L	Ω (omega)
Reluctance	S	/H
Resistance	R	Ω (omega)
Resistivity	ρ (rho)	Ω m
Slip	s	—
Speed, rotational	n, n_r	rev/s
	N, N_r	rev/min
Speed, synchronous	n_1	rev/s
	N_1	rev/min
Temperature coefficient	α (alpha)	/°C
Temperature difference	θ (theta)	°C
Time	t	s
Torque	T	N m
Work	W	J
Velocity, angular	ω (omega)	rad/s
Velocity, linear	u	m/s
Volume	V	m^3

Preface

Periphrastic methods spurning,
To this audience discerning,
I admit this show of learning
Is the fruit of steady cram!

W. S. Gilbert

The most important development in technical education in the last decade has been the introduction of new courses under the guidance of the Business and Technician Education Council (BTEC). These courses take into account the educational requirements of technicians and will fulfil their needs possibly for the remainder of the century.

The aim of this book is to cover the studies of a student presenting himself for the Electrical and Electronics Principles 3 unit of BTEC. The text is based on this programme and is liberally illustrated by means of worked examples.

The content of that BTEC unit bears a strong resemblance to the Principles 3 unit of the SCOTEC Certificate in Electrical and Electronic Engineering, hence additional sections have been included to cover most of that SCOTEC unit as well. However, this book does not cover those parts of the unit dealing with magnetic circuits and with power and power factor in series a.c. circuits, as these have already been dealt with in *Electrical and electronic principles Level 2.*

The International System of Units (SI) is used throughout the text. The student may therefore concern himself with that system only, it being the system generally recognised in engineering.

The symbols and abbreviations conform to those recommended in BS1991, BS3939 and PD5686, all published by the British Standards Institute. Use is made also of the current recommendations of the Institution of Electrical Engineers. To avoid confusion, symbols and abbreviations are not mixed.

Finally the author wishes to thank his wife for checking the manuscript, helping with its duplication and generally putting up with the trials and tribulations of its being typed. Also he wishes to thank his friends and colleagues both in Stow College and in the Institution of Electrical and Electronics Technician Engineers for their advice and assistance.

Milngavie,
January 1979

Chapter 1

Circuit theorems

In our previous studies of electrical circuitry, it has been assumed that sources of electrical energy, such as batteries and generators, have supplied current at a constant voltage. For instance, it has been assumed that a 12-V battery maintains that terminal voltage whether supplying 10 A or 20 A or even no current at all. In practice the supply voltage varies with the current. This variation may be small enough to be negligible, but in those instances where the variation is significant, we require to modify our representation of the circuit.

To help with these changes, there are two circuit theorems which are important to us, being Thévenin's theorem and the maximum power transfer theorem. However we must start by introducing the concept of internal resistance.

1 E.M.F. and internal resistance of a cell

Any source of electrical energy is required to provide an e.m.f. to a circuit in order that a current may be induced to flow in the circuit. If a circuit is such that a cell causes a current to flow in the circuit, the current experiences a resistance in passing through the cell. This resistance encountered by the current as it flows through the cell from the negative terminal to the positive terminal is termed the internal resistance of the cell.

As in any other electrical conductor, the cell resistance depends on the materials used, the length and the cross-sectional area of the current path and the temperature at which the system is operating. It follows that the area and the spacing of the electrodes in the cell both affect the internal resistance of the cell even though the e.m.f. developed may be independent of these factors. Let r be the internal resistance of a cell as indicated in the circuit shown in Fig. 1.

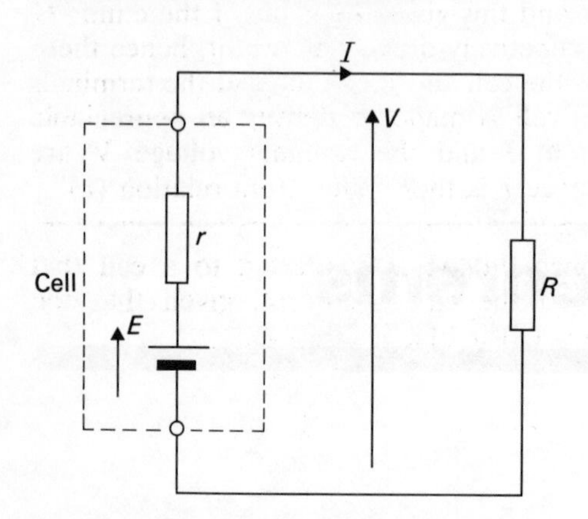

Fig. 1 Cell with internal resistance

The current in the circuit is given by

$$E = V + Ir$$

where $V = IR$

and $$V = E - Ir \tag{1}$$

The voltage V is that which would be measured at the terminals of the cell. It follows that, if the internal resistance r were zero, the voltage measured at the terminals would be the e.m.f. of the cell.

Under open-circuit conditions, the terminal voltage of a cell is equal to the e.m.f. developed and this may be measured by any device which measures voltage and is connected to the terminals of the cell without taking any current. A potentiometer is such a device. In practice, it may be acceptable to use a high-resistance voltmeter because such an instrument takes very little current, especially if the voltmeter had a figure of merit in excess of 10 kΩ/V.

Under load conditions, when a cell is delivering current, some of the e.m.f. is lost in overcoming the internal resistance of the cell with the result that the terminal voltage is lower than the cell e.m.f. The greater the current, the greater is the volt drop within the cell and the lower is the terminal voltage.

From relation (1), the terminal voltage is given by

$V = E - Ir$

From this relation, the internal resistance is given by

$$r = \frac{E - V}{I} \qquad (2)$$

The values for this relation may be obtained from a two-part test as follows. First the open-circuit voltage of the cell is measured by means of a high-resistance voltmeter and this gives the value of the e.m.f. E. You will note that the meter effectively draws no current, hence there is a negligible volt drop within the cell and the voltage at the terminals equals the e.m.f. Second, the cell is made to deliver an appreciable current I and both the current I and the terminal voltage V are measured. The internal resistance r is then found from relation (2).

Example 1 A coil of resistance 0·04 Ω is connected to a cell that develops an e.m.f. 1·5 V. Find the circuit current, given that the internal resistance of the cell is (*a*) 0·1 Ω, (*b*) 0·02 Ω.

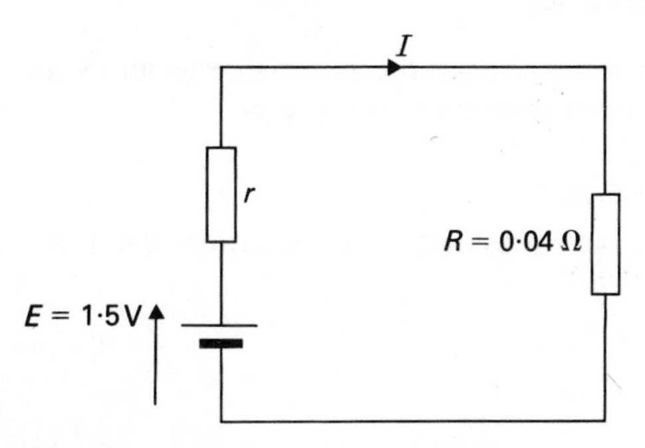

Fig. 2

(*a*) $I = \frac{E}{R + r} = \frac{1 \cdot 5}{0 \cdot 04 + 0 \cdot 1} = 10 \cdot 7 \text{ A}$

(*b*) $I = \frac{E}{R + r} = \frac{1 \cdot 5}{0 \cdot 04 + 0 \cdot 02} = 25 \cdot 0 \text{ A}$

Example 2 A cell has an internal resistance 0·2 Ω and produces an e.m.f. of 1·5 V. Assuming that the cell is connected to a resistance 5·3 Ω calculate:

(*a*) the current delivered by the cell;
(*b*) the terminal potential difference of the cell;
(*c*) the power produced by the cell;

(*d*) the power output of the cell.

$$I=\frac{E}{R+r}=\frac{1\cdot5}{5\cdot3+0\cdot2}=0\cdot273\ \text{A}=\underline{273\ \text{mA}}$$

$$V=E-Ir=1\cdot5-(0\cdot273\times0\cdot2)=\underline{1\cdot445\ \text{V}}$$

$$P_i=EI=1\cdot5\times0\cdot273=\underline{0\cdot41\ \text{W}}$$

$$P_0=VI=1\cdot445\times0\cdot273=\underline{0\cdot39\ \text{W}}$$

Most cells only produce an e.m.f. of about 1·3 – 2·1 V and, as most devices require to operate at higher voltages, it is usual to connect a number of cells in series to make a battery.

Example 3 Sixteen 1·5-V cells are connected in series to a load of 44 Ω. The internal resistance of each cell is 0·25 Ω. Calculate:

(*a*) the circuit current;
(*b*) the volt drop in the battery due to the internal resistance;
(*c*) the p.d. at the battery terminals.

The cells are connected in series; hence the internal resistances are effectively in series, and the total internal resistance is

$$r=16\times0\cdot25=4\cdot0\ \Omega$$

The e.m.f.s of the cells are also in series and the total battery e.m.f. is

$$E=16\times1\cdot5=24\cdot0\ \text{V}$$

Thus the circuit current is

$$I=\frac{E}{R+r}=\frac{24}{44+4}=\underline{0\cdot5\ \text{A}}$$

$$Ir=0\cdot5\times4=\underline{2\ \text{V}}$$

$$V=E-Ir=24-2=\underline{22\ \text{V}}$$

In Example 3, the sixteen cells were effectively reduced to a voltage source (of 24 V) connected in series with a resistor (of 4 Ω) representing the e.m.f. and internal resistance respectively. It follows that the same general form of representation could be applied to any battery, and this would take the equivalent representation shown in Fig. 3.

In the equivalent circuit, the constant-voltage source develops the same e.m.f. regardless of the current passing through it. At the same time, the volt drop due to the internal resistance causes the terminal voltage of the battery to drop as the current increases.

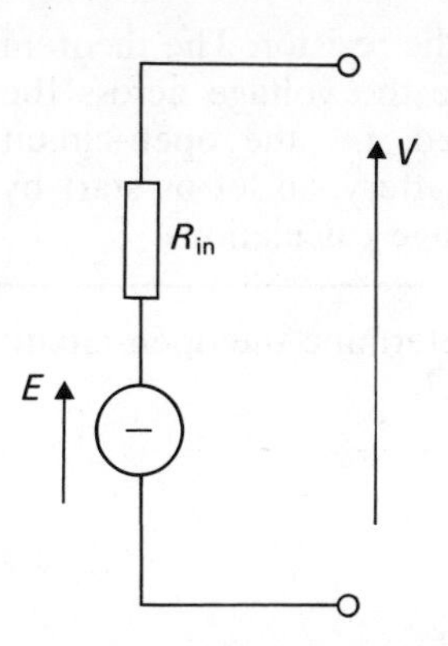

Fig. 3 Equivalent circuit of a battery

2 Thévenin's theorem

Our observations concerning the equivalent circuit of a battery can be extended to take in any other form of voltage source. This generality was first observed by Thévenin, who stated the following theorem:

> The current which flows in any branch of a circuit is the same as that which would flow in the branch if it were connected across a source of electrical energy, the e.m.f. of which is equal to the potential difference that would appear across the branch if it were open-circuited, and the internal resistance of which is equal to the resistance that appears across the open-circuited branch terminals.

When read for the first time, this theorem appears to be almost impossible to understand. However it is really quite a simple theorem if we work up to it. The first point to note is that it is dealing with any source of electrical energy. Thus the source could be a simple cell or a battery, it could be one or more generators connected in parallel, or it could be a number of such energy sources feeding through a network to an energy outlet. This most complicated instance is that with which we are all familiar and is the electric socket outlet, which is a simple enough pair of terminals yet is supplied from all the generators feeding in to the National Grid.

Whatever the arrangement within the network, we are going to look at it from the point of view of the outlet terminals. Thévenin's theorem is therefore dealing with any network with two outlet terminals and containing one or more sources of electrical energy. The theorem then states that no matter how complicated the network is, it can be replaced by an equivalent circuit consisting of a constant-voltage source in series with a resistor. This equivalent circuit is the same as that used for the battery.

For the equivalent circuit, we require to determine the e.m.f. of the constant-voltage source and the resistance of the resistor. The theorem says that to determine the e.m.f., we require the voltage across the terminals when the external load is removed, i.e. the open-circuit voltage. Again this is the same as with the battery, so let us start by looking at a similar form of open-circuit voltage calculation.

Example 4 For the circuit shown in Fig. 4, determine the open-circuit voltage that would appear across terminals *AB*.

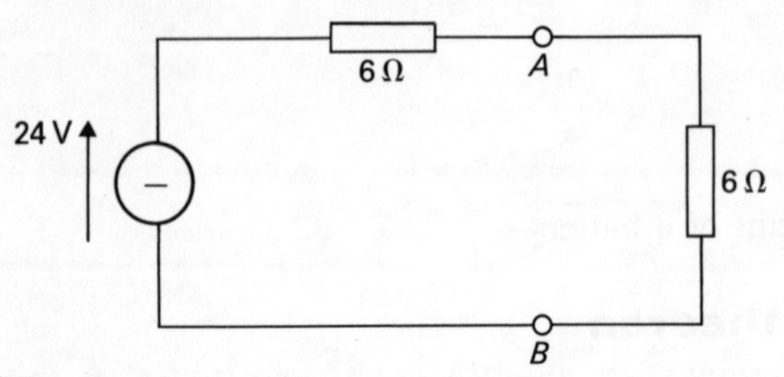

Fig. 4

Under open-circuit conditions, the load across terminals *AB* is removed. Rather than imagine what part of the circuit arrangement remains, let us redraw the circuit for this new condition as shown in Fig. 5.

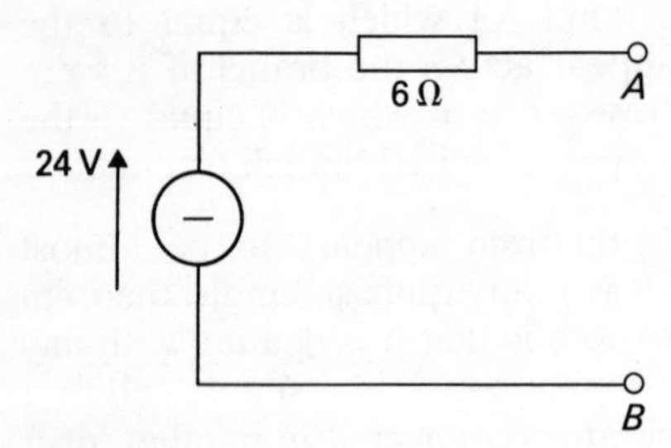

Fig. 5

From Fig. 5, it is now clear that a circuit no longer remains and therefore no current can flow. There can be no volt drops across the remaining resistances in the arrangement, and the source e.m.f. appears across terminals *AB*, thus

$V_{AB_{oc}} = \underline{24\ \text{V}}$

It is surprising how often students will give the open-circuit voltage in this example as being 12 V. This arises because they did not think it necessary to redraw the circuit for open-circuit conditions, and hence forgot that there could be no current in the load. The moral of the story is – always redraw the circuit diagram showing the load removed.

Example 5 emphasises this point when a slightly more complicated arrangement is considered.

Example 5 For the circuit shown in Fig. 6, determine the open-circuit voltage that would appear across the terminals *XY*.

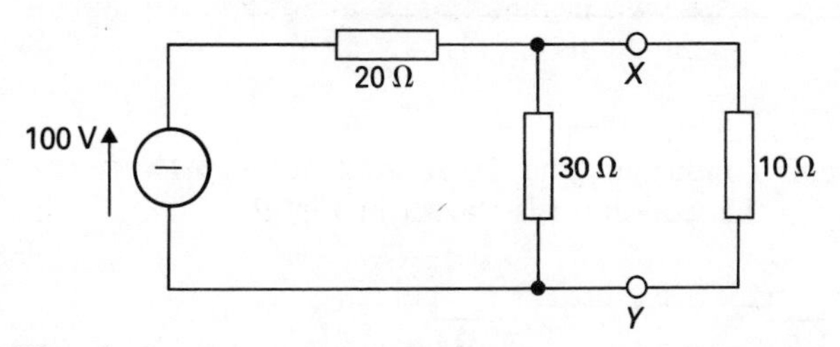

Fig. 6

Under open-circuit conditions, the 10-Ω load across *XY* is removed. The circuit for this condition is shown in Fig. 7.

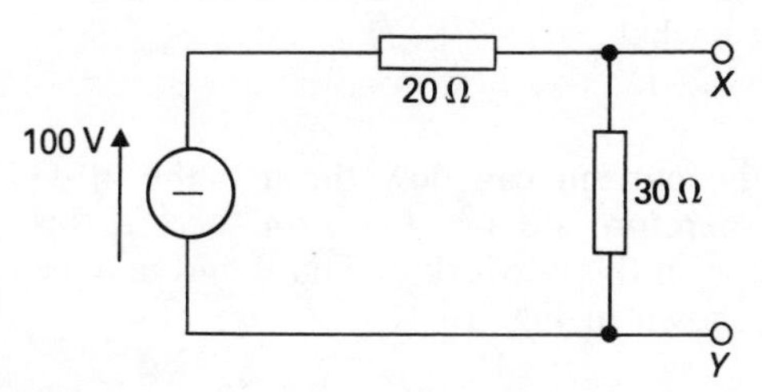

Fig. 7

From Fig. 7, it is clear that a circuit remains and for this condition,

$$V_{XY}=\frac{30}{20+30}\times 100=60\text{ V}$$

However the voltage V_{XY} under the given open-circuit conditions is the voltage appearing at the terminals *XY*, hence

$V_{XY_{oc}}=\underline{60\text{ V}}$

One form of open-circuit voltage calculation which can give students some problems arises when the network introduces a mixture of the two forms of calculation so far demonstrated. A network of this type is given in Example 6.

Example 6 For the network shown in Fig. 8, determine the open-circuit voltage that would appear across terminals *MN*.

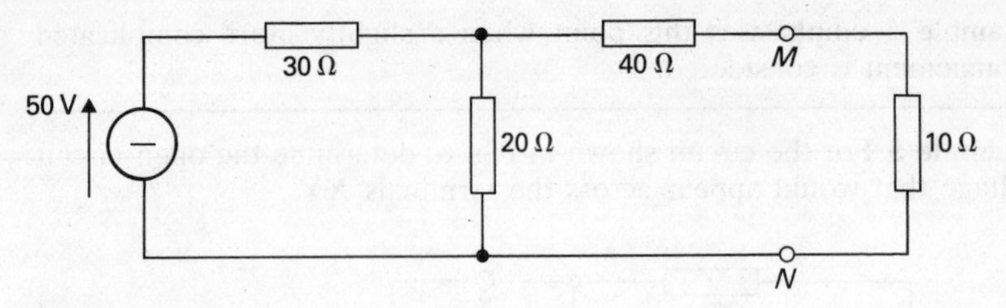

Fig. 8

Under open-circuit conditions, the 10-Ω load across *MN* is removed. The circuit for this condition is shown in Fig. 9.

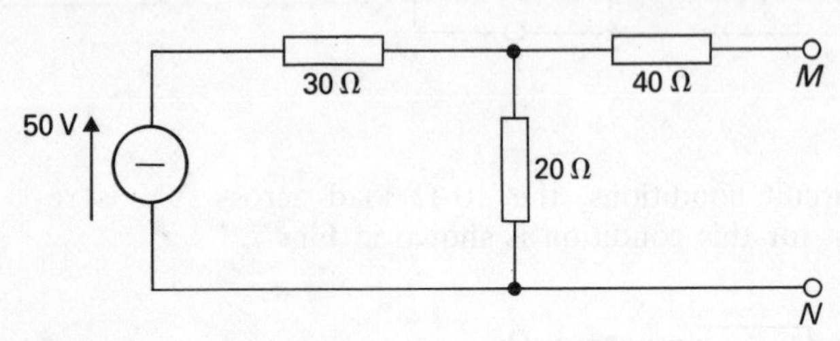

Fig. 9

From Fig. 9, it is clear that no current can flow through the 40-Ω resistor and that there can therefore be no volt drop across this component. It serves no function in the network of Fig. 9 and may be omitted from the diagram, as shown in Fig. 10

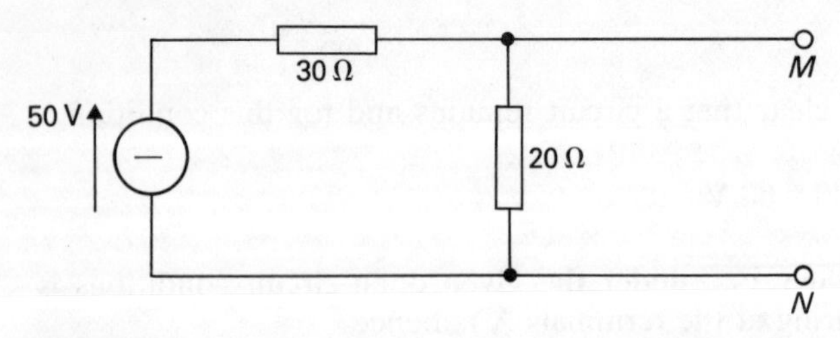

Fig. 10

From Fig. 10, the open-circuit voltage across *MN* is given by

$$V_{MN_{oc}} = \frac{20}{20+30} \times 50 = \underline{20\ \text{V}}$$

This method of reducing a network by omitting components which serve no electrical purpose can make circuit analysis very much easier, but do remember to redraw the diagram each time. You may feel that this is a waste of time but it makes sure that you come to the correct conclusion. Most mistakes occur when students try to imagine the changes without redrawing the circuits.

The equivalent circuit of a network as defined by Thévenin's theorem requires a constant-voltage source and a series resistance. We have now seen how to obtain the open-circuit voltage which will be the voltage of the constant-voltage source. It remains to find the resistance. The value of that resistance is that which is presented at the output terminals when all the sources of e.m.f. have been replaced by their internal resistances. Two examples will make this statement more clear.

Example 7 For the circuit shown in Fig. 11, determine the internal resistance of the part of the circuit supplying terminals *AB*. It may be assumed that the internal resistance of the battery is negligible.

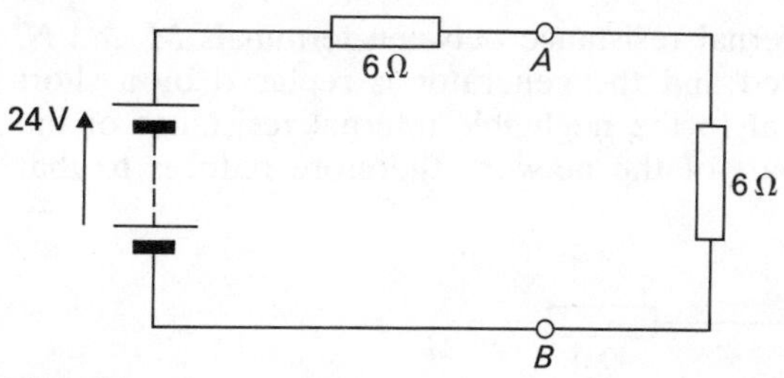

Fig. 11

To determine the internal resistance between terminals *A* and *B*, the 6-Ω load is removed and the battery is replaced by a short circuit which is equivalent to the negligible internal resistance of the battery. The supply part of the circuit therefore reduces to that shown in Fig. 12.

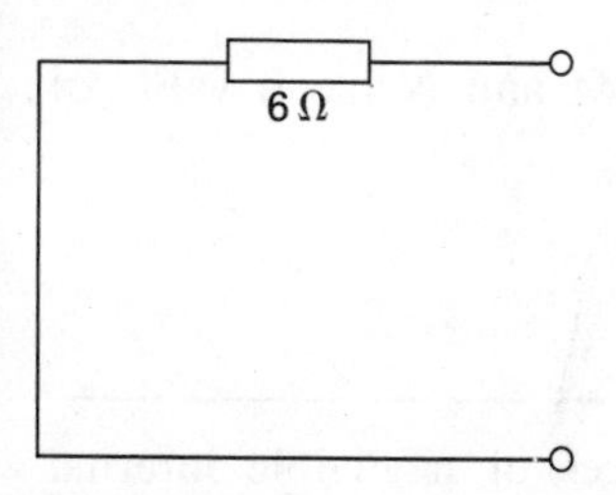

Fig. 12

From Fig. 12, it can be seen that the internal resistance of the part of the circuit supplying terminals *AB* is

$R_{in} = \underline{6\ \Omega}$

Example 8 For the network shown in Fig. 13, determine the internal resistance of the part of the network supplying terminals *MN*. It may be assumed that the internal resistance of the 50-V generator is negligible.

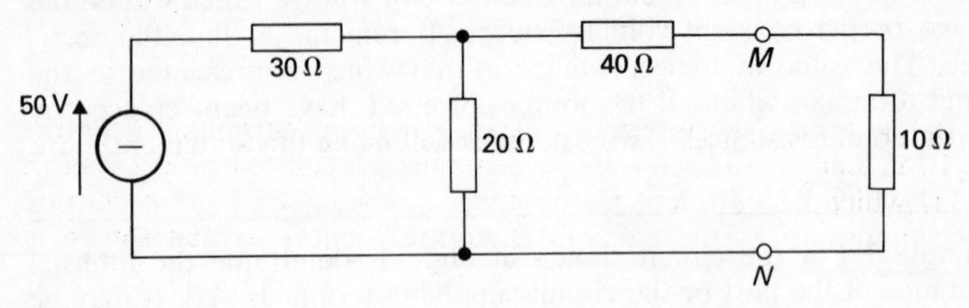

Fig. 13

To determine the internal resistance between terminals *M* and *N*, the 10-Ω load is removed and the generator is replaced by a short circuit which is equivalent to the negligible internal resistance of the generator. The supply part of the network therefore reduces to that shown in Fig. 14.

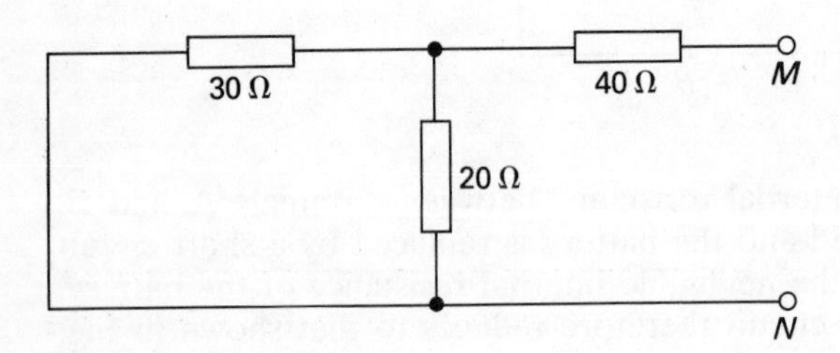

Fig. 14

The remaining network between terminals *M* and *N* has a value of resistance given by

$$R_{MN} = 40 + \frac{30 \times 20}{30 + 20} = 40 + 12 = \underline{52\ \Omega}$$

Both the examples have involved sources of negligible internal resistance. Example 9 illustrates the procedure adopted when the source of e.m.f. has internal resistance.

Example 9 For the network shown in Fig. 15, determine the internal resistance of the part of the network supplying terminals *PS*. The internal resistance of the generator is 3 Ω.

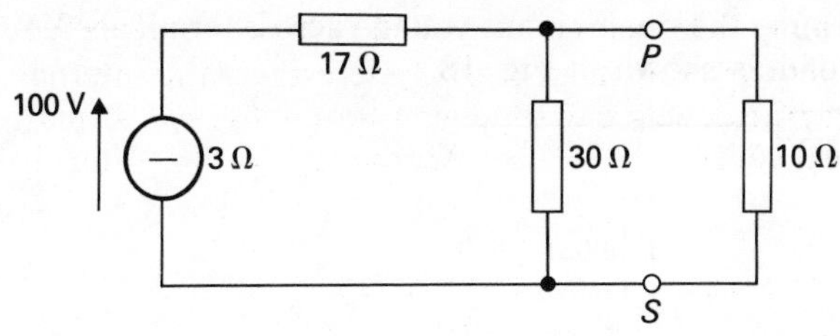

Fig. 15

To determine the internal resistance between terminals P and S, the 10-Ω load is removed and the generator is replaced by a resistance of 3 Ω which is equivalent to the internal resistance of the generator. The supply part of the network therefore reduces to that shown in Fig. 16.

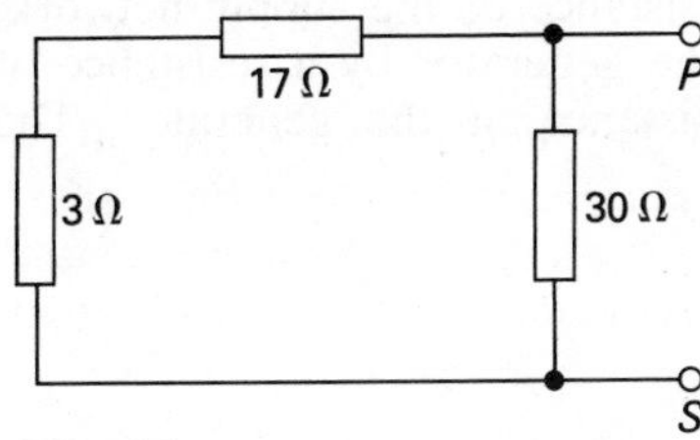

Fig. 16

The remaining network between terminals P and S has a value of resistance given by

$$R_{in} = R_{PS} = \frac{(3+17)\times 30}{(3+17)+30} = \underline{12\ \Omega}$$

From these examples, we have now seen how to obtain the value of the e.m.f. of the constant-voltage source and the resistance internal to the equivalent circuit required for Thévenin's theorem. By putting these two processes together, we can proceed to find the current in a branch of a network by means of that theorem.

Example 10 For the network shown in Fig. 17, determine the current in the 3·6-Ω resistor by means of Thévenin's theorem.

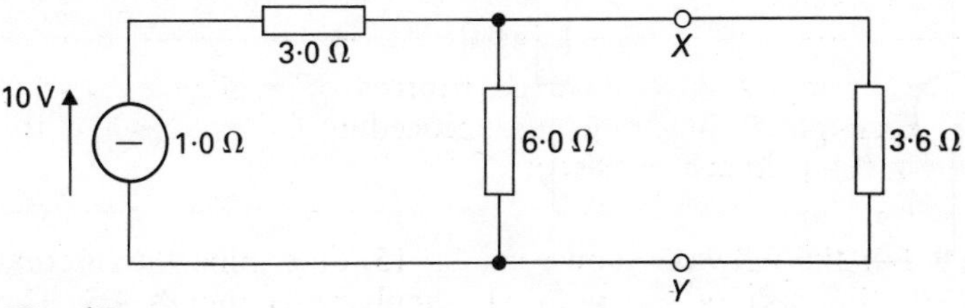

Fig. 17

 In order to determine the open-circuit voltage across terminals XY, remove the 3·6-Ω load as shown in Fig. 18.

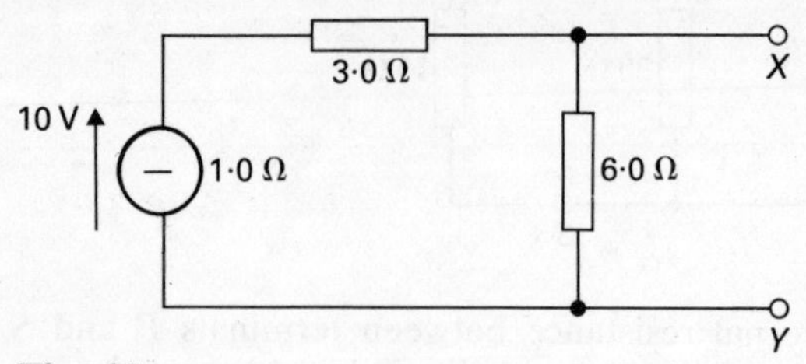

Fig. 18

$$V_{XY_{oc}} = \frac{6\cdot0}{1\cdot0 + 6\cdot0 + 3\cdot0} \times 10 = 6\cdot0 \text{ V}$$

In order to determine the internal resistance of the supply network, remove the 3·6-Ω load and replace the generator by a resistance of 1·0 Ω equivalent to the internal resistance of the generator. The resulting network is shown in Fig. 19.

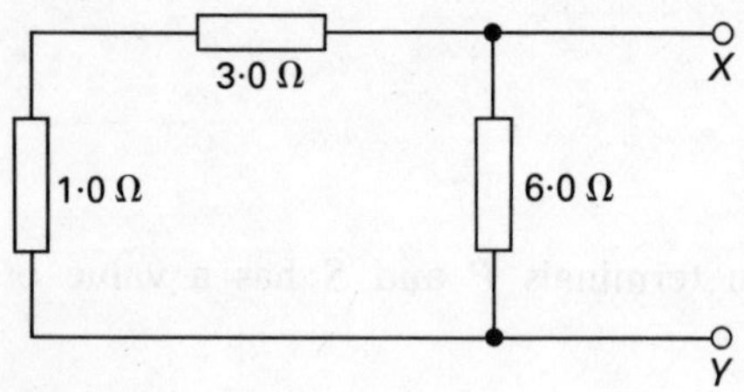

Fig. 19

$$R_{in} = R_{XY} = \frac{(1\cdot0 + 3\cdot0) \times 6\cdot0}{1\cdot0 + 3\cdot0 + 6\cdot0} = 2\cdot4\ \Omega$$

The equivalent circuit supplying terminals XY is therefore as shown in Fig. 20.

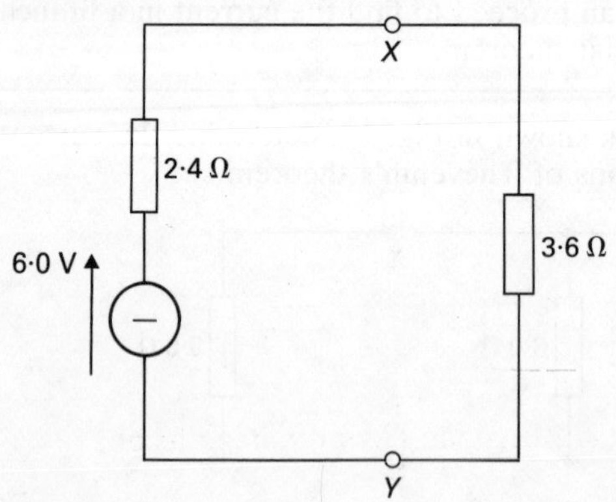

Fig. 20

From the circuit shown in Fig. 20, the current is given by

$$I=\frac{6\cdot0}{2\cdot4+3\cdot6}=\underline{1\cdot0\ \text{A}}$$

Example 11 For the network shown in Fig. 21, determine the power dissipated by the 5-Ω resistor.

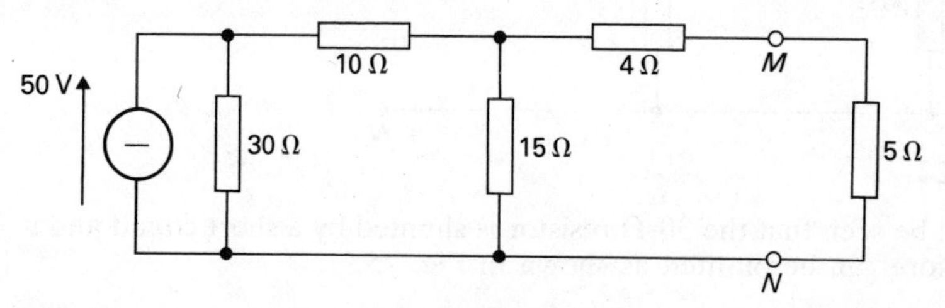

Fig. 21

To determine the open-circuit voltage across terminals *MN*, remove the 5-Ω resistor, as shown in Fig. 22.

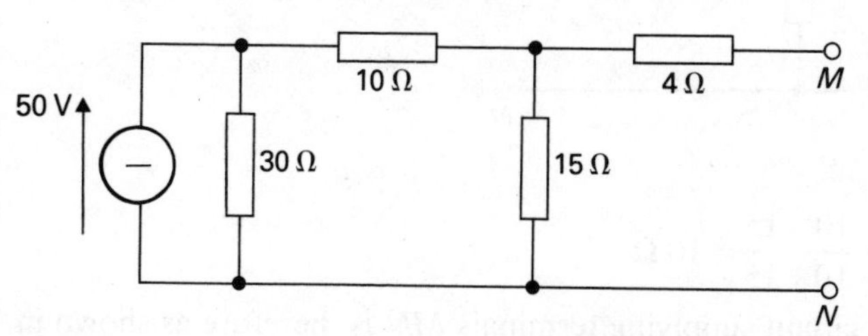

Fig. 22

We can now see that there can be no current in the 4-Ω resistor, which may be omitted as shown in Fig. 23.

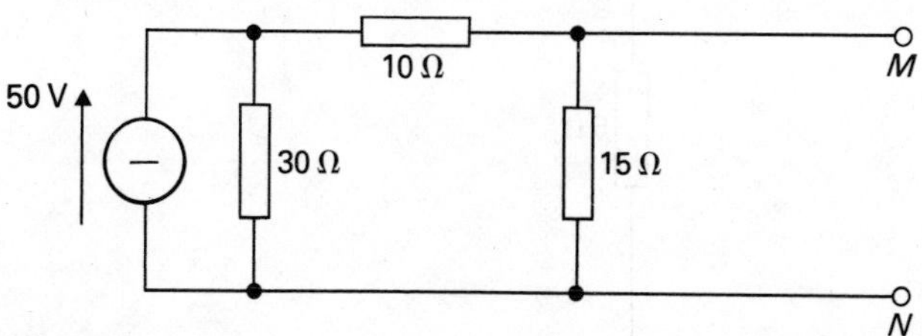

Fig. 23

From Fig. 23, we can determine the open-circuit voltage to be

$$V_{MN_{oc}}=\frac{15}{10+15}\times50=30\ \text{V}$$

To determine the internal resistance of the supply network, remove the 5-Ω resistor and replace the generator by a short circuit. Since no internal resistance of the voltage source is given, it is usual to assume it to be negligible. The resulting network is shown in Fig. 24.

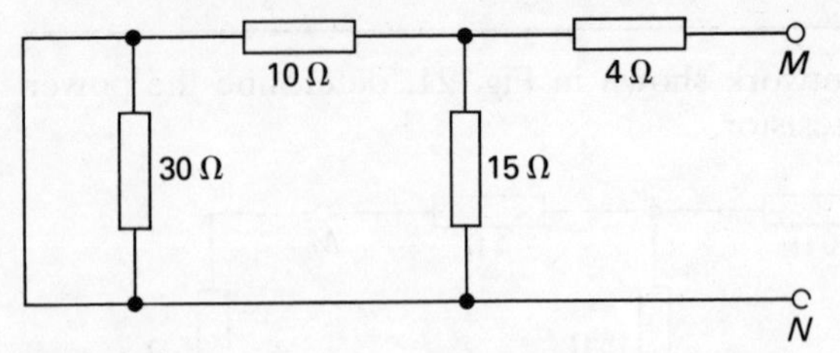

Fig. 24

It will be seen that the 30-Ω resistor is shunted by a short circuit and it therefore can be omitted as shown in Fig. 25.

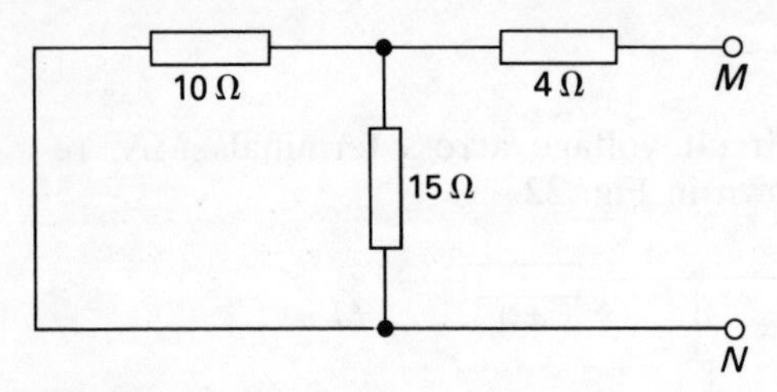

Fig. 25

$$R_{in} = R_{MN} = 4 + \frac{10 \times 15}{10 + 15} = 10\ \Omega$$

The equivalent circuit supplying terminals MN is therefore as shown in Fig. 26.

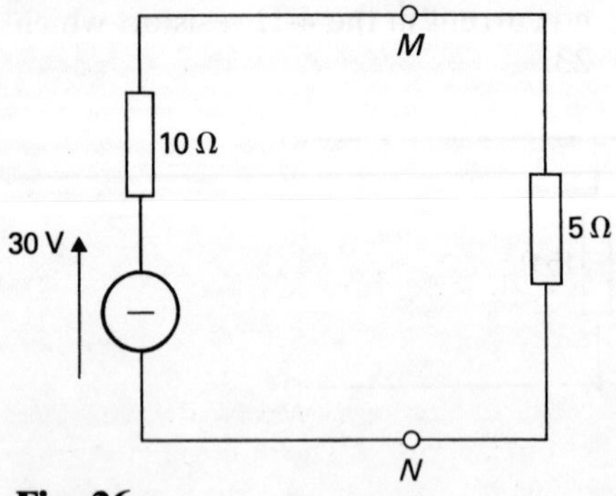

Fig. 26

$$I = \frac{30}{10 + 5} = 2\ \text{A}$$

The power dissipated by the 5-Ω resistor is therefore

$P = I^2R = 2^2 \times 5 = \underline{20\ \text{W}}$

Example 12 For the network shown in Fig. 27, determine, by means of Thévenin's theorem, the current in the 2·0-Ω resistor.

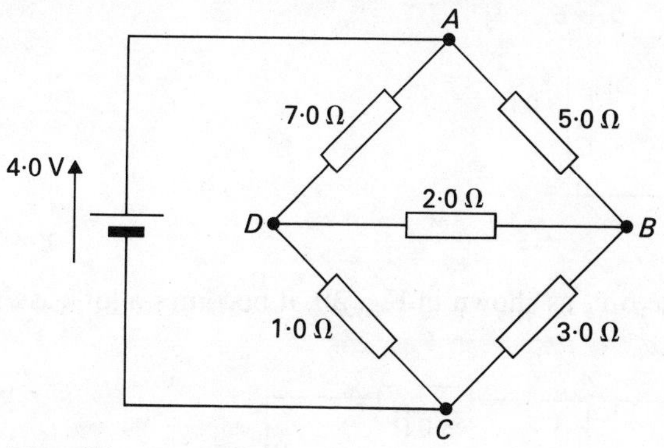

Fig. 27

To determine the open-circuit voltage across the 2·0-Ω resistor, remove that resistor and redraw the network as shown in Fig. 28.

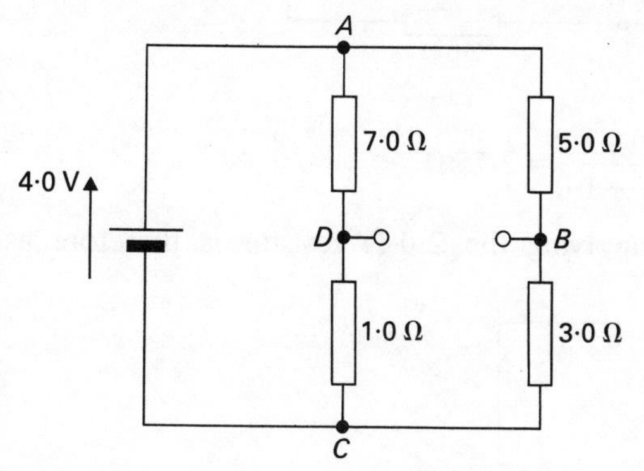

Fig. 28

$$V_{BD_{oc}} = V_{BC} - V_{DC} = \frac{3{\cdot}0}{3{\cdot}0 + 5{\cdot}0} \times 4{\cdot}0 - \frac{1{\cdot}0}{1{\cdot}0 + 7{\cdot}0} \times 4{\cdot}0$$

$$= 1{\cdot}0\ \text{V}$$

To determine the internal resistance of the supply network, remove the 2·0-Ω resistor and replace the battery by a short circuit. Again since no internal resistance for the battery is given, we may assume it to be negligible. The resulting network is shown in Fig. 29.

With the network drawn in this form, you have difficulty in determining the resistance between terminals B and D. However, if we

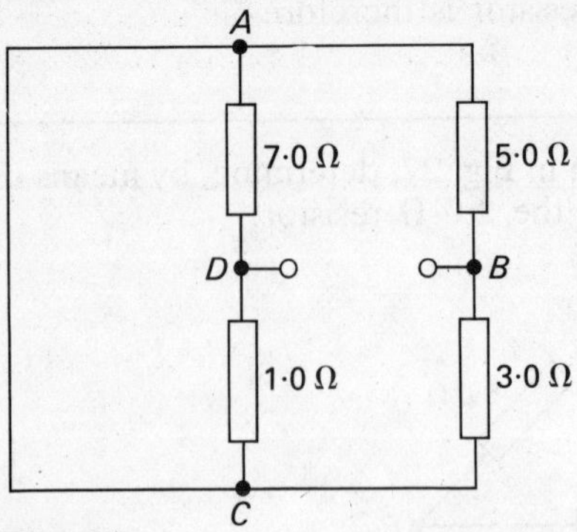

Fig. 29

turn the diagram inside out, as shown in Fig. 30, it becomes a lot easier to understand.

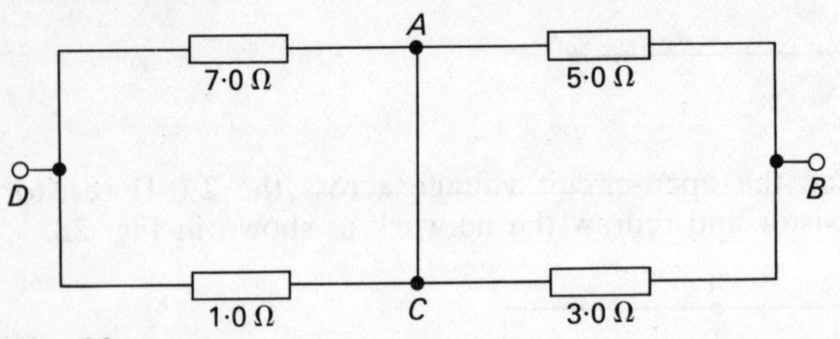

Fig. 30

$$R_{\text{in}} = R_{BD} = \frac{5{\cdot}0 \times 3{\cdot}0}{5{\cdot}0 + 3{\cdot}0} + \frac{7{\cdot}0 \times 1{\cdot}0}{7{\cdot}0 + 1{\cdot}0} = 2{\cdot}75\ \Omega$$

The equivalent circuit supplying the 2·0-Ω resistor is therefore as shown in Fig. 31.

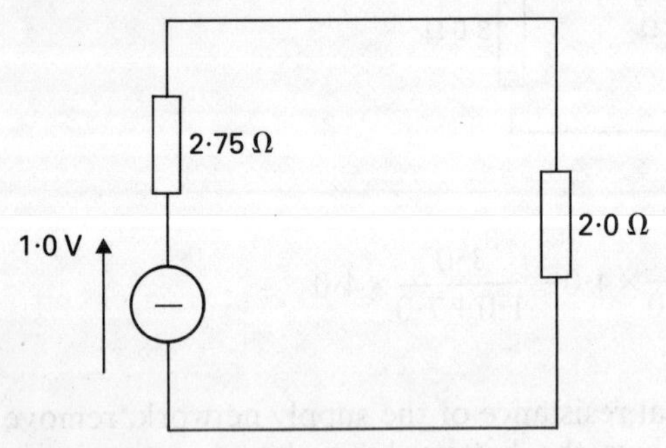

Fig. 31

$$I = \frac{1{\cdot}0}{2{\cdot}75 + 2{\cdot}0} = 0{\cdot}21\ \text{A}$$

The various examples of the application of Thévenin's theorem to network analysis have been limited to sources of direct current. However Thévenin's theorem can also be applied to a.c. networks.

For the simple problems considered, you may feel that there was no advantage in using this form of solution compared with the application of Kirchhoff's laws. Nevertheless when we have to deal with larger and more complicated networks, the principle of replacing the supply network by a constant-voltage generator in series with a resistance or an impedance is very useful. In particular, power engineers find this a convenient method of representing the supply system to their installation, say for the purpose of determining the volt drops within a factory, or the fault currents that might occur.

3 Maximum power transfer theorem

Consider the circuit shown in Fig. 32, in which a 12-V d.c. source of internal resistance 1 Ω is connected to a variable-resistance load. Let the load resistance be varied and determine the corresponding values of power dissipated by the load. These values of power are shown in the power/resistance characteristic, from which we observe that the greatest power dissipation occurs when the load resistance is equal to the internal resistance of the source. This observation is known as the maximum power transfer theorem.

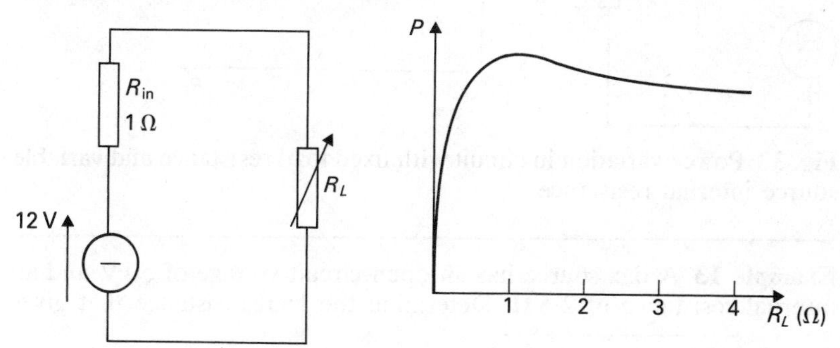

Fig. 32 Power variation in circuit with variable load resistance

The maximum power transfer theorem states that in a circuit of fixed source internal resistance, the power dissipated in the load is a maximum if the load resistance is made equal to the internal resistance of the source. It can be added that if the load variation is insufficient to permit the load resistance to equal the source internal resistance, then the maximum power dissipated by the load will arise when the load resistance is made as near as possible equal to the internal resistance.

For the circuit given in Fig. 32, the maximum power dissipated by the load occurs when R_L is 1 Ω, hence $P = 36$ W. However if R_L were only $\frac{1}{2}R_{in}$, then $P = 32$ W, and if R_L were $2R_{in}$, then $P = 32$ W again. This shows that it is not too important that the load resistance be adjusted to be exactly equal to the internal resistance. If it is adjusted to be almost equal, then similar values of power transfer can be obtained.

The process of varying the load resistance to be equal or almost equal to the source internal resistance is termed load matching. In d.c. circuits, load matching can only be achieved by adjusting the load resistance, but other techniques can be applied to a.c. circuits. A method of load matching by means of a transformer is described in Section 8 of Chapter 6.

The maximum power transfer theorem applies to circuits in which the load is varied to match the source. The theorem does not apply to circuits in which the load resistance is fixed and the source internal resistance can be varied. This situation is given in Fig. 33 and we can see that the maximum power dissipation in the load occurs when the internal resistance of the source is a minimum.

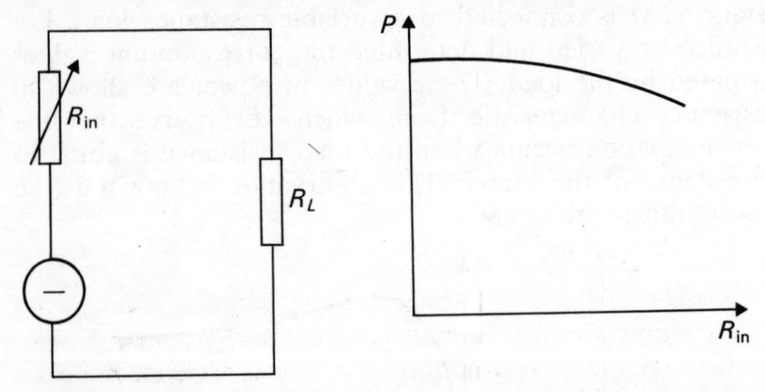

Fig. 33 Power variation in circuit with fixed load resistance and variable source internal resistance

---

Example 13 A d.c. source has an open-circuit voltage of 50 V and an internal resistance of 2·5 Ω. Determine the load resistance that gives

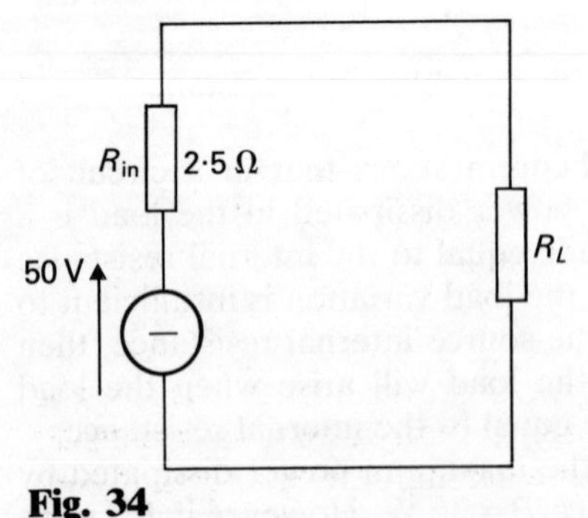

Fig. 34

maximum power dissipation, and the value of that power.

For maximum power, $R_L = R_{in} = \underline{2{\cdot}5\ \Omega}$

$$I = \frac{V}{R_L + R_{in}} = \frac{50}{2{\cdot}5 + 2{\cdot}5} = 10\ \text{A}$$

$$P = I^2 R_L = 10^2 \times 2{\cdot}5 = \underline{250\ \text{W}}$$

Example 14 For the network shown in Fig. 35, determine the resistance of load R_L that gives maximum power dissipation and the value of that power.

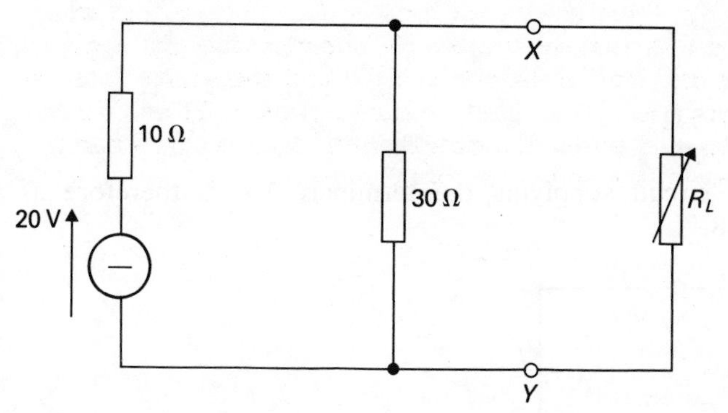

Fig. 35

In cases involving complicated sources supplying the load, it is necessary to determine the equivalent constant-voltage source by means of Thévenin's theorem. To find the open-circuit voltage, remove the load R_L as shown in Fig. 36.

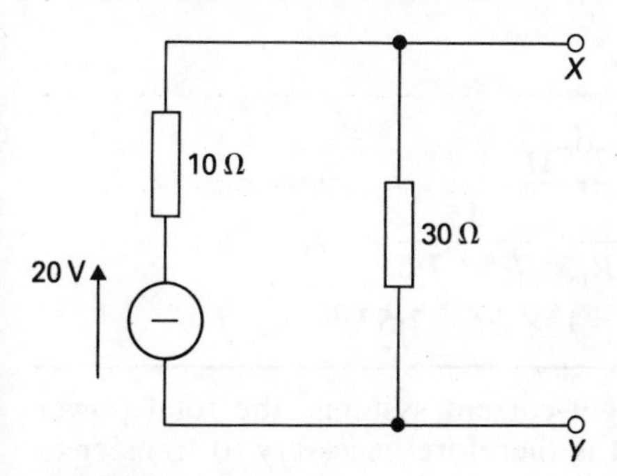

Fig. 36

To find the internal resistance of the source network, remove the load R_L and replace the generator by a short-circuit, assuming its internal resistance to be negligible, the resulting network being shown in Fig. 37.

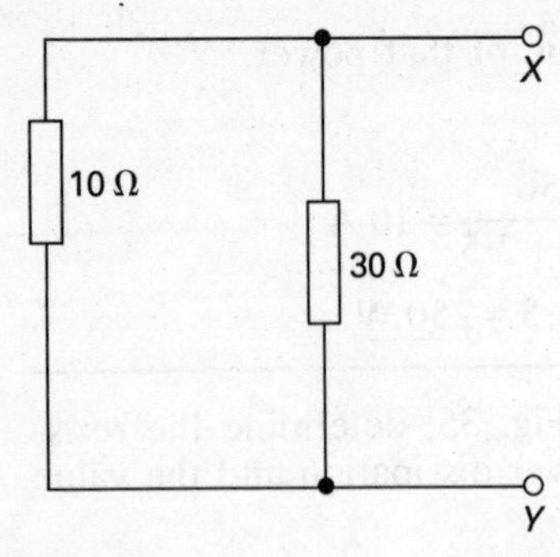

Fig. 37

$$R_{in} = \frac{10 \times 30}{10 + 30} = 7{\cdot}5\ \Omega$$

The equivalent circuit supplying the terminals XY is therefore as shown in Fig. 38.

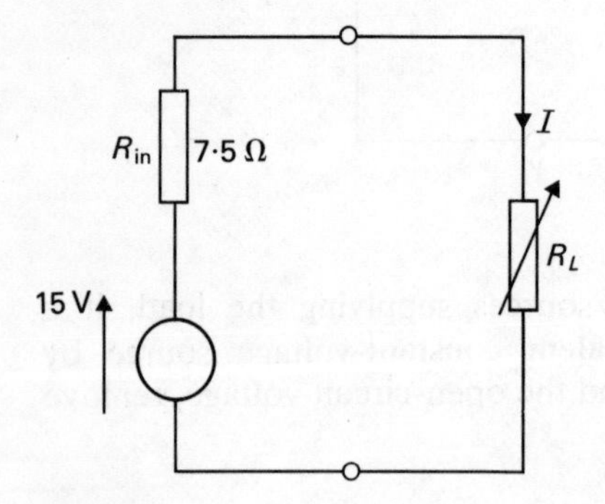

Fig. 38

For maximum power, $R_L = R_{in} = \underline{7{\cdot}5\ \Omega}$

$$I = \frac{V}{R_L + R_{in}} = \frac{15}{7{\cdot}5 + 7{\cdot}5} = 1\ \text{A}$$

$$P = I^2 R_L = 1^2 \times 7{\cdot}5 = \underline{7{\cdot}5\ \text{W}}$$

When we are dealing with light-current systems, the total power available is generally small and it is therefore necessary to transfer as much as possible to the load. The sources of the power are often either amplifiers or lines, and in either case, the internal resistance (or impedance) is fixed; thus it is necessary to match the load to the source in order to obtain maximum power transfer. With such small power levels, it is unimportant that the efficiency of power transfer is low, so long as the power to the load is maximised.

In systems involving high values of current and power, usually the supply voltage to the load is fixed and the load power is determined by its application. It follows that the load resistance (or impedance) is fixed, and the best operating condition is obtained by reducing the source internal resistance (or impedance) as much as possible. In this circumstance, the efficiency of the supply is much higher than the 50 per cent value obtained by matching the load to the source. Such an improvement is essential to an electricity supply authority which could not contemplate operating at low levels of efficiency, such would be the cost of the lost energy.

4 Constant-current generator

In the section dealing with Thévenin's theorem, it was shown that a source of electrical energy could be represented by a source of constant voltage in series with a resistance. This is not the only possible form of representation. Again consider a source feeding a load R_L as shown in Fig. 39.

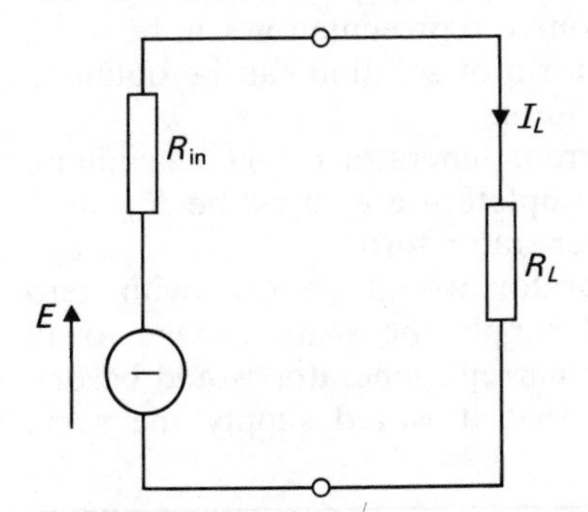

Fig. 39 Energy source feeding load

The load current I_L is given by

$$I_L = \frac{E}{R_{in}+R_L} = \frac{E/R_{in}}{(R_{in}+R_L)/R_{in}}$$

If a short circuit were applied to the source terminals, the current I_{sc} that would flow in the short circuit would be given by E/R_{in}, hence

$$I_L = \frac{I_{sc}}{(R_{in}+R_L)/R_{in}} = \frac{R_{in}}{R_{in}+R_L} \times I_{sc} \tag{3}$$

This relation takes a form generally recognised as the division of current between two parallel resistors, sometimes called the current sharing rule. From relation (3), it can be seen that, when viewed from the load, the supply appears as a source of current I_{sc} which is divided between the internal resistance R_{in} and the load resistance R_L connected in parallel. This arrangement is shown in Fig. 40.

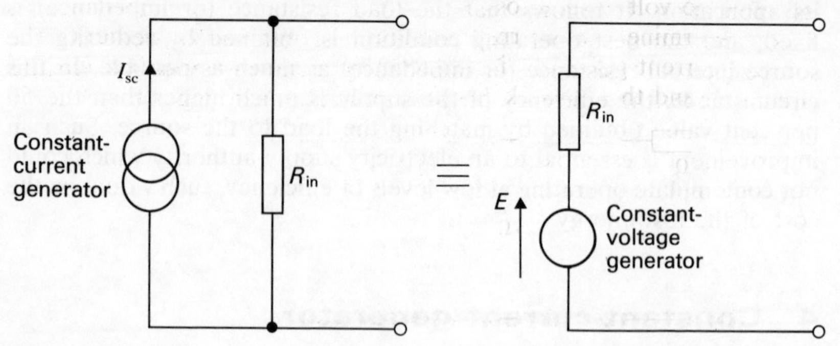

Fig. 40 Constant-current generator and constant-voltage generator forms

To distinguish the current generator from the more common voltage generator, the symbol used is that of two intersecting circles. For the solution of problems, either the constant-voltage generator or the constant-current generator form of source representation can be used, but in a number of cases, an easier form of solution can be obtained from the constant-current generator form.

The resistance of the constant-current generator is taken as infinite since the internal resistance of the complete source must be R_{in}, as is obtained with the constant-voltage generator form.

The ideal constant-voltage generator would be one with zero internal resistance, so that it would supply the same voltage to all loads. Conversely, the ideal constant-current generator would be one with infinite internal resistance, so that it would supply the same current to all loads.

Example 15 Represent the network shown in Fig. 41 by a source of e.m.f. in series with a resistance.

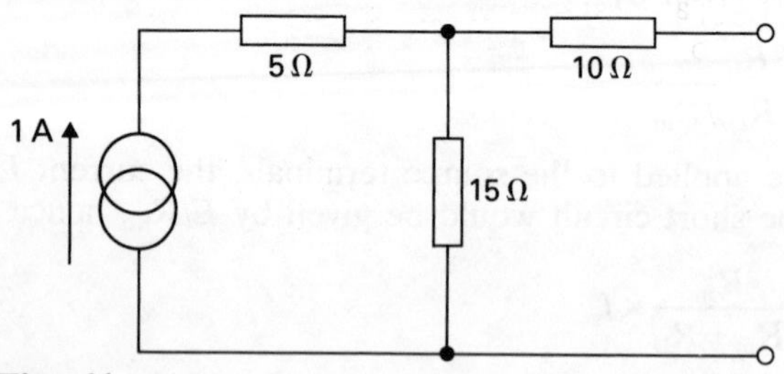

Fig. 41

The terminals are experiencing an open-circuit condition, hence

$V_{oc} = 1 \times 15 = 15 \text{ V}$

Notice that the 10-Ω resistance plays no part in the open-circuit voltage calculation since no current is passing through it and there is therefore no volt drop across it.

To determine the internal resistance of the supply network, the constant-current generator is replaced by an open-circuit (i.e. infinite resistance) and the network becomes that shown in Fig. 42.

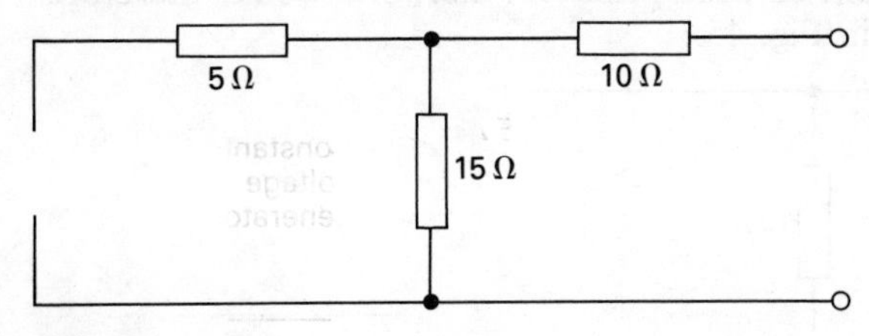

Fig. 42

$R_{in} = 10 + 15 = 25\ \Omega$

The network can therefore be represented as shown in Fig. 43.

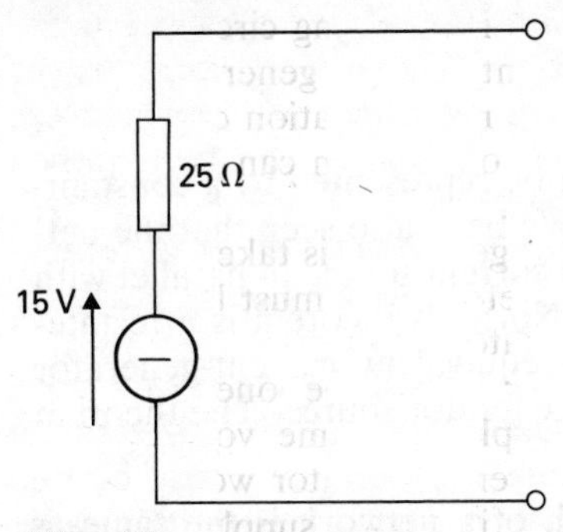

Fig. 43

Example 16 A supply network is represented in constant-voltage generator form by the circuit shown in Fig. 44. Determine the equivalent network incorporating a constant-current generator.

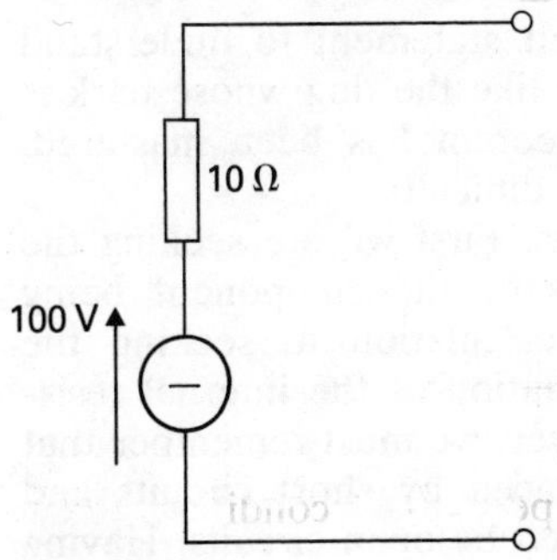

Fig. 44

For this circuit, if the terminals were short-circuited, then the short-circuit current would be

$$I_{sc} = \frac{100}{10} = 10\ \text{A}$$

The equivalent constant-current generator network would therefore take the form shown in Fig. 45.

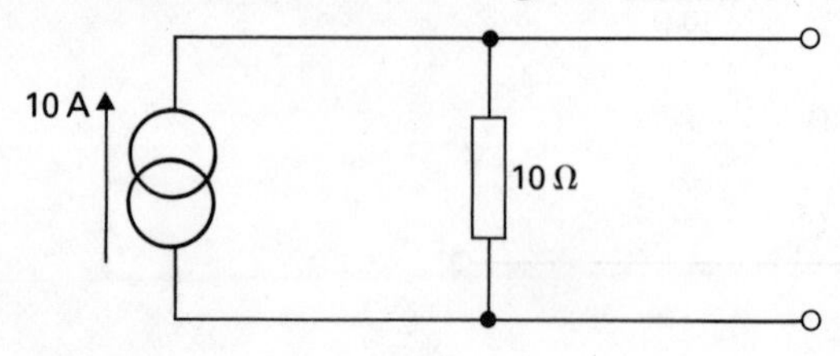

Fig. 45

5 Norton's theorem

We have seen that a supply network can be represented by a constant-voltage source in series with a resistor. We have also seen that the network could be represented by a constant-current source in parallel with the same resistor. Following from this, Norton's theorem is a restatement of Thévenin's theorem using an equivalent current-generator source instead of an equivalent voltage-generator source. The theorem may be stated as follows:

> The current that flows in any branch of a network is the same as that which would flow in the branch if it were connected across a source of electrical energy, the short-circuit current of which is equal to the current that would flow in a short circuit across the branch, and the internal resistance of which is equal to the resistance which appears across the open-circuited branch terminals.

Like Thévenin's theorem, this is a difficult statement to understand when met for the first time but again it is like the dog whose bark is worse than its bite. Once Thévenin's theorem has been mastered, Norton's theorem presents much less of a difficulty.

We need only bear in mind two points. First we are seeking the short-circuit current across the terminals of the component being investigated, and this is a comparable calculation to seeking the open-circuit voltage. Second, the determination of the internal resistance is exactly the same as before. However, we must remember that voltage sources which are ideal are replaced by short circuits and current sources which are ideal are replaced by open circuits. Having made these points, we can proceed directly to an example.

Example 17 Calculate the current in the 5-Ω resistor in the network shown in Fig. 46.

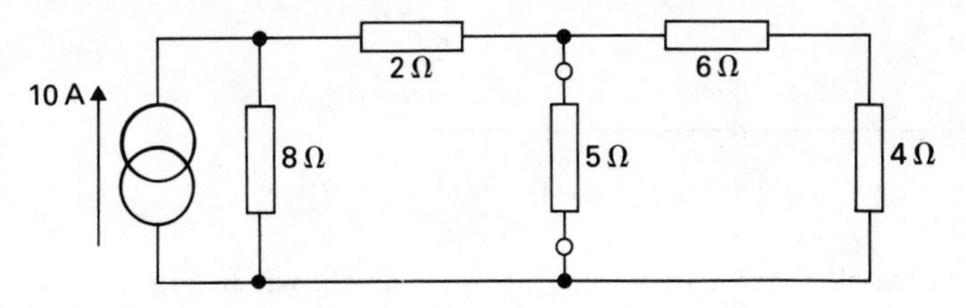

Fig. 46

To obtain the short-circuit current, apply across the terminals of the 5-Ω resistor a short circuit as shown in Fig. 47.

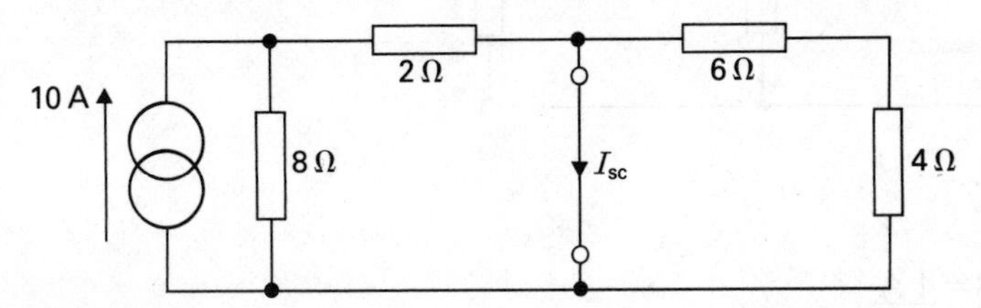

Fig. 47

The short circuit bypasses the 6-Ω resistor and the 4-Ω resistor which can be removed from the network as shown in Fig. 48.

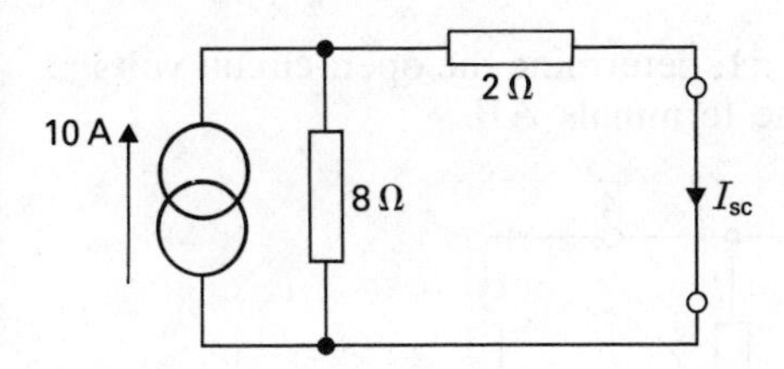

Fig. 48

$$I_{sc} = \frac{8}{8+2} \times 10 = 8 \text{ A}$$

To obtain the internal resistance of the supply network, the 5-Ω resistor is removed and the current generator replaced by an open circuit as shown in Fig. 49.

$$R_{in} = \frac{(8+2) \times (6+4)}{(8+2) + (6+4)} = 5 \ \Omega$$

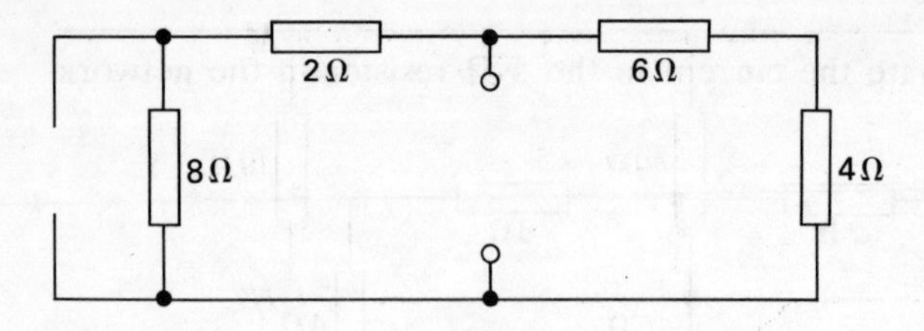

Fig. 49

The supply network there can be represented as shown in Fig. 50.

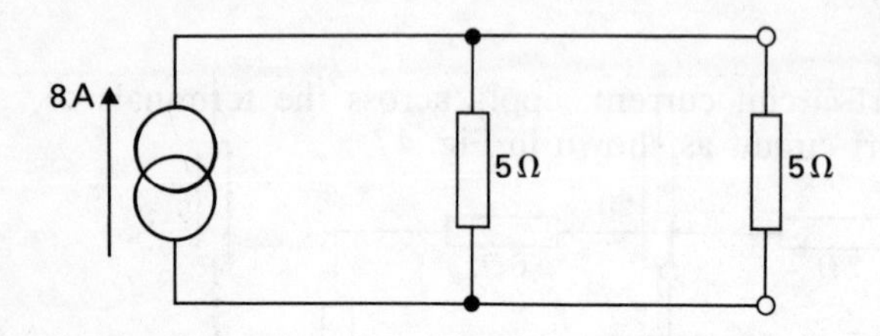

Fig. 50

$$I=\frac{5}{5+5}\times 8=\underline{4\ \text{A}}$$

Problems

1. For the circuit shown in Fig. 51, determine the open-circuit voltage that would appear across the terminals AB.

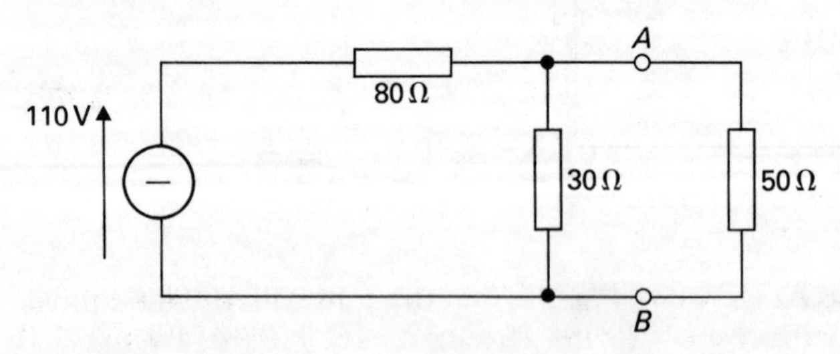

Fig. 51

2. For the circuit shown in Fig. 52, determine the internal resistance of the part of the network supplying terminals MN. It may be assumed that the internal resistance of the 50-V generator is negligible.

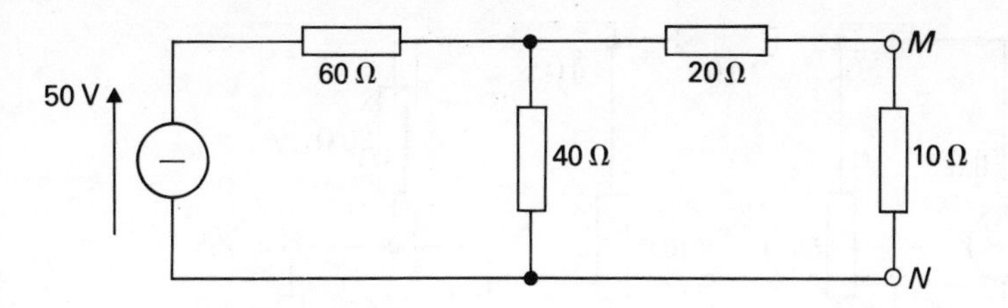

Fig. 52

3. For the network shown in Fig. 53, determine the power dissipated by the 5-Ω resistor.

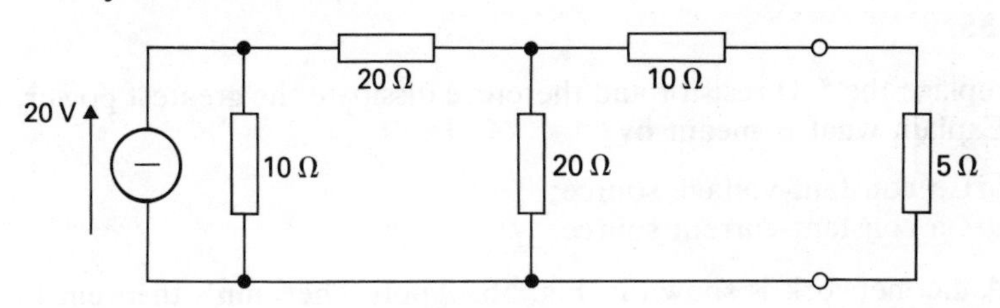

Fig. 53

4. By applying Thévenin's theorem, find the current in the 8-Ω resistor connected between *AB* of the circuit shown in Fig. 54.

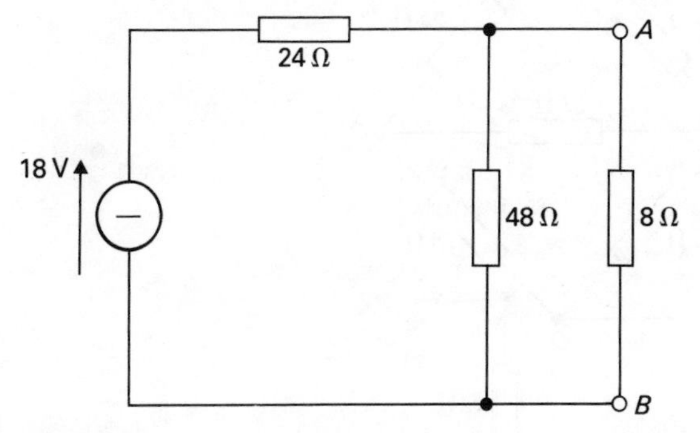

Fig. 54

5. For the network shown in Fig. 55, find the constant-voltage equivalent circuit with respect to the terminals *AB*. Hence, for the 5-Ω resistor connected across *AB*, determine:

   (*a*) the voltage across it;
   (*b*) the current;
   (*c*) the power dissipated.

6. For the network shown in Fig. 55, state the resistance which could

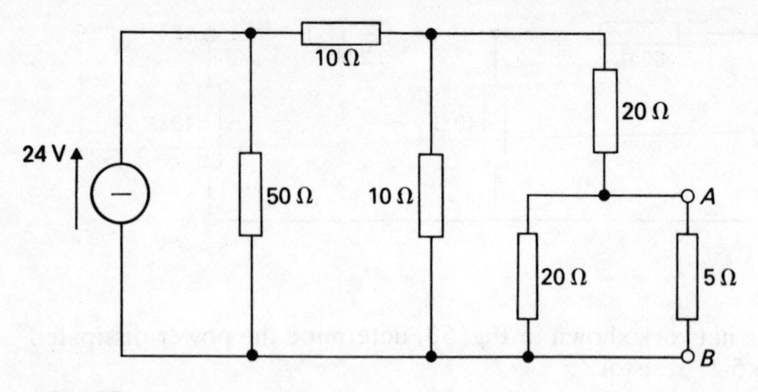

Fig. 55

replace the 5-Ω resistor and therefore dissipate the greatest power.

7. Explain what is meant by

   (*a*) a constant-voltage source;
   (*b*) a constant-current source.

   A.d.c. network is shown in Fig. 56. Apply Thévenin's theorem to find the constant-voltage equivalent circuit with respect to the branch *AC* and hence find the current in this branch.

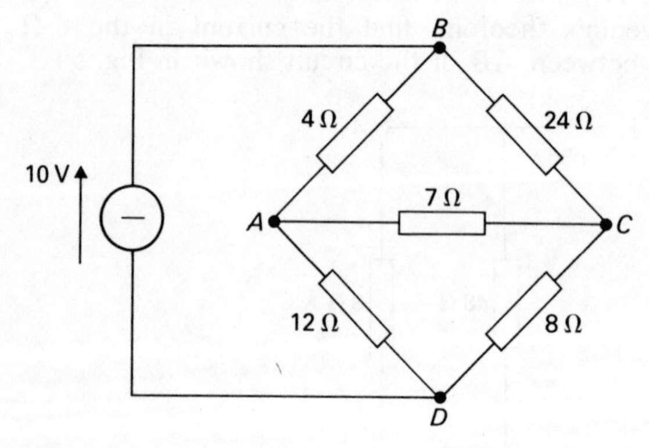

Fig. 56

8. Figures 57(a) and (b) show two battery networks. What, in each case, is the power dissipated in the load and the power supplied by the battery when a 4-Ω resistor is connected across the terminals *XY*?

   Convert each of the networks into equivalent sources (a) of the Thévenin type, (b) of the Norton type. Hence show in what way the two networks are equivalent, and in what way they are not equivalent, with respect to the terminals *XY*.

9. For the circuit shown in Fig. 55, determine the Norton constant-current equivalent circuit with respect to terminals AB.

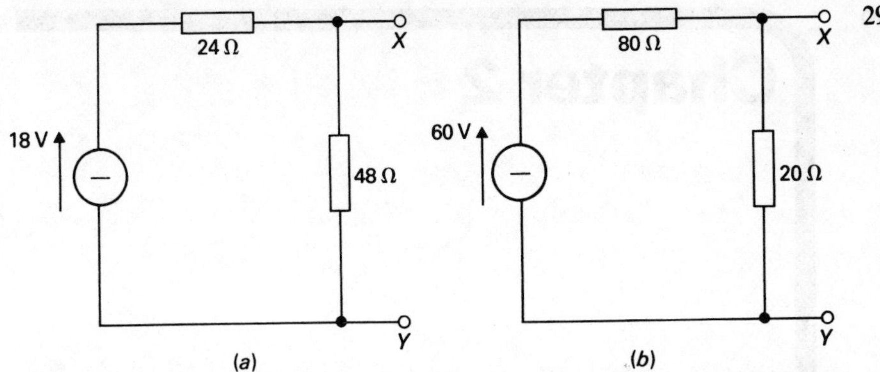

Fig. 57

10. A generator has an open-circuit voltage V_{oc} and a short-circuit current I_{sc}. Draw the Norton and Thévenin equivalent circuits, showing values associated with all components in the equivalent circuits in terms of V_{oc} and I_{sc}.

    Determine for the network shown in Fig. 58 the constant-voltage equivalent circuit relative to terminals *AB*. Hence find the voltage across, and the current in, a 5-Ω resistor connected across *AB*.

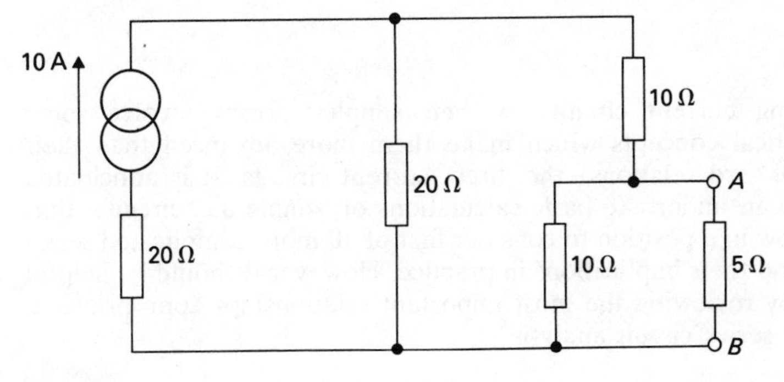

Fig. 58

Answers

1. 30 V
2. 44 Ω
3. 0·8 W
4. 0·5 A
5. 11·1 Ω; 5·33 V; 1·65 V; 0·33 A; 0·54 W
6. 11·1 Ω
7. 9 Ω; 5 V; 0·33 A
8. 1·44 W; 11·7 W; 43·2 W; 12 V; 16 Ω; 0·75 A; 16 Ω
9. 0·48 A; 11·1 Ω
10. 7·5 Ω; 50 V; 20 V; 4·0 A

Chapter 2

Alternating current circuits

Alternating current circuits in their simplest forms involve some mathematical concepts which make them more advanced than their straightforward relations – the direct current circuits. It is anticipated that you can undertake basic calculations on simple a.c. circuits; thus we are now in a position to consider first of all more complicated series circuits and their implications in practice. However it should be helpful to start by reviewing the most important relationships appropriate to basic a.c. series circuit analysis.

1 Basic series a.c. circuits

There are three possible constituent components to an a.c. circuit. These are resistance R, inductance L and capacitance C. Their individual characteristics are summarised in Fig. 1.

Example 1 A 500-μH inductor is connected to a 1·0-V, 5-kHz sinusoidal supply. Assuming the inductor to have negligible resistance,

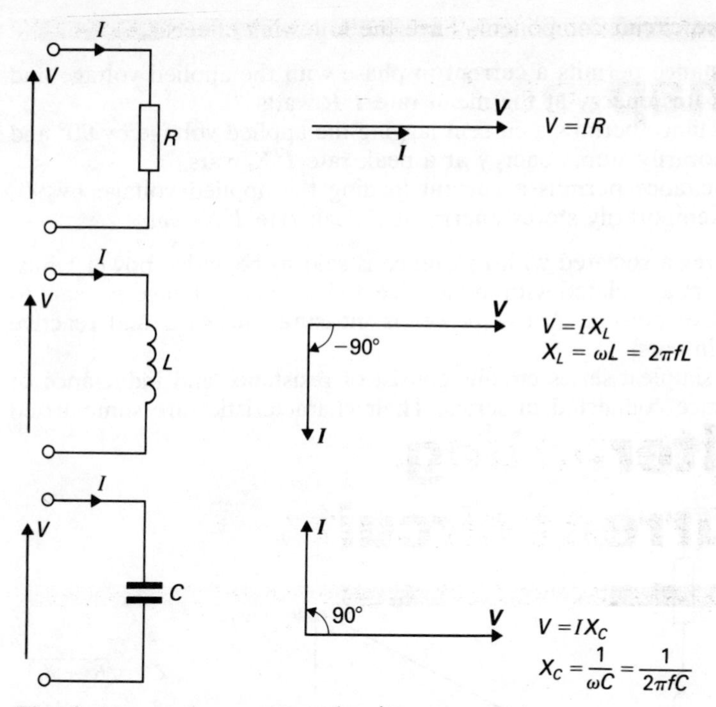

Fig. 1 Alternating current circuit components

determine the circuit current and hence calculate the capacitance of a capacitor that would take the same current from the same source.

$X_L = 2\pi fL = 2\pi \times 5000 \times 500 \times 10^{-6} = 15{\cdot}7\ \Omega$

$$I = \frac{V}{X_L} = \frac{1{\cdot}0}{15{\cdot}7} = 0{\cdot}0637\ \text{A}$$

$$= \underline{63{\cdot}7\ \text{mA}}$$

For the same current to flow,

$$X_C = X_L$$

$$= \frac{1}{2\pi fC}$$

$$C = \frac{1}{2\pi fX_C} = \frac{1}{2\pi \times 5000 \times 15{\cdot}7} = 2{\cdot}03 \times 10^{-6}\ \text{F}$$

$$= \underline{2{\cdot}03\ \mu\text{F}}$$

The three circuit components have the following effects:

1. Resistance permits a current in phase with the applied voltage and dissipates energy at the mean rate I^2R watts.
2. Inductance permits a current lagging the applied voltage by 90° and temporarily stores energy at a peak rate I^2X_L vars.
3. Capacitance permits a current leading the applied voltage by 90° and temporarily stores energy at a peak rate I^2X_C vars.

The power associated with resistance is said to be active power whilst the powers associated with inductance and with capacitance are said to be reactive powers. Active power is measured in watts and reactive powers in vars.

The simplest series circuits consist of resistance and inductance or capacitance connected in series. Their characteristics are summarised in Fig. 2.

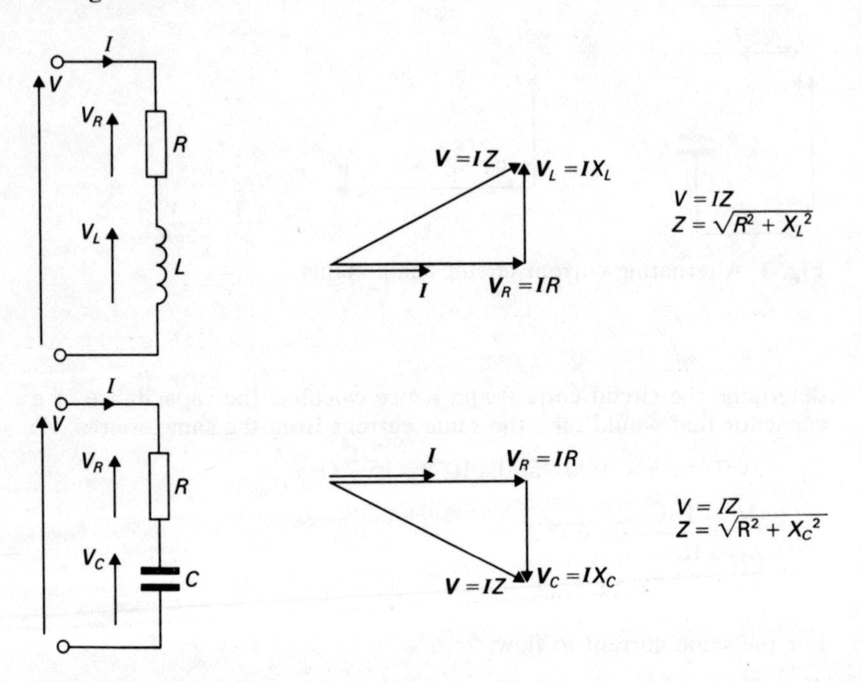

Fig. 2 *R–L* and *R–C* series circuits

A phasor is denoted in print by italic bold-faced or Clarendon type, e.g. ***I***. In handwriting, an over-dotted symbol is used, e.g. $\dot{I}$. Note that ***I*** represents a current phasor whilst *I* represents only the magnitude of the current.

Example 2 A pure inductance of 5·0 mH is connected in series with a pure resistance of 69 Ω. The circuit is supplied from a 4-kHz sinusoidal source and the voltage across the 69-Ω resistor is found to be 1·5 V. Calculate the supply voltage.

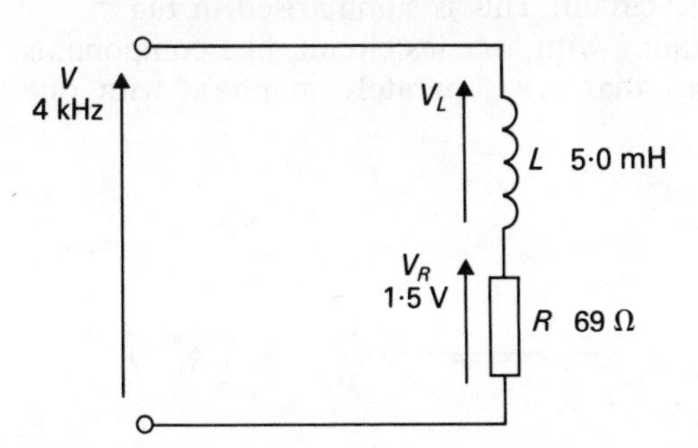

Fig. 3

$V_R = 1{\cdot}5\,V$

$$I = \frac{V_R}{R} = \frac{1{\cdot}5}{69} = 0{\cdot}0217\text{ A}$$

$$X_L = 2\pi fL = 2\pi \times 4000 \times 5 \times 10^{-3} = 125{\cdot}7\ \Omega$$

$$Z = (R^2 + X_L^2)^{1/2} = (69^2 + 125{\cdot}7^2)^{1/2} = 143{\cdot}4\ \Omega$$

$$V = IZ = 0{\cdot}0217 \times 143{\cdot}4 = \underline{3{\cdot}1\text{ V}}.$$

In each of the combinations shown, the phase relationship between voltage and current lies between 0° and 90°. In the *R–L* circuits, the phase angle is a lagging one and in the *R–C* circuit, the phase angle is a leading one.

Inductance is introduced into a circuit usually by the inclusion of a coil. The conductor forming the coil has resistance which may be negligible but is never zero. This resistance may be added to that of a series resistance to obtain the total circuit resistance as indicated in Fig. 4.

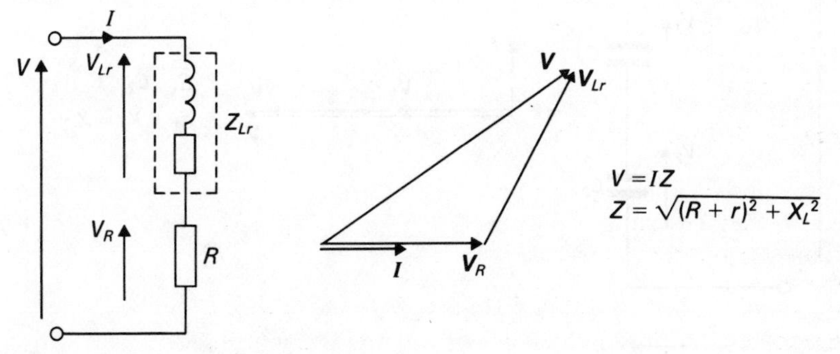

Fig. 4 *R–L* series circuit with resistive coil

This is a particular instance of a more general rule that like components in a series circuit may be added to one another in order to obtain the total effect of the circuit. This is summarised in Fig. 5.

Provided that we are dealing with a series circuit, like components are associated with voltages that are separately in phase with one

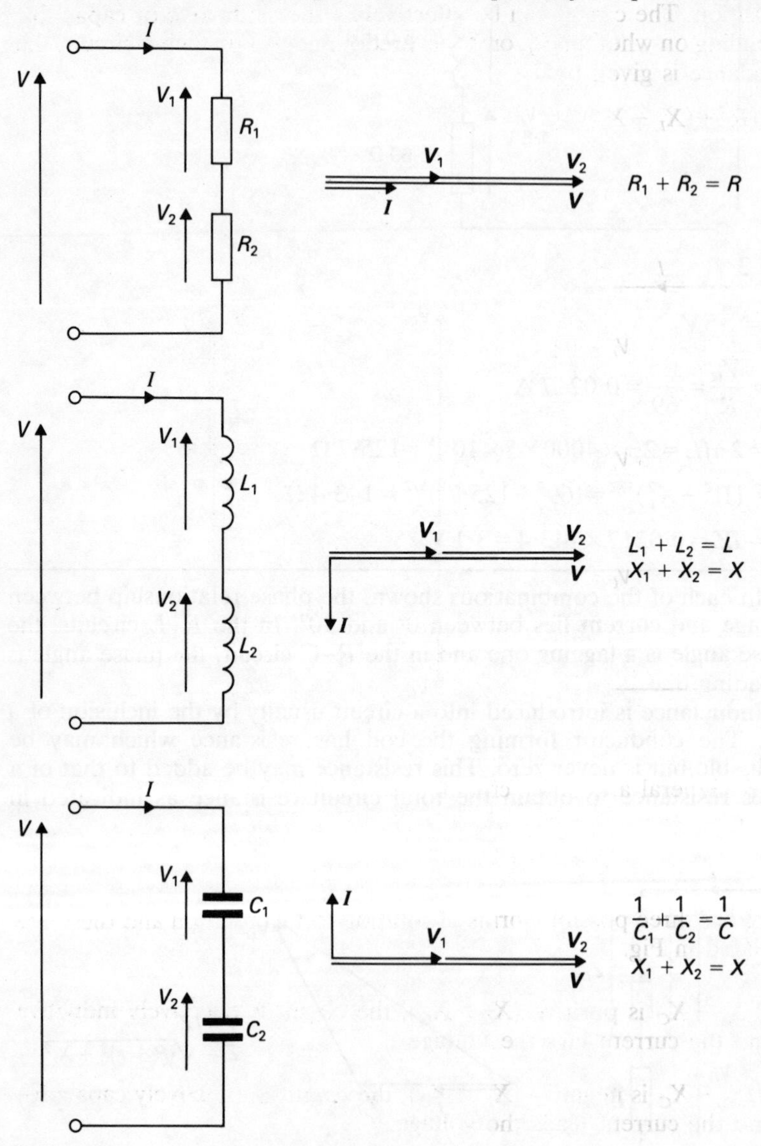

Fig. 5 Series connection of like components

another. It follows that the voltages associated with like series-connected components may be added arithmetically.

Finally there is the general a.c. series circuit containing resistance, inductance and capacitance. In such a circuit, the voltage $\mathbf{V}_L$ across the inductance and the voltage $\mathbf{V}_C$ across the capacitance are in phase opposition. The circuit can be effectively either inductive or capacitive depending on whether $\mathbf{V}_L$ or $\mathbf{V}_C$ is predominant. For such a circuit, the impedance is given by

$$X=(R^2+(X_L-X_C)^2)^{1/2}$$

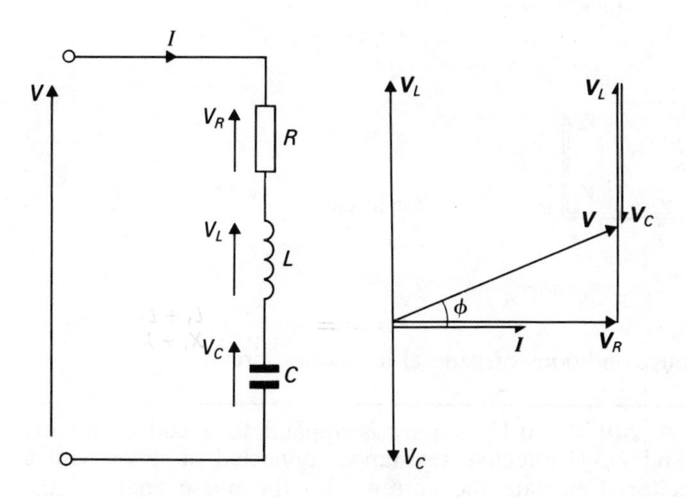

Fig. 6 General a.c. series circuit

There are three possible forms of solution to this relation and these are tabulated in Fig. 7.

1. If X_L-X_C is positive ($X_L>X_C$), the circuit is effectively inductive and the current lags the voltage
2. If X_L-X_C is negative ($X_L<X_C$), the circuit is effectively capacitive and the current leads the voltage.
3. If X_L-X_C is zero ($X_L=X_C$), the circuit is effectively resistive and it can be said to be in resonance.

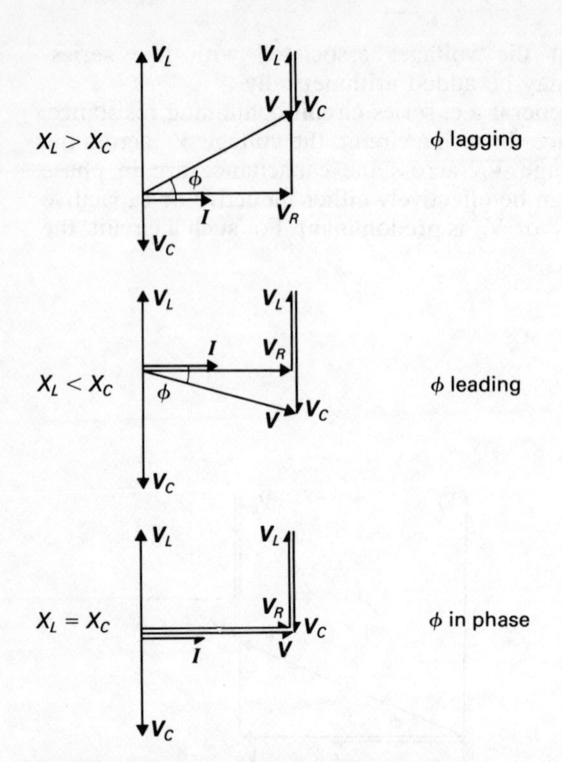

Fig. 7 Various conditions of general a.c. series circuit

Example 3 A 240-V, 50-Hz supply is applied to a coil of 60 mH inductance and 2·5 Ω effective resistance connected in series with a 68-μF capacitor. Calculate the current and the phase angle of the circuit. Also calculate the voltage across each of the circuit components.

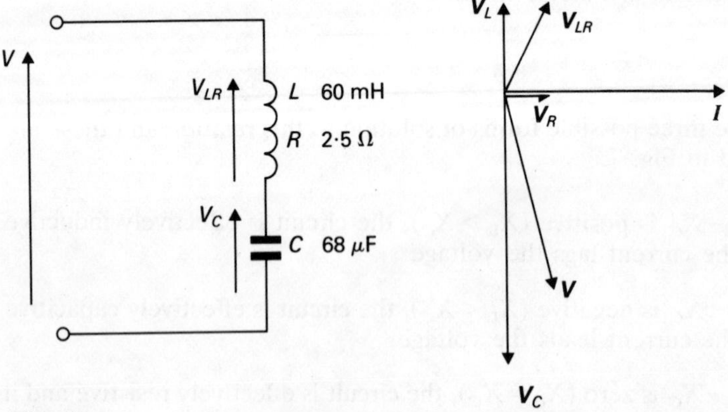

Fig. 8

$$X_L = 2\pi fL = 2\pi \times 50 \times 60 \times 10^{-3} = 18{\cdot}85\ \Omega$$

$$X_C = \frac{1}{2\pi fC} = \frac{1}{2\pi \times 50 \times 68 \times 10^{-6}} = 46{\cdot}80\ \Omega$$

$$X = X_L - X_C = 18{\cdot}85 - 46{\cdot}80 = -27{\cdot}95\ \Omega$$

$$Z = (R^2 + X^2)^{1/2} = (2{\cdot}5^2 + 27{\cdot}95^2)^{1/2} = 28{\cdot}06\ \Omega$$

$$I = \frac{V}{Z} = \frac{240}{28{\cdot}06} = \underline{8{\cdot}55\ \text{A}}$$

$$\cos\phi = \frac{R}{Z} = \frac{2{\cdot}5}{28{\cdot}06} = 0{\cdot}0356$$

$$\phi = \underline{88^\circ\ \text{lead}}$$

$$Z_{LR} = (R^2 + X_L^2)^{1/2} = (2{\cdot}5^2 + 18{\cdot}85^2)^{1/2} = 19{\cdot}1\ \Omega$$

$$V_{LR} = IZ_{LR} = 8{\cdot}55 \times 19{\cdot}1 = \underline{163\ \text{V}}$$

$$V_C = IX_C = 8{\cdot}55 \times 46{\cdot}8 = \underline{400\text{V}}$$

---

2 Impedance

Our understanding of an a.c. circuit is based on the concept of a source and a load which comprises resistance, inductance and capacitance in series. However, in more general terms, we have really considered the impedance Z of a load as seen from its terminals. This understanding is that illustrated in Fig. 9.

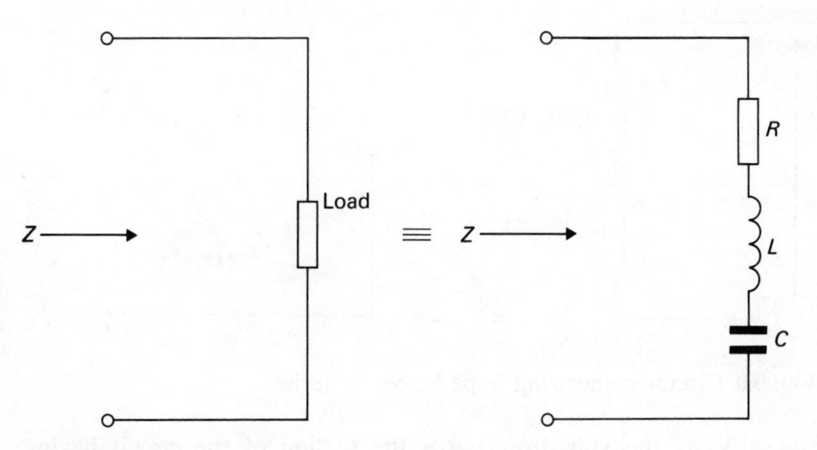

Fig. 9 Circuit impedance

Impedance is the property of a part of a circuit relating the volt drop V across it to the current I, i.e.

$$Z=\frac{V}{I}$$

The term impedance is also used to describe a part of a circuit, thus if we have two or more distinct parts of a circuit connected in series, then we may say that we have series-connected impedances. This is a short method of saying that there are loads, each of particular impedance, connected in series. We ought to be careful to remember that impedance is the property of a part of a circuit and is not the part of the circuit itself. In the same way, resistance is the property of a resistor and is not the resistor itself.

Just as we can have resistances in series, or in parallel, so we can have impedances in series, or in parallel. Let us consider the first of these possibilities further.

3 Series-connected impedances

Series circuits are more complicated when they comprise two or more impedances. Consider two impedances connected in series as shown in Fig. 10. As usual in series-circuit analysis, the phasor diagram is drawn with the current I as reference, and is completed by applying the following relations:

$$V_1=IZ_1$$
$$V_2=IZ_2$$
$$\mathbf{V}=\mathbf{V}_1+\mathbf{V}_2$$

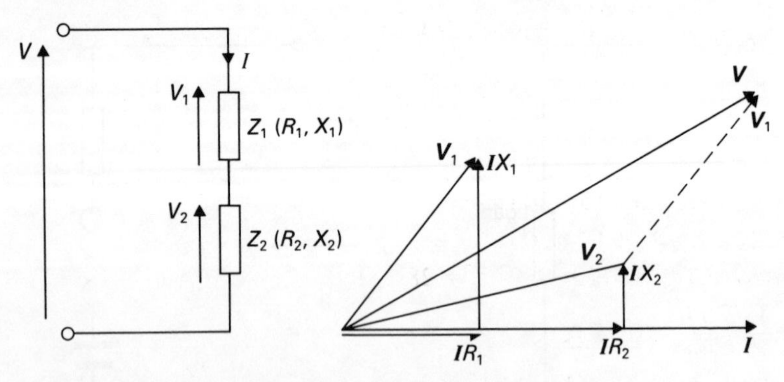

Fig. 10 Circuit comprising impedances in series

where V_1 is the volt drop across the section of the circuit having impedance Z_1, and V_2 is the volt drop across the section of the circuit having impedance Z_2. Remember that the current is the same in all parts of the circuit and therefore can be simply defined as I. By the geometry of the phasor diagram,

$$\begin{aligned} V &= ((IR_1+IR_2)^2+(IX_1+IX_2)^2)^{1/2} \\ &= I((R_1+R_2)^2+(X_1+X_2)^2)^{1/2} \\ &= IZ \end{aligned}$$

where

$$Z=((R_1+R_2)^2+(X_1+X_2)^2)^{1/2} \qquad (1)$$

Thus the circuit impedance is found by collecting the separate resistances and reactances into like groups; this was anticipated by our observations in Section 1. The complicated series circuit is now simplified into one effective impedance equivalent in every way when observed from the circuit terminals.

Example 4 Two coils are connected in series across a 5·0-V sinusoidal a.c. supply. Coil A has resistance 5 Ω and reactance 8 Ω, and coil B has resistance 15 Ω and reactance 12 Ω. Calculate the circuit current and the volt drops across each of the coils.

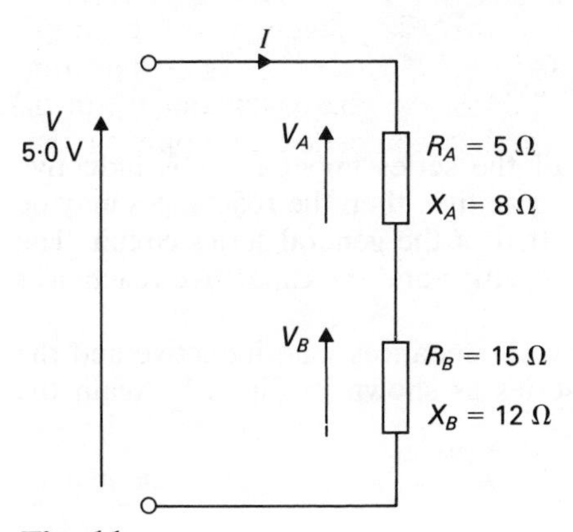

Fig. 11

$$\begin{aligned} R &= R_A+R_B=5+15=20\ \Omega \\ X &= X_A+X_B=8+12=20\ \Omega \\ Z &= (R^2+X^2)^{1/2}=(20^2+20^2)^{1/2}=28{\cdot}3\ \Omega \\ I &= \frac{V}{Z}=\frac{5{\cdot}0}{28{\cdot}3}=\underline{0{\cdot}18\ \text{A}} \\ Z_A &= (R_A^2+X_A^2)^{1/2}=(5^2+8^2)^{1/2}=9{\cdot}4\ \Omega \\ V_A &= IZ_A=0{\cdot}18\times 9{\cdot}4=\underline{1{\cdot}69\ \text{V}} \\ Z_B &= (R_B^2+X_B^2)^{1/2}=(15^2+12^2)^{1/2}=19{\cdot}2\ \Omega \\ V_B &= IZ_B=0{\cdot}18\times 19{\cdot}2=\underline{3{\cdot}46\ \text{V}} \end{aligned}$$

You will observe that if you were to arithmetically add V_A and V_B, you would obtain a total voltage of 5·15 V, which is greater than 5·0 V. This is because the volt drops must be added together by means of a phasor diagram making due allowance for their phase relations with the current. The phasor addition appropriate to Example 4 is shown in Fig. 12.

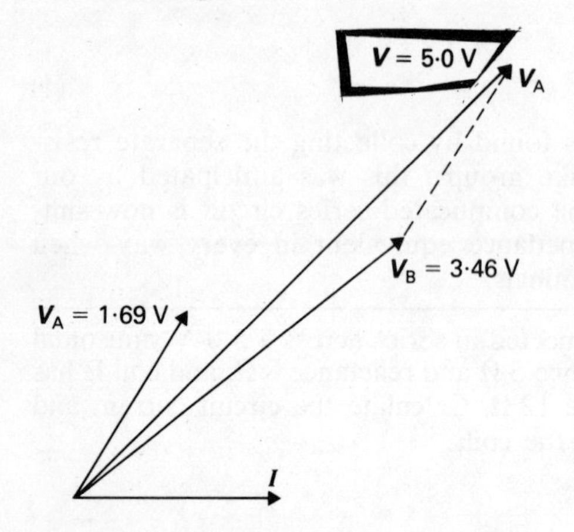

Fig. 12 Addition of voltage phasors

In cases where at least one of the series impedances is inductive and at least one of the others is capacitive, then the reactances may be combined in a similar manner to that of the general series circuit. The inductive reactances are taken as positive and the capacitive reactances as negative.

Let us consider the case of two impedances, one inductive and the other capacitive, connected in series as shown in Fig. 13. Again the impedance is given by

$$Z=((R_1+R_2)^2+(X_1+X_2)^2)^{1/2}$$

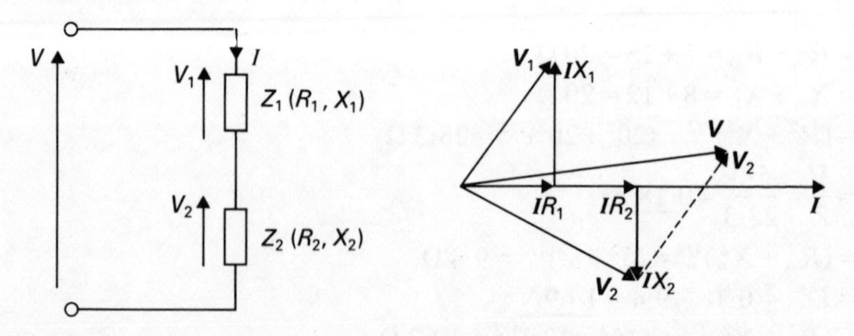

Fig. 13 Circuit comprising dissimilar impedances in series

This may make you wonder that the capacitive reactance X_2 did not give rise to a term $X_1 - X_2$. X_2 has a negative value; thus when we substitute in the term $X_1 + X_2$, we might typically obtain, say, $X_1 + X_2 = 10 - 15 = -5\ \Omega$, the negative sign of the answer indicating that the reactance is predominantly capacitive.

The polarity of the reactance can be indicated more directly when the type of reactance is known; thus in a series circuit containing both inductance and capacitance, the total reactance is given by $X_L - X_C$. In this case, we are letting X_L and X_C be the magnitudes of their respective reactances and the negative sign already indicates the effect of a capacitive reactance.

Thus, to summarise, when X represents a reactance, its value in ohms is positive if the reactance is inductive and negative if it is capacitive. However, when the respective forms of reactance and capacitance are represented by X_L and X_C, the polarity can be expressed separately and only the magnitude of the reactance need concern us.

Example 5 Three impedances are connected in series across a 20-V, 50-kHz a.c. supply. The first impedance is a 10-Ω resistor, the second a coil of 15 Ω inductive reactance and 5 Ω resistance, and the third consists of a 15-Ω resistor in series with a 25-Ω capacitor. Calculate:

(*a*) the circuit current;
(*b*) the circuit phase angle;
(*c*) the impedance volt drops.

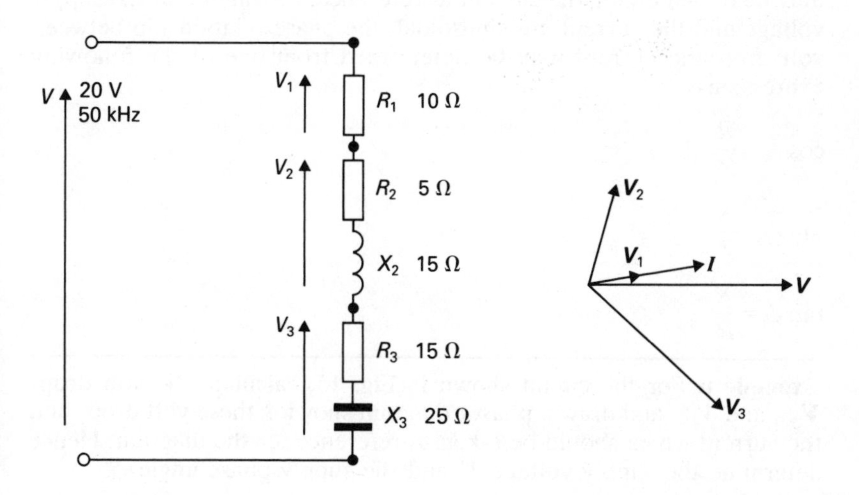

Fig. 14

$$R = R_1 + R_2 + R_3 = 10 + 5 + 15 = 30\ \Omega$$

$$X = X_2 + X_3 = 15 - 25 = -10\ \Omega$$

$$Z = (R^2 + X^2)^{1/2} = (30^2 + 10^2)^{1/2} = 31{\cdot}6\ \Omega$$

$$I = \frac{V}{Z} = \frac{20}{31{\cdot}6} = 0{\cdot}63\ \text{A}$$

$$\cos\phi = \frac{R}{Z} = \frac{30}{31{\cdot}6} = 0{\cdot}95$$

$$\underline{\phi = 18{\cdot}5^\circ\ \text{lead}}$$

$$V_1 = IR_1 = 0{\cdot}63 \times 10 = \underline{6.3\ \text{V}}$$

$$V_2 = IZ_2 = I(R_2^2 + X_2^2)^{1/2} = 0{\cdot}63(15^2 + 5^2)^{1/2} = \underline{10{\cdot}0\ \text{V}}$$

$$V_3 = IZ_3 = I(R_3^2 + X_3^2)^{1/2} = 0{\cdot}63(15^2 + 25^2)^{1/2} = \underline{18{\cdot}4\ \text{V}}$$

It should be stressed that in the last two examples, the determination of the volt drops across each of the component parts was derived from their respective impedances. The impedance between any two points in a circuit is the relation of the voltage to the current, and it is determined by the combination of the complete resistance and the complete reactance between the points of measurement. Only resistances may be added together and reactances may be added together but impedances must not be added arithmetically. The key to the analysis of series-connected impedance circuits lies in the manipulation of adding resistances and adding reactances.

It has already been mentioned that the volt drops across different parts of a circuit may be achieved by means of a phasor diagram. This may be drawn taking the current as reference. Provided that the supply voltage and the current are sinusoidal, the phase relationship between volt drop and current may be determined from one of the following expressions:

$$\cos\phi = \frac{R}{Z}$$

$$\sin\phi = \frac{X}{Z}$$

$$\tan\phi = \frac{X}{R}$$

Example 6 For the circuit shown in Fig. 15, calculate the volt drops $\mathbf{V}_{AB}$ and $\mathbf{V}_{BC}$ and draw a phasor diagram showing these volt drops and the current which should be taken as reference for the diagram. Hence determine the supply voltage V and the supply phase angle ϕ.

$$V_{AB} = IZ_{AB} = I(R_{AB}^2 + X_{AB}^2)^{1/2} = 2(10^2 + 35^2)^{1/2} = 2 \times 36{\cdot}4$$

$$= \underline{72{\cdot}8\ \text{V}}$$

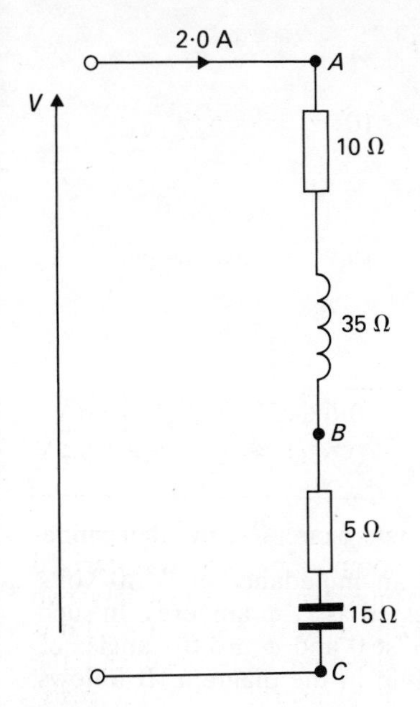

Fig. 15

$$\cos\phi_{AB}=\frac{R_{AB}}{Z_{AB}}=\frac{10}{36\cdot4}=0\cdot27$$

$$\phi_{AB}=74\cdot3°$$

Thus $\mathbf{V}_{AB}$ leads $\boldsymbol{I}$ by 74·3°.

$$V_{BC}=IZ_{BC}=I(R_{BC}^2+X_{BC}^2)^{1/2}=2(5^2+15^2)^{1/2}=2\times15\cdot8$$
$$=\underline{31\cdot6\text{ V}}$$

$$\cos\phi_{BC}=\frac{R_{BC}}{Z_{BC}}=\frac{5}{15\cdot8}=0\cdot32$$

$$\phi_{BC}=71\cdot6°$$

Thus $\mathbf{V}_{BC}$ lags $\boldsymbol{I}$ by 71·6°.

The phasor diagram using this information is shown in Fig. 16. By adding $\mathbf{V}_{AB}$ and $\mathbf{V}_{BC}$, we find that the supply voltage is 50 V and the circuit phase angle is a lagging one of 45°.

---

4 Impedance and polar notation

Sometimes it is necessary to define both the magnitude and also the phase angle of a quantity. By means of polar notation, we may express

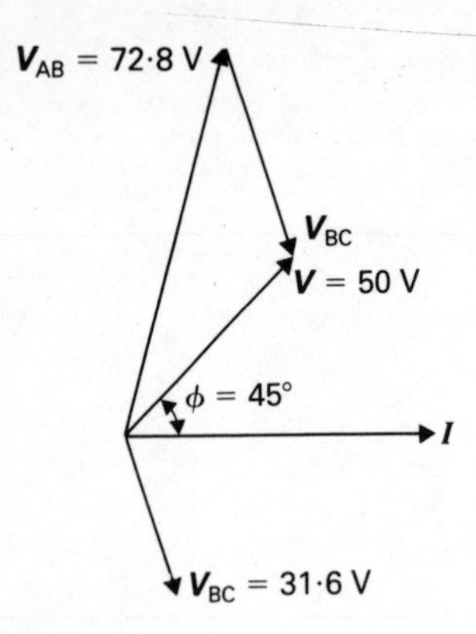

Fig. 16

the voltage, taken as reference, across an impedance as $V\angle 0$ volts whilst the corresponding current is expressed as $I\angle\phi$ amperes. In such an instance, V and I are magnitudes whilst 0 and ϕ are the angles of rotation of the phasors relative to a datum in the diagram. It follows that the impedance is given by

$$\frac{V\angle 0}{I\angle\phi}$$

To simplify this expression, the magnitudes are operated arithmetically thus V/I becomes the magnitude of the impedance, i.e. Z ohms. The angles are subtracted to give $\angle(0-\phi)=\angle-\phi$. Thus the impedance is given by $Z\angle-\phi$ ohms.

Example 7 An impedance passes a current $2{\cdot}5\angle 30°$ A when a voltage $50\angle-15°$ V is applied across it. Determine the impedance in similar form.

$$Z=\frac{\mathbf{V}}{\mathbf{I}}=\frac{50\angle-15°}{2{\cdot}5\angle 30°}=\underline{20\angle-45°\ \Omega}$$

We see that the phase angle is reversed, which is appropriate to earlier assumptions. Thus for a capacitive circuit, the current leads the voltage and ϕ is positive, yet the angle of the reactance is negative and X_C is taken as negative, as was done in Section 3. Conversely the inductive circuit has the current lagging the voltage, in which case ϕ is negative and the reactance becomes positive, which again we have accepted.

Impedance may therefore be expressed in the form $Z\angle-\phi$ ohms. Let us look at this applied to a series impedance circuit as shown in Fig. 17. In this instance, the current has been taken as reference and is

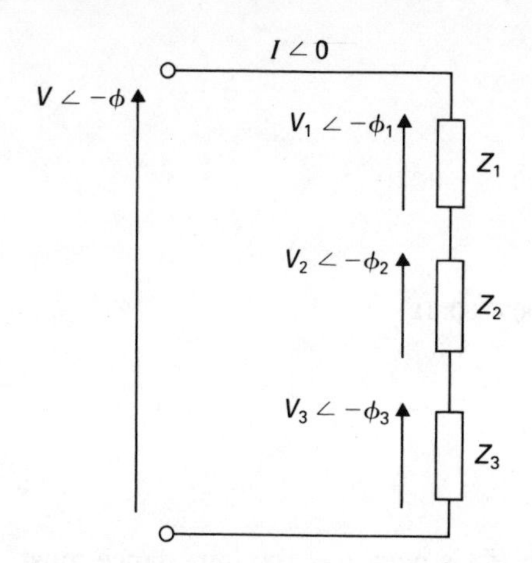

Fig. 17 Polar impedance in series

defined as $I\angle 0$. The phase angle ϕ usually indicates the phase angle of the current relative to the voltage, therefore the voltage has a phase angle $-\phi$ relative to the current. In phasor notation,

$$\mathbf{V} = \mathbf{V}_1 + \mathbf{V}_2 + \mathbf{V}_3$$

In polar notation

$$V\angle -\phi = V_1\angle -\phi_1 + V_2\angle -\phi_2 + V_3\angle -\phi_3$$

$$I\angle 0 \cdot Z\angle -\phi = I\angle 0 \cdot Z_1\angle -\phi_1 + I\angle 0 \cdot Z_2\angle -\phi_2 + I\angle 0 \cdot Z_3\angle -\phi_3$$

$$Z\angle -\phi = Z_1\angle -\phi_1 + Z_2\angle -\phi_2 + Z_3\angle -\phi_3 \quad (2)$$

However from our knowledge of impedance triangles, $R = Z \cdot \cos \phi$ and from relation (2) it may be observed that

$$Z \cdot \cos \phi = Z_1 \cos \phi_1 + Z_2 \cos \phi_2 + Z_3 \cos \phi_3$$

thus $\quad R = R_1 + R_2 + R_3 \quad (3)$

which agrees with our previous understanding of series impedance circuits. Similarly

$$X = X_1 + X_2 + X_3 \quad (4)$$

Example 8 Two impedances $20\angle -45°\ \Omega$ and $30\angle 30°\ \Omega$ are connected in series across a supply and the resulting current is found to be 10 A. Determine the supply voltage and the supply phase angle.

$$R_1 = Z_1 \cos \phi_1 = 20 \cdot \cos -45° = 14{\cdot}1\ \Omega$$
$$R_2 = Z_2 \cos \phi_2 = 30 \cdot \cos 30° = 26{\cdot}0\ \Omega$$
$$R = R_1 + R_2 = 14{\cdot}1 + 26{\cdot}0 = 40{\cdot}1\ \Omega$$

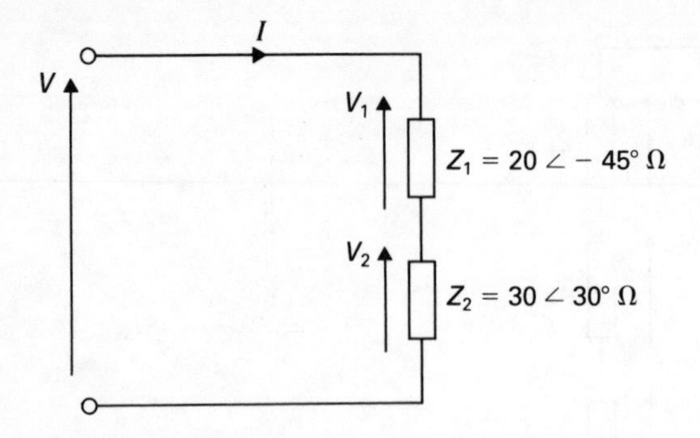

Fig. 18

Because the angle associated with Z_1 is negative, the impedance must be capacitive, thus

$$X_1 = Z_1 \sin \phi_1 = 20 \cdot \sin -45° = -14{\cdot}1\ \Omega$$

$$X_2 = Z_2 \sin \phi_2 = 30 \cdot \sin 30° = 15{\cdot}0\ \Omega$$

$$X = X_1 + X_2 = -14{\cdot}1 + 15{\cdot}0 = 0{\cdot}9\ \Omega$$

$$Z = (R^2 + X^2)^{1/2} = (40{\cdot}1^2 + 0{\cdot}9^2)^{1/2} = 40{\cdot}1\ \Omega$$

$$V = IZ = 10 \times 40{\cdot}1 = \underline{401\ \text{V}}$$

$$\tan \phi = \frac{X}{R} = \frac{0{\cdot}9}{40{\cdot}1} = 0{\cdot}022$$

$$\phi = \underline{1{\cdot}3°}$$

The total reactance is positive and therefore the circuit is inductive. The current therefore lags the supply voltage and the phase angle is $-1{\cdot}3°$.

5 Resonance in a series circuit

When the inductive reactance in a general series circuit is numerically equal to the capacitive reactance, $V_L = V_C$ and consequently V and I are in phase with one another. This is an instance of a condition that is termed resonance – in this case, series resonance because it concerns a series circuit. At resonance, the circuit is said to resonate. The phasor diagram for a series resonant circuit is shown in Fig. 19.

The condition of resonance is used extensively in electronic and communications networks but, prior to explaining the phenomenon of resonance, it is worth further investigation of the a.c. general series circuit in terms of reactances and their respective volt drops being equal.

The a.c. circuits discussed in this chapter have generally had the

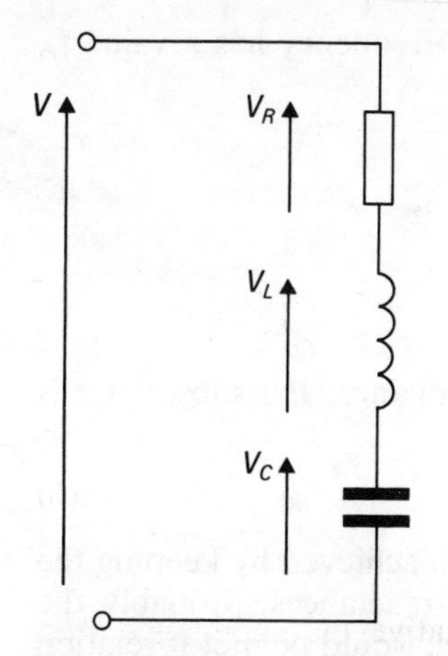

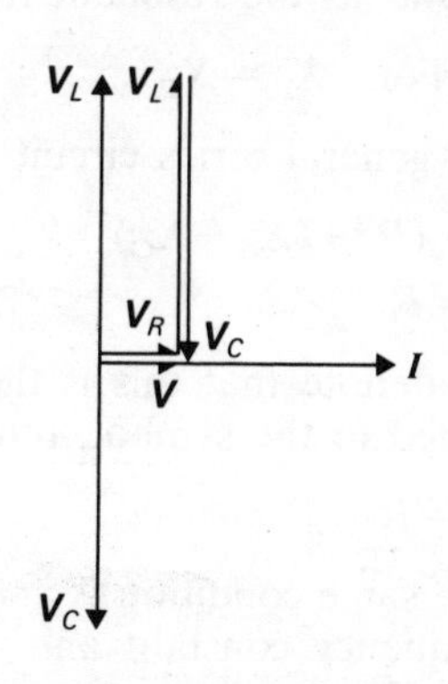

Fig. 19 Series resonant circuit

condition of constant supply frequency. However, many circuits – again particularly in electronic and communications networks – are supplied from variable-frequency sources or from sources supplying a number of frequencies. It is therefore pertinent to consider the effect of variation of supply frequency on the general series circuit.

When the frequency is zero, i.e. corresponding to a d.c. supply, $X_L = 2\pi fL = 0$ and $X_C = 1/2\pi fC = \infty$. As the frequency increases, X_L increases in direct proportion, whilst X_C decreases inversely with increase of frequency. For resonance:

$$V_L = V_C$$
$$IX_L = IX_C$$
$$X_L = X_C \qquad (5)$$

From the graph of reactance against frequency shown in Fig. 20, it is

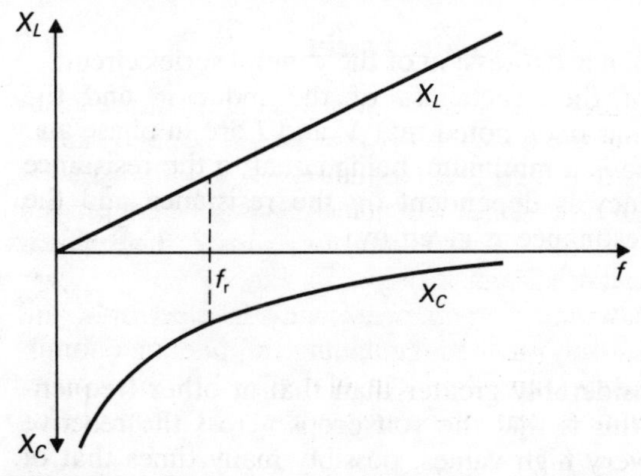

Fig. 20 Variation of reactance with frequency

seen that this condition is satisfied when the frequency has a value f_r, known as the resonant frequency.

For $X_L = X_C$

the general series circuit impedance is given by

$$Z = (R^2 + (X_L - X_C)^2)^{1/2}$$
$$= R$$

To denote that this is the impedance at resonance, the subscript r is added to the symbol, thus

$$Z_r = R \qquad (6)$$

The same condition $X_L = X_C$ could have been achieved by keeping the frequency constant and varying one of the reactances, probably the capacitance. Again the condition of resonance would be met if relation (5) were satisfied.

Thus the two methods of achieving resonance in a series circuit are:

(*a*) vary the frequency until resonance occurs;
(*b*) vary one of the components until resonance occurs.

The resonant frequency can be related to the inductance and the capacitance of the circuit as follows:

$$X_L = X_C$$

$$2\pi f_r L = \frac{1}{2\pi f_r C}$$

$$f_r^2 = \frac{1}{(2\pi)^2 LC}$$

$$f_r = \frac{1}{2\pi (LC)^{1/2}} \qquad (7)$$

This is the resonant frequency expression of the general series circuit.

For the condition of the reactances of the inductor and the capacitor being equal, it has been noted that V and I are in phase and that the circuit impedance is a minimum, being equal to the resistance R. The resonant frequency is dependent on the resistance and the value of the current at resonance is given by

$$I_r = \frac{V}{R}$$

This current may be considerably greater than that at other frequencies. A consequence of this is that the volt drops across the reactive components may reach very high values, possibly many times that of the supply voltage, and Fig. 19 is drawn to illustrate this increase of voltage. It may be added that the power dissipation in the circuit is consequently a maximum, whilst the peak rates of energy storage for the reactions become equal and maximum. The variation of circuit values is illustrated in Fig. 21 by the current and impedance characteristics.

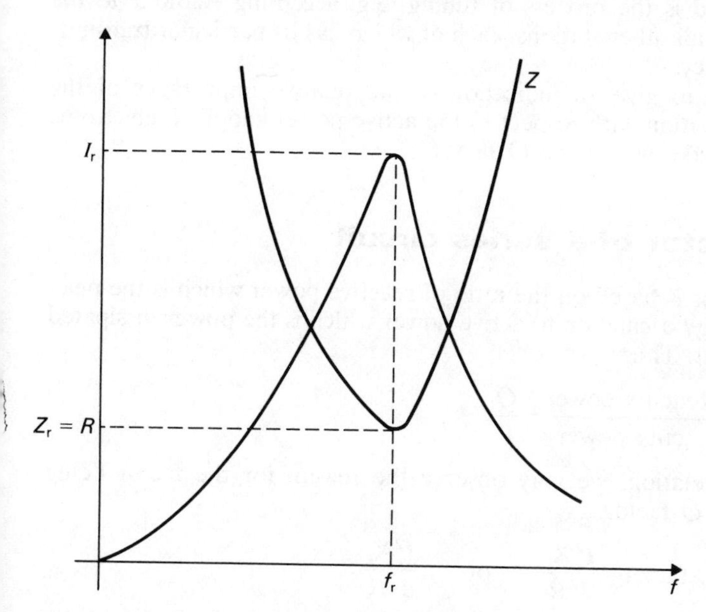

Fig. 21 Series resonance characteristics

Because the peak rates of energy storage in the reactors are appreciable at resonance, we must ask how this relates with the current being in phase with the supply voltage, which implies that the circuit has no reactive power but only the active power, which is the rate of energy dissipation by the circuit resistance.

Effectively the energy stored by the reactors is constant and it oscillates between the electric and magnetic modes of storage, i.e. first it is stored in the inductor then it transfers to the capacitor and then it transfers back again. The energy therefore remains with the reactors and it is this oscillation of energy between them that makes the circuit appear to be purely resistive.

A circuit is said to resonate whenever this effect of energy oscillation predominates; this is usually interpreted to be when the peak rate of energy storage in each of the reactors – given by I^2X – is at least ten times the power associated with the circuit resistance. If this condition is not met, then the circuit is a power-orientated arrangement which just happens to have the current in phase with the voltage.

To the electronics and communications engineers, the degree of predominance of the energy oscillation is important, since the better the ratio of predominance, the better the circuit is able to accept current (and power) at the resonant frequency to the exclusion of other frequencies. This selection of frequency is used in radio and television receivers and is the process of tuning, e.g. accepting Radio 1 to the exclusion of all other stations, each of which has its particular transmitting frequency.

In order to give an indication of the relative importance of the energy oscillation with respect to the active power supplied, electronic engineers make use of the *Q* factor.

6 Q factor of a series circuit

The *Q* factor is based on the ratio of reactive power which is the peak rate of energy oscillation to active power which is the power dissipated in the circuit. Thus

$$Q \text{ factor} = \frac{\text{Reactive power}}{\text{Active power}} = \frac{Q}{P}$$

From this relation, we may observe the reason for the factor being termed the *Q* factor.

$$\text{Thus} \quad Q \text{ factor} = \frac{I^2 X_L}{I^2 R} \quad \text{or} \quad \frac{I^2 X_C}{I^2 R}$$

$$= \frac{X_L}{R} \quad \text{or} \quad \frac{X_C}{R}$$

$$= \frac{\omega_r L}{R} \quad \text{or} \quad \frac{1}{\omega_r CR}$$

At resonance, $X_L = X_C$

and $$Q \text{ factor} = \frac{\omega_r L}{R} = \frac{1}{\omega_r CR} \qquad (8)$$

For a simple *RLC* series circuit, as shown in Fig. 22, the circuit current at resonance is given by

$$I = \frac{V}{R}$$

The voltage across the inductance is given by

$$V_L = IX_L = I\omega_r L$$

$$= \frac{V}{R} \cdot \omega_r L = \frac{\omega_r L}{R} \cdot V$$

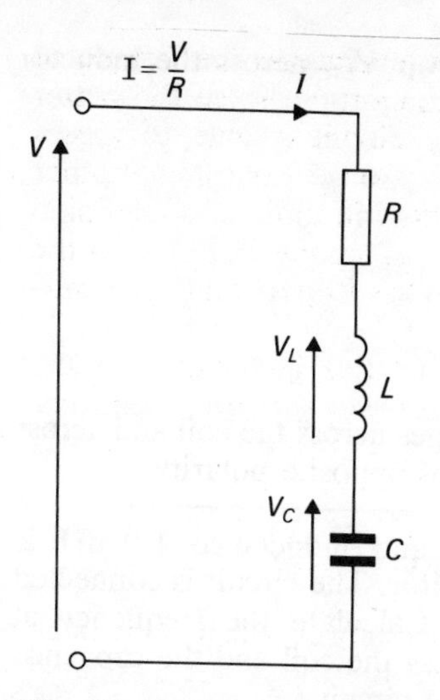

Fig. 22 *RLC* series circuit

But the Q factor is given by

$$Q = \frac{\omega_r L}{R}$$

thus $\quad V_L = QV \qquad (9)$

Q is therefore a factor of magnification of the supply voltage, the voltage across the induction being many times that across the circuit. Similarly in the case of the capacitor,

$$V_C = IX_C = I \cdot \frac{1}{\omega_r C}$$

$$= \frac{V}{R} \cdot \frac{1}{\omega_r C} = \frac{1}{\omega_r CR} \cdot V$$

But the Q factor is given by

$$Q = \frac{1}{\omega_r CR}$$

thus $\quad V_C = QV \qquad (10)$

Q may be very large, especially if R is only the resistance of the inductor coil, and values of Q can be many hundreds. In such cases, the voltage across the capacitor will be many hundreds of times that of the supply voltage. The voltage across the inductor will also be many hundreds of times that of the supply voltage, but generally we cannot observe V_L but rather the overall volt drop V_{LR} across the inductor coil, which is due to the coil impedance.

$$V_{LR} = IZ_{LR} = \frac{V}{R}(R^2+\omega_r^2L^2)^{1/2}$$
$$= V\left(1+\frac{\omega_r^2L^2}{R^2}\right)^{1/2}$$
$$= V(1+Q^2)^{1/2}$$

When Q is much greater than 1, the voltages across the coil and across the capacitor tend to be equal although of opposite polarity.

Example 9 A coil of resistance 5·0 Ω and inductance 1·0 mH is connected in series with an 0·2-μF capacitor. The circuit is connected to a 2·0-V, variable frequency supply. Calculate the frequency at which resonance occurs, the voltages across the coil and the capacitor at this frequency, and the Q factor of the circuit.

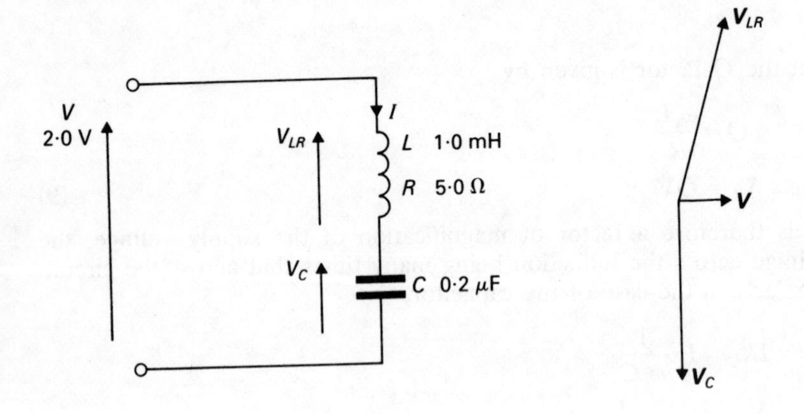

Fig. 23

$$f_r = \frac{1}{2\pi(LC)^{1/2}} = \frac{1}{2\pi(1\times10^{-3}\times0{\cdot}2\times10^{-6})^{1/2}}$$
$$= 11\,250\text{ Hz} = \underline{11{\cdot}25\text{ kHz}}$$

$$I_r = \frac{V}{R} = \frac{2{\cdot}0}{5{\cdot}0} = 0{\cdot}4\text{ A}$$

$$X_L = X_C = 2\pi f_r L = 2\pi\times11\,250\times1\times10^{-3} = 70{\cdot}7\ \Omega$$
$$Z_{LR} = (R^2+X_L^2)^{1/2} = (5{\cdot}0^2+70{\cdot}7^2)^{1/2} = 71{\cdot}0\ \Omega$$
$$V_{LR} = I_rZ_{LR} = 0{\cdot}4\times71{\cdot}0 = \underline{28{\cdot}4\text{ V}}$$
$$V_C = I_rX_C = 0{\cdot}4\times70{\cdot}7 = \underline{28{\cdot}3\text{ V}}$$

The Q factor is the ratio of the voltage across the capacitor to the supply voltage at resonance, thus

$$Q \text{ factor} = \frac{V_C}{V} = \frac{28{\cdot}3}{2{\cdot}0} = \underline{14{\cdot}15}$$

The magnification of the supply voltage by the circuit at resonance is therefore just over 14 times. This value could also have been obtained from

$$Q \text{ factor} = \frac{\omega_r L}{R} = \frac{70{\cdot}7}{5{\cdot}0} = \underline{14{\cdot}15}$$

---

7 Series resonance in practice

In order that the validity of the deductions concerning series resonance may be observed, it is possible to arrange an investigation using the circuit shown in Fig. 24.

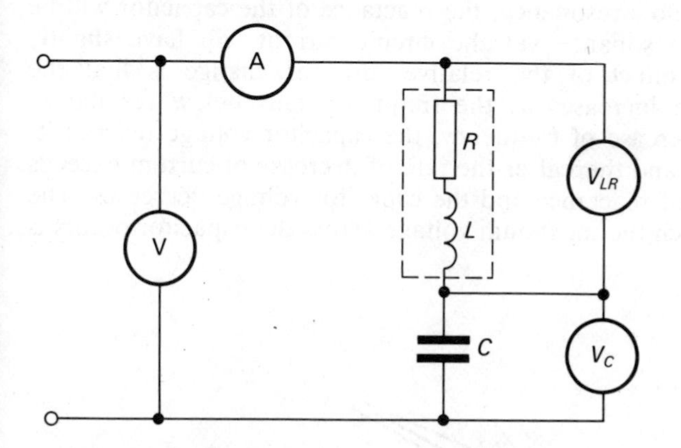

Fig. 24 Investigation of series resonance

Let the supply voltage be constant and the resistance, inductance and capacitance of the circuit be fixed. It remains possible to vary the supply frequency and to measure the variations of current and of the voltages across the coil and across the capacitor. The results may be plotted to give the characteristics shown in Fig. 25.

For a circuit with a high value of Q (say over 100), the maximum volt drop across the coil and the maximum volt drop across the capacitor coincide with the maximum circuit current at the resonant frequency f_r. However, if a circuit of low Q (say under 10) is used, we observe that the maximum voltage across the capacitor occurs at a frequency less than f_r, whilst the maximum voltage across the coil

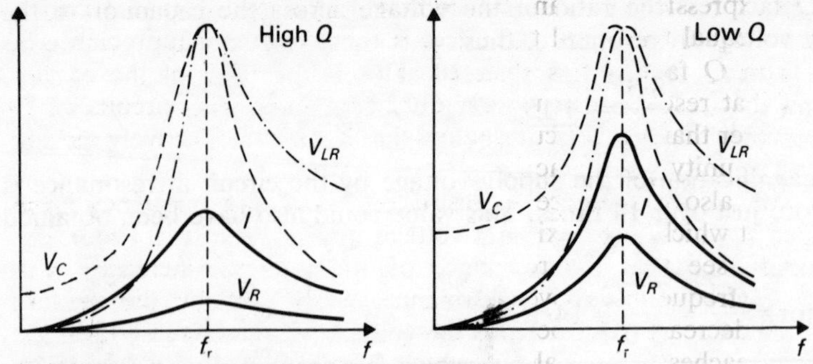

Fig. 25 Series resonance characteristics

occurs at a higher frequency than f_r. The maximum current however still occurs at the resonant frequency.

In order to explain this phenomenon, it is necessary to consider again the variation of reactance and impedance with frequency as shown in Fig. 26. The reactance of the capacitor decreases rapidly with increase of frequency, at least up to the resonant frequency. Thus at a frequency just below resonance, the reactance of the capacitor will be greater than at resonance yet the circuit current will have slightly decreased. The effect of the relative rates of change is that the capacitor voltage increases as the frequency falls below resonance. After a small decrease of frequency, the capacitor voltage reaches its maximum value, and thereafter the rate of decrease of current exceeds that of increase of reactance and the capacitor voltage decreases. The frequency at which the maximum voltage across the capacitor occurs is given by

$$f = f_r\left(1 - \frac{1}{2Q^2}\right)^{1/2}$$

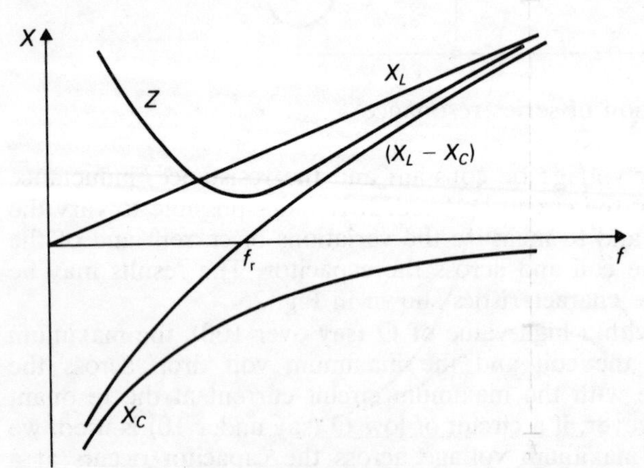

Fig. 26 Reactance and impedance variation with frequency

From this expression, we can deduce that if Q is greater than 10, then f is almost equal to f_r, and the effect is therefore only appreciable in circuits with Q factors less than 10. This is the basis of the earlier comment that resonance is usually only connected with circuits of Q factors greater than 10 – circuits below that value are effectively merely operating at unity power factor.

The coil also experiences a displacement of the frequency from resonance at which the maximum voltage appears across it. From Fig. 31 you can see that the reactance of the inductor increases with increase of frequency above resonance, whilst initially the current shows little decrease. This permits the voltage V_{LR} to increase until the frequency reaches a level above which further increase in frequency causes the rate of decrease in current to outweigh the rate of increase of reactance. It will be noted however that it is also affected by the coil resistance and this in turn depends on the Q factor.

Thus for series circuits in which the Q factor would be less than about 10, greater magnification of the supply voltage appears across the capacitor and across the coil at frequencies other than that of resonance, and slightly higher values of magnification than Q are therefore experienced. This is associated with circuits however that have such poor values of Q that they are more accurately circuits operating at unity power factor – a condition more appropriate to power systems than to electronics and communications systems.

Now consider how the voltage magnification varies with frequency. Because we have already obtained the characteristic of V_C/f shown in Fig. 25 from an investigation in which the supply voltage V was constant, it is therefore possible to produce the characteristic of voltage magnification (V_C/V) against frequency as shown in Fig. 27.

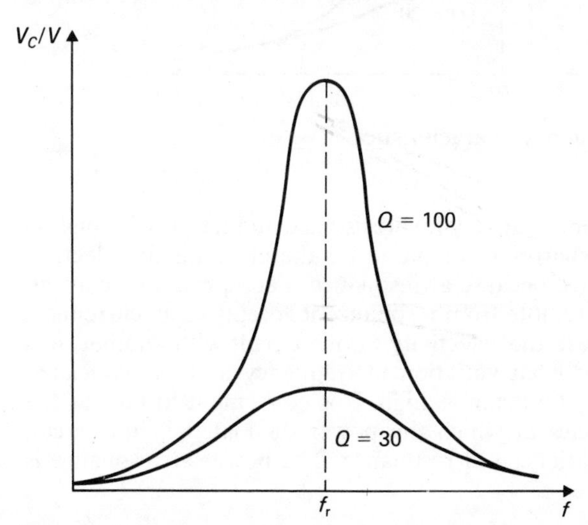

Fig. 27 Voltage magnification/frequency characteristics

By changing the resistance of the circuit yet maintaining the same values of inductance and capacitance, we can determine other characteristics as shown. With a high value of Q, it is seen that the voltage magnification is very much greater at and near to the resonant frequency than at other values. The *RLC* series circuit therefore appears to be selective in that it responds better to a narrow band of frequencies about the resonant frequency than to frequencies outwith that band.

In producing the characteristics of Fig. 27, we could also have produced similar (although not identical in shape) characteristics of the variation of current with frequency as shown in Fig. 28. At resonance, the maximum value of current is determined by the circuit resistance, but at frequencies remote from resonance, the current is decreasingly dependent on the resistance.

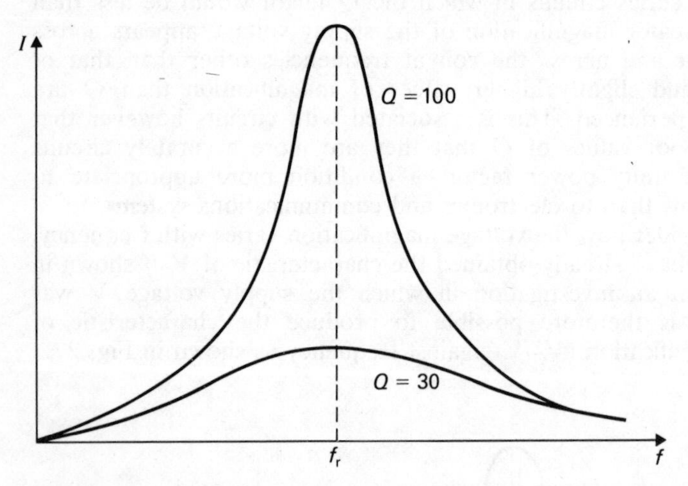

Fig. 28 Current/frequency characteristics

In circuits with comparatively low resistance and hence with high Q factor, the circuit is sharply resonant and is therefore highly selective. It is said to be selective because at resonance it accepts a high current, whilst at frequencies remote from resonance it accepts small currents.

In order to compare the selectivity of one circuit with another, it is necessary to consider what variation of frequency may be associated with current that may be taken as high. It is convenient to take as the limits those frequencies at which the power dissipated in the circuit falls to half that experienced at resonance. The power at resonance is given by

$$P_r = I_r^2 R$$

For the power to be half, then

$$\tfrac{1}{2}P_r = I^2R$$

$$\tfrac{1}{2}I_r^2R = I^2R$$

$$I = \frac{1}{\surd 2} \cdot I_r$$

Let the frequencies at which $I = (1/\surd 2)I_r$ be f_1 and f_2 as shown in Fig. 29. The range of frequencies given by $f_2 - f_1 = \Delta f$ is termed the bandwidth and it may be shown that

$$Q \text{ factor} = \frac{f_r}{\Delta f}$$

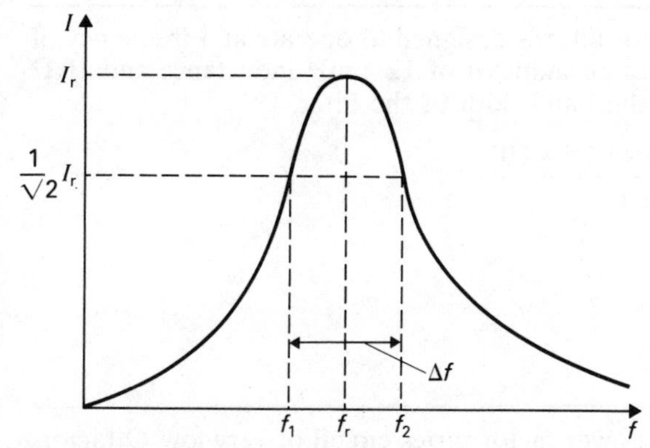

Fig. 29 Derivation of bandwidth

Thus the higher the Q factor, the less is the bandwidth and the more selective is the circuit. This property is most useful to communications engineers in particular in a circuit that can be arranged to accept current within a small range of frequency and effectively reject current at all other frequencies. A circuit that makes such a selection is termed a filter and, because this filter accepts current within the bandwidth, it is termed an acceptor filter.

In communications systems, many signals can pass along a transmission line, each signal at a different frequency. An acceptor filter accepts a signal at a given frequency and rejects all others because their frequencies lie outside the bandwidth. As previously noted, this is part of the process of accepting Radio 1 and rejecting all other radio signals.

Before leaving the use of resonance in communications and other light current applications, it is worth noting why the point of half power has been accepted as the criterion for bandwidth. It has been found that in reproducing sound, we can make it louder or quieter, as we can readily appreciate. However if we were to record three sounds at frequencies f_r, f_1 and f_2 respectively and all originally of equal power, and we were to then reproduce the sounds such that the power reproduction of the sounds at f_1 and f_2 were only half that of the sound at f_r, we would find that remarkably few people would notice the difference in reproduction. This result is rather surprising but it also indicates the quality required of radios, tape recorders, etc. We can also observe that if another originally identical sound in terms of power were recorded at a frequency f_3 lying outside of the bandwidth, then the reproduced sound would be at a power less than half that of the sound at f_r. The result would be noticed by almost everyone, thus confirming that variations of reproduction must lie within the limits of half power. These observations were made in the earliest days of the telephone and the criterion remains one of the most important.

Example 10 An acceptor filter is designed to operate at a frequency of 15 kHz and incorporates an inductor of 12·5 mH inductance and 10 Ω resistance. Determine the bandwidth of the filter.

$$Q = \frac{\omega_r L}{R} = \frac{2\pi \times 15\,000 \times 12{\cdot}5 \times 10^{-3}}{10}$$

$$= 118$$

$$= \frac{f_r}{\Delta f} = \frac{15\,000}{\Delta f}$$

$$\Delta f = \underline{127 \text{ Hz}}$$

Returning to the unity power factor series circuit of very low Q factor, we find that voltage magnification can also be a nuisance as well as an aid. In power applications, a power factor approaching unity minimises the size of cable required and minimises the supply losses. However if we have a circuit that would have even a Q factor of 1·5, this would result in the voltage across the capacitor being 1·5 times that of the supply voltage and therefore the conductor joining the capacitor to the inductive coil requires extra insulation. Power engineers must therefore be aware of this problem and bear in mind the need for such extra insulation. One common instance of extra insulation occurs in capacitor-start induction motors used in domestic appliances such as washing machines; the wire joining the capacitor to the motor generally requires extra insulation even although the circuit is not quite at unity power factor but is near enough to experience voltage magnification.

Example 11 The start winding of an induction motor has an effective resistance of 200 Ω and inductance 1·27 H and is connected in series with an 8-μF capacitor. Given that the circuit is supplied from a 240-V, 50-Hz source, determine the rating of the insulation for the capacitor.

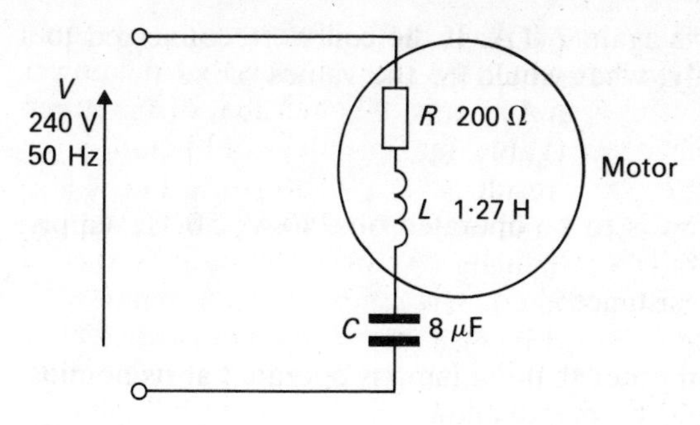

Fig. 30

$$X_L = 2\pi fL = 2\pi \times 50 \times 1{\cdot}27 = 400\ \Omega$$

$$X_C = \frac{1}{2\pi fC} = \frac{1}{2\pi \times 50 \times 8 \times 10^{-6}} = 400\ \Omega$$

It follows that the circuit operates at unity power factor, thus

$$I = \frac{V}{R} = \frac{240}{200} = 1{\cdot}2\ \text{A}$$

$$V_C = IX_C = 1{\cdot}2 \times 400 = 480\ \text{V}$$

Thus the capacitor must be insulated for 480 V.

Problems

1. A ferromagnetic-cored coil connected to a 100-V, 50-Hz sinusoidal supply is found to take a current 5·0 A and to dissipate 200 W. Find:
   (*a*) the impedance;
   (*b*) the effective resistance;
   (*c*) the inductance;
   (*d*) the power factor of the circuit.
2. A coil takes a current 10·0 A at a power factor 0·6 lagging when connected to a 200-V, 50-Hz supply. Calculate the inductance and resistance of the coil. What would be the power factor if the resistance of the coil were doubled?
3. A coil is connected in series with a non-reactive resistor of 30 Ω across a 240-V, 50-Hz supply. The indication on a voltmeter connected across the coil is 180 V and across the resistance is 130 V. Calculate the power dissipated in the coil.
4. When a coil is connected to a 20-V d.c. source, it takes a current of 2·0 A. When the same coil is connected to a 150-V, 60-Hz

supply, the current is again 2·0 A. If the coil were connected to a 240-V, 50-Hz supply, what would be the values of:
(*a*) the current;
(*b*) the power;
(*c*) the power factor?

5. A 110-V, 60-W lamp is to be operated on 240-V, 50-Hz supply. Find what values of:
(*a*) non-inductive resistance;
(*b*) pure capacitance
would be required in order that the lamp is operated at its nominal voltage. Which would be preferable?
6. A load consists of a capacitor and a resistor connected in series and has an impedance of 50 Ω and a power factor of 0·707 leading. The load is connected in series with a 40-Ω resistor across an a.c. supply and the resulting circuit current is 3·0 A. Determine the supply voltage and the circuit phase angle.
7. An inductor of inductance 0·2 H and resistance 15 Ω is connected in series with another inductor of inductance 0·05 H and resistance 30 Ω. The circuit is supplied from a 240-V, 50-Hz supply. Determine:
(*a*) the voltage across each induction;
(*b*) the power dissipated by each induction;
(*c*) the circuit power factor.
8. Explain the meaning of the term 'power factor' of an a.c. circuit. An inductor of inductance 2·5 H and resistance 50 Ω is connected in series with a 150-Ω resistor to a 50-Hz supply. Determine the power factor of the circuit.
9. A series circuit comprises two impedances Z_A and Z_B. The circuit current is $2{\cdot}0\angle 0°$ A and the voltages across the impedances are $V_A = 20\angle 0°$ V and $V_B = 15\angle -90°$ V. By means of a phasor diagram drawn to scale, determine the supply voltage.
If the circuit operates at 50 kHz, determine the form and value of each of the impedances and express the circuit total impedance in the form $Z\angle -\phi$ ohms.
10. An impedance A takes a current $5{\cdot}0\angle 45°$ A when connected to a source of $110\angle 0°$ V. Another impedance B takes a current $2{\cdot}5\angle -60°$ A when connected to the same supply. The impedances are now connected in series to the supply. Calculate the supply current and express your answer in the polar form used above.
11. A 3180-pF capacitor is connected in series with a coil of inductance 63·6 mH and resistance 1·0 kΩ. A 10-kHz sinusoidally-varying voltage 100 mV is applied across the circuit. Calculate:
(*a*) the current taken from the supply;
(*b*) the voltage across the coil;
(*c*) the phase angle of the coil voltage relative to the supply current;
(*d*) the phase angle of the current.

12. Draw and explain the phasor diagram for a series circuit containing resistance, inductance and capacitance when connected to a sinusoidal supply, and hence derive an expression for the impedance of such a circuit.

    A coil is connected in series with a 150-μF capacitor across a 240-V, 50-Hz supply. Determine the resistance and inductance of the coil if the circuit current is 10·0 A lagging the voltage by 37°.
13. An inductor of inductance 0·6 H and resistance 40 Ω is connected in series with a resistor of resistance 80 Ω to a 10-V, 60-Hz supply. Determine the power factor and the power dissipated in the circuit.

    What component connected in series with the circuit would change the power factor to 0·8, and what would be the rating of the component?
14. A circuit takes an active power 4·2 kW at a power factor 0·6 lagging. Find the values of the apparent power and the reactive power.
15. An alternating current flowing through an inductive circuit consists of an active component 7·2 A and a reactive component 5·4 A. The supply voltage is 200 V. Find:
    (*a*) the value of the supply current;
    (*b*) the power factor;
    (*c*) the power dissipated.
16. A circuit comprises 20-Ω resistance, 150-mH inductance and 100-μF capacitance connected in series to a 240-V, 50-Hz supply. Calculate:
    (*a*) the active and reactive components of the current;
    (*b*) the power factor;
    (*c*) the current.
17. A capacitor of negligible resistance, when connected to a 220-V, variable-frequency supply takes a current of 10·0 A when the supply frequency is 50 Hz. A non-inductive resistor when connected to the same supply takes a current of 12·0 A. Given that the two are connected in series and placed across the same supply, calculate:
    (*a*) the current taken and its phase angle when the supply frequency is 50 Hz;
    (*b*) the supply frequency when the circuit current is 8·0 A.
18. Two identical coils each have self inductance 2·4 mH and resistance 10 Ω. Their fields are mutually linked with a coupling coefficient of 0·7. The effect of the mutual inductance is additive and the coils are connected in series to a 15-V, 4-kHz supply. Determine the current and the power factor of the circuit.

    If the coils were connected such that the mutual effects were in opposition, what would be the current and power factor?
19. Deduce an expression for the resonant frequency f_r in terms of L and C for a series RLC circuit. Draw for the resonant condition a

phasor diagram to show the phase relation between the current I, the sinusoidal supply voltage V and the voltages V_R, V_L and V_C across R, L and C respectively.

An RLC series circuit consists of a coil of inductance L and resistance C in series with a 10-μF capacitor. Connected to a 200-V variable-frequency sinusoidal supply, the circuit takes a maximum current of 20 A when the supply frequency is 1 kHz. Determine:

(a) the voltage across the capacitor;
(b) the resistance and inductance of the coil.

20. An air-cored coil of inductance L and resistance R is connected in series with a 100-μF capacitor across a 200-V, variable-frequency sinusoidal supply. The current in the circuit is a maximum when the supply frequency is 80 Hz and is 9·0 A at 60 Hz. Determine the values of L and R.
21. An RLC series circuit comprises inductance 0·5 H and capacitance 0·5 μF and is supplied from a variable-frequency source. Given that the frequency varies from 0 to 1 kHz, draw, on the same axes, the reactance/frequency characteristics for the circuit and hence explain series resonance.

    The resistance of the circuit is 16 Ω. Calculate the resonant frequency and hence determine the Q factor of the circuit at resonance.
22. An RLC series circuit is connected to a 0·2-V, variable-frequency source. The maximum current in the circuit is obtained when the supply frequency is 5 kHz, and the current is then 5·0 mA. It is also found that the Q factor of the circuit under these conditions is 120. Calculate:
    (a) the voltage across the capacitor;
    (b) the values of R, L and C.
23. A resistive coil is connected in series with a variable capacitor across a constant-voltage, 100-kHz supply. Maximum current in the circuit is obtained when the capacitance of the capacitor is 700 pF and the current is reduced to 0·707 of the maximum value when the capacitance is 665 pF. Find:
    (a) the inductance and resistance of the coil;
    (b) the Q factor of the coil at 100 kHz.
24. A circuit consisting of a coil of resistance 100 Ω and inductance 0·15 H in series with a 30-μF capacitor operates at unity power factor when connected to a 250-V sinusoidal supply. Determine:
    (a) the supply current;
    (b) the voltage across each component;
    (c) the energy dissipated by the circuit in 5 min.
25. A coil whose resistance is not negligible is connected in series with a capacitor across a variable-frequency supply, the frequency being adjusted until the current has a maximum value 0·5 A. The supply voltage is 20 V and the frequency is 318 Hz. Calculate the

resistance and inductance of the coil given that the voltage across the capacitor is 50 V.

If the frequency is doubled, calculate the change in capacitance required to keep the current at 0·5 A, the supply voltage remaining constant.

Answers

1. 20 Ω; 8 Ω; 58 mH; 0·4 lag
2. 12 Ω; 51 mH; 0·83 lag
3. 138 W
4. 3·8 A; 147 W; 0·16 lag
5. 239 Ω; 8·13 μF
6. 250 [illegible]°
7. 171 V; 90 V; 105 W; 211 W; 0·5 lag
8. 0·2[illegible]
9. 25∠−37° V; 0·42 μF; 12·5∠−37° Ω
10. 2·36∠-[illegible] A
11. 70 μA; 288 mV; −76°; +45°
12. 19·2 Ω; 113 mH
13. 0·47 lag; 0·18 W; 19·5 μF; 0·6 var
14. 7·0 kVA; 5·6 kvar
15. 9·0 A; 0·8 lag; 1440 W
16. 7·6 A; 5·8 A; 0·79 lag; 9·5 A
17. 7·7 A; 50°; 53·6 Hz
18. 73 mA; 0·10 lag; 362 mA; 0·48 lag
19. 318 V; 10 Ω; 2·53 mH
20. 39·6 mH; 19 Ω
21. 316 Hz; 62
22. 24 V; 40 Ω; 153 mH; 6620 pF
23. 3·6 mH; 120 Ω; 18·9
24. 2·5 A; 306 V; 177 V; 187·5 kJ
25. 40 Ω; 50 mH; 3·75 μF

Chapter 3

Simple alternating current networks

In practice, most alternating current systems consist of a number of circuits connected in parallel to form a network. The branches of the parallel network may each consist of one component or, possibly, two or more components connected in series. So far, we are able to deal with any such circuit, from the simple one with only one component to the complicated series circuit with several components. In order to progress to parallel connected systems, it is important to appreciate that a parallel-connected network merely consists of circuits connected in parallel and that, as a consequence, all we need to do is to add the currents together in order to obtain an understanding of parallel-connected networks.

1 Simple circuits in parallel

There are three possible arrangements of simple circuits in parallel and we require to analyse these. The three arrangements are R in parallel with L, R in parallel with C and finally L in parallel with C.

The network for resistance and inductance in parallel is shown in

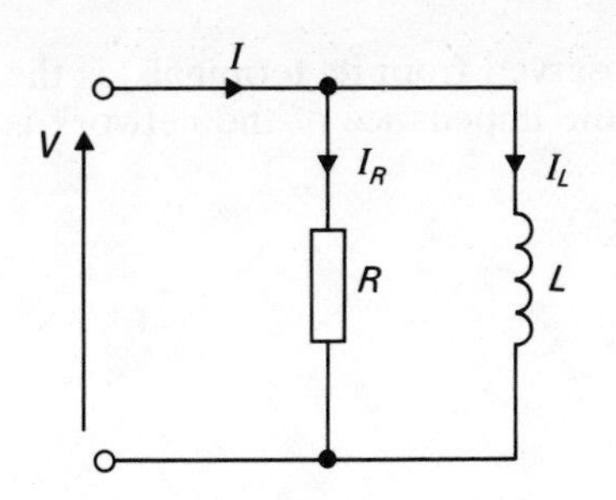

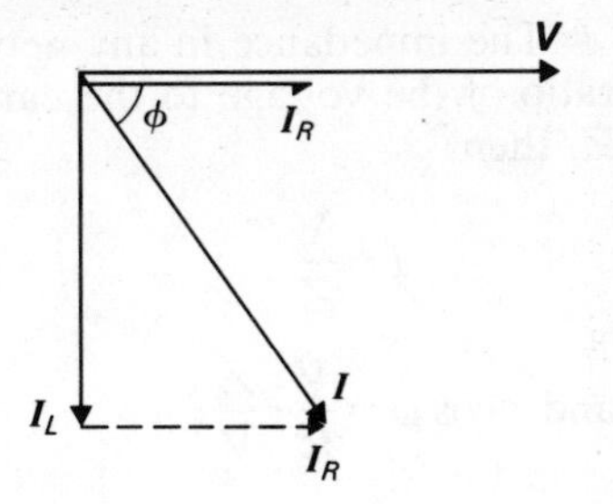

Fig. 1 Resistance and inductance in parallel

Fig. 1. In the resistive branch, the current is given by

$$I_R = \frac{V}{R}$$

where I_R and V are in phase.

In the inductive branch, the current is given by

$$I_L = \frac{V}{X_L}$$

where $\boldsymbol{I}_L$ lags $\boldsymbol{V}$ by 90°.

Unlike the series circuits in which the current was common to all parts of the circuit, the parallel network must take the supply voltage as the common reference since it is applied to each of the network branches. Thus taking the voltage which is common to both branches as reference, we can construct the phasor diagram shown in Fig. 1. The total supply current $\boldsymbol{I}$ is obtained by adding the branch currents on the phasor diagram, i.e.

$$\boldsymbol{I} = \boldsymbol{I}_R + \boldsymbol{I}_L \tag{1}$$

It can be seen from the phasor diagram that the phase angle ϕ is a lagging angle such that

$$\tan\phi = \frac{I_L}{I_R} \tag{2}$$

$$= \frac{V}{X_L} \cdot \frac{R}{V}$$

$$= \frac{R}{X_L} = \frac{R}{2\pi fL} \tag{3}$$

Also $$\cos\phi = \frac{I_R}{I} \tag{4}$$

The impedance in any network, observed from its terminals, is the ratio of the voltage to the current. If the impedance of the network is Z, then

$$I = \frac{V}{Z}$$

and $$\cos\phi = \frac{V}{R} \cdot \frac{Z}{V}$$

$$= \frac{Z}{R} \quad (5)$$

Particularly relation 5 makes us aware that there are problems in changing from series circuits to parallel networks. After all, $\cos\phi$ for the series circuit is given by R/Z, yet for the parallel network it is given by Z/R. Thus at an early stage in our studies of parallel arrangements, we learn that the relations of the previous chapter must be used in connection with series circuits only.

Another similar difficulty may be seen from the following analysis:

From the phasor diagram in Fig. 1

$$I = (I_R^2 + I_L^2)^{1/2}$$

$$= \left(\left(\frac{V}{R}\right)^2 + \left(\frac{V}{X_L}\right)^2\right)^{1/2}$$

$$= V\left(\frac{1}{R^2} + \frac{1}{X_L^2}\right)^{1/2}$$

$$Z = \frac{V}{I}$$

$$= \frac{1}{\left(\frac{1}{R^2} + \frac{1}{X_L^2}\right)^{1/2}} \quad (6)$$

Clearly this expression is quite different from that for the impedance of an R–L series circuit, and also it is considerably more difficult to manipulate. From this observation, we can make two conclusions which are important if we are to make parallel-network calculations easy;

1. The relation $Z = (R^2 + X_L^2)^{1/2}$ applies only to series circuits and should not be used for parallel networks.
2. The expression for the impedance of a parallel arrangement is rather complex and it is therefore much easier to give our attention to manipulating the addition of the branch currents. The network impedance is always easily derived from the ratio of V to I.

Example 1 A 20-Ω resistor is connected in parallel with an inductor of 318-μH inductance and negligible resistance to a 1·0-V, 5-kHz supply. Determine:

(*a*) the supply current;
(*b*) the network impedance;
(*c*) the supply phase angle.

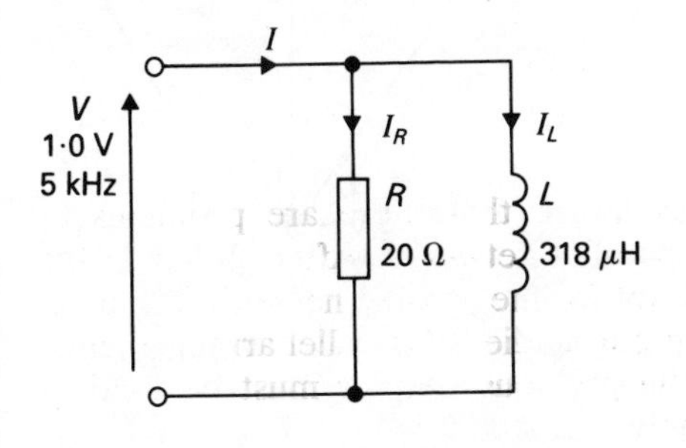

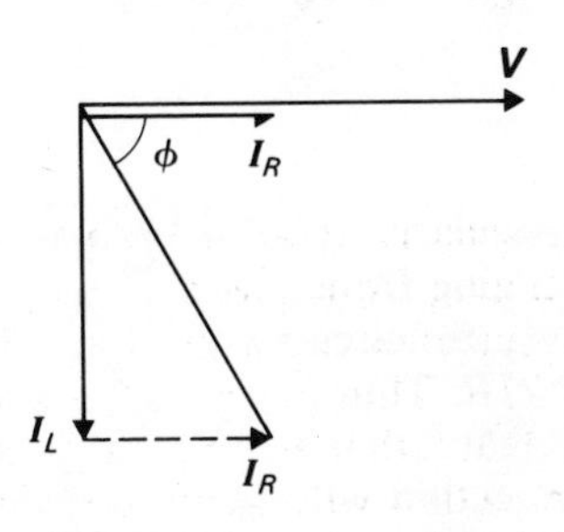

Fig. 2

$$X_L = 2\pi fL = 2\pi \times 5 \times 10^3 \times 318 \times 10^{-6} = 10\ \Omega$$

$$I_L = \frac{V}{X_L} = \frac{1{\cdot}0}{10} = 0{\cdot}10\text{ A}$$

$$I_R = \frac{V}{R} = \frac{1{\cdot}0}{20} = 0{\cdot}05\text{ A}$$

$$I = (I_R^2 + I_L^2)^{1/2} = (0{\cdot}05^2 + 0{\cdot}10^2)^{1/2} = \underline{0{\cdot}11\text{ A}}$$

$$Z = \frac{V}{I} = \frac{1{\cdot}0}{0{\cdot}11} = \underline{8{\cdot}9\ \Omega}$$

$$\cos\phi = \frac{I_R}{I} = \frac{0{\cdot}05}{0{\cdot}11} = 0{\cdot}47$$

$$\underline{\phi = 62^\circ\text{ lag}}$$

In Example 1, the solution to the problem was performed by applying trigonometry to the phasor diagram. However we could also have used a phasor diagram drawn to scale. Let us look at another problem and solve it by means of a scale diagram.

Example 2 A 50-Ω resistor is connected in parallel with a coil of negligible resistance to a 100-V, 50-Hz supply and the supply current is found to be 3·5 A. By means of a phasor diagram drawn to scale, find the inductance of the coil.

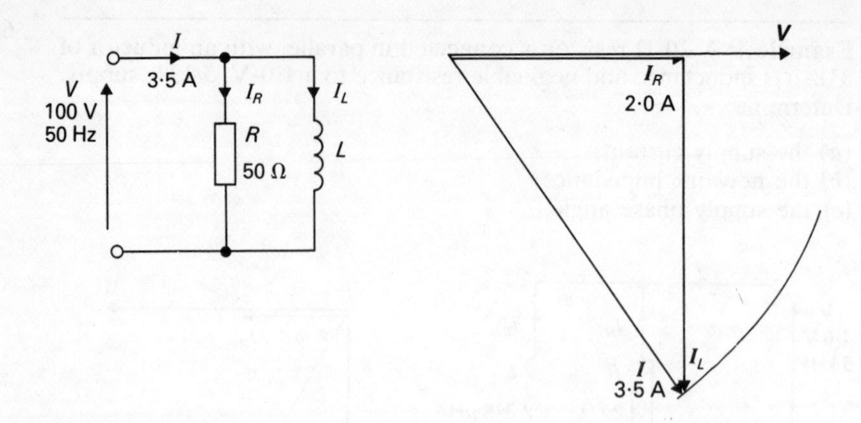

Fig. 3

The current in the 50-Ω resistor is in phase with the supply voltage and its magnitude is given by

$$I_R = \frac{V}{R} = \frac{100}{50} = 2{\cdot}0\ \text{A}$$

The current in the coil lags the supply voltage by 90°. Thus we can construct the phasor diagram shown in Fig. 3, taking the supply voltage as reference. A circle of radius equivalent to 3·5 A is drawn and we know that the current phasor must lie on it at some point. However the phasor I_L, when added to I_R, must give the supply current I; thus when we project a line from I_R, as shown, to meet the circle, then the point of intersection will be appropriate to the condition

$$\boldsymbol{I} = \boldsymbol{I}_R + \boldsymbol{I}_L$$

From the diagram, we can find that $I_L = 2{\cdot}87$ A, thus

$$X_L = \frac{V}{I_L} = \frac{100}{2{\cdot}87} = 35{\cdot}8\ \Omega$$

$$L = \frac{X_L}{2\pi f} = \frac{35{\cdot}8}{2\pi \times 50} = \underline{0{\cdot}11\ \text{H}}$$

Before leaving the R–L parallel network, it can be observed that the preferred methods of network solution depend on the addition (or subtraction) of the currents, and that this may either be achieved by analysing the vertical and horizontal components of the phasor diagram, or by drawing the phasor diagram to scale. In the light of these observations, let us now proceed to the other configurations.

In the case of resistance and capacitance in parallel, as shown in

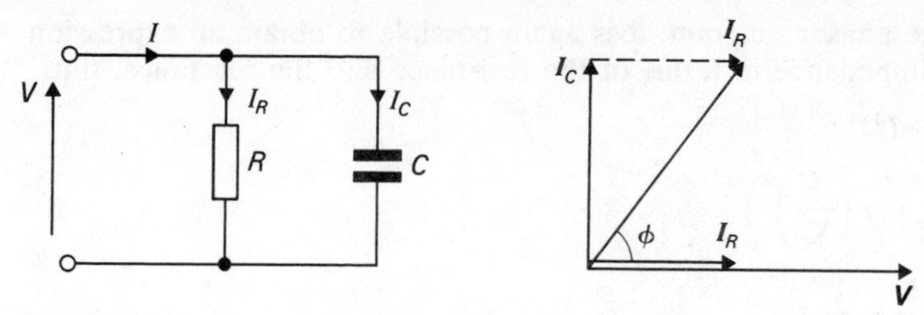

Fig. 4 Resistance and capacitance in parallel

Fig. 4, the current in the resistive branch is again given by

$$I_R = \frac{V}{R}$$

where I_R and V are in phase.

In the capacitive branch, the current is given by

$$I_C = \frac{V}{X_C}$$

where I_C leads V by 90°. The phasor diagram is constructed again taking the voltage as reference and on the basis of

$$\boldsymbol{I} = \boldsymbol{I}_R + \boldsymbol{I}_C \tag{7}$$

It can be seen from the phasor diagram in Fig. 4 that the phase angle ϕ is a leading angle. It follows that parallel networks behave in a similar fashion to series circuits in that the combination of resistance with inductance produces a lagging circuit whilst the combination of resistance with capacitance gives rise to a leading circuit.

$$\tan \phi = \frac{I_C}{I_R} \tag{8}$$

$$= \frac{V}{X_C} \cdot \frac{R}{V}$$

$$= \frac{R}{X_C} = 2\pi fCR \tag{9}$$

Also $$\cos \phi = \frac{I_R}{I} \tag{10}$$

$$= \frac{V}{R} \cdot \frac{Z}{V}$$

$$= \frac{Z}{R} \tag{11}$$

From the phasor diagram, it is again possible to obtain an expression for the impedance in terms of the resistance and the reactance, thus

$$I=(I_R^2+I_C^2)^{1/2}$$

$$=\left(\left(\frac{V}{R}\right)^2+\left(\frac{V}{X_C}\right)^2\right)^{1/2}$$

$$=V\left(\frac{1}{R^2}+\frac{1}{X_C^2}\right)^{1/2}$$

$$Z=\frac{V}{I}$$

$$=\frac{1}{\left(\frac{1}{R^2}+\frac{1}{X_C^2}\right)^{1/2}} \tag{12}$$

Example 3 A network consists of a 120-Ω resistor in parallel with a 40-μF capacitor and is connected to a 240-V, 50-Hz supply. Calculate:

(*a*) the branch currents and the supply current;
(*b*) the supply phase angle;
(*c*) the network impedance.

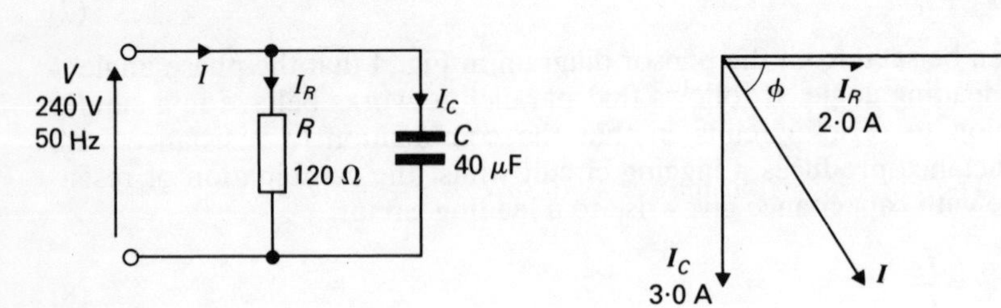

Fig. 5

$$I_R=\frac{V}{R}=\frac{240}{120}=\underline{2\cdot 0\text{ A}}$$

$$X_C=\frac{1}{2\pi fC}=\frac{1}{2\pi\times 50\times 40\times 10^{-6}}=80\ \Omega$$

$$I_C=\frac{V}{X_C}=\frac{240}{80}=\underline{3\cdot 0\text{ A}}$$

$$I^2=I_R^2+I_C^2=(2\cdot 0^2+3\cdot 0^2)=13$$

$$I=(13)^{1/2}=\underline{3\cdot 6\text{ A}}$$

$$\cos\phi = \frac{I_R}{I} = \frac{2\cdot 0}{3\cdot 6} = 0\cdot 56$$

$$\phi = \underline{56\cdot 3^\circ \text{ lead}}$$

$$Z = \frac{V}{I} = \frac{240}{3\cdot 6} = \underline{66\cdot 7\ \Omega}$$

Finally, in the case of capacitance connected in parallel with inductance, as shown in Fig. 6, the current in the capacitive branch is given by

$$I_C = \frac{V}{X_C}$$

where I_C leads V by 90°.

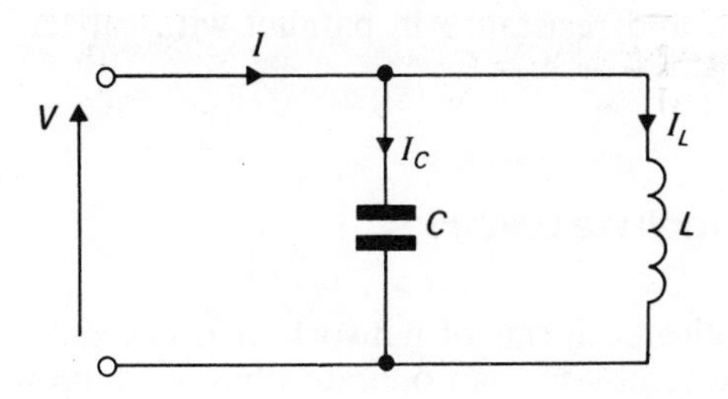

Fig. 6 Capacitance and inductance in parallel

The current in the inductive branch is given by

$$I_L = \frac{V}{X_L}$$

where I_L lags V by 90°.

The phasor diagram is constructed in the usual manner from the relation

$$\boldsymbol{I} = \boldsymbol{I}_C + \boldsymbol{I}_L \qquad (13)$$

The phasor diagram however can take one of two forms, depending on whether I_C is greater or less than I_L. Both forms of diagram are shown in Fig. 7.

From the phasor diagrams, we see that ϕ can be either +90° or −90° depending on whether the capacitance or the inductance predominates. However if I_C is equal to I_L, there is the peculiar condition that I is zero. Electrically this would be the equivalent of perpetual motion whereby the energy would continually move around the C–L circuit without loss and therefore continue indefinitely. This cannot be,

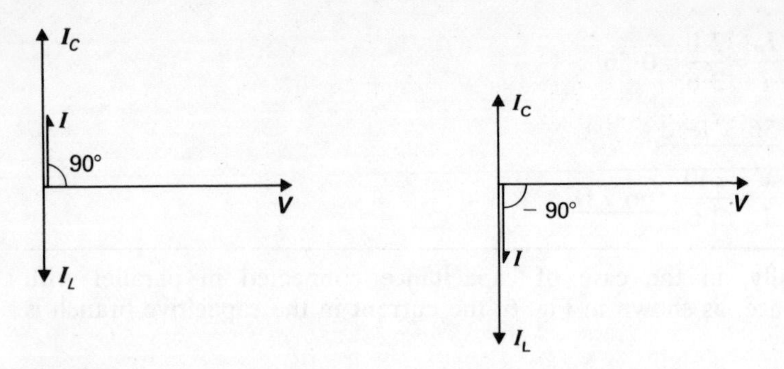

Fig. 7 Phasor diagrams for *C–L* parallel network

due to circuit resistance, and for this reason we need to introduce resistance into the arrangement. This is most easily done by not considering the resistance of the inductor coil to be negligible, thus the network comprises inductance and resistance in parallel with capacitance.

2 Parallel impedance networks

Before looking at more complicated forms of network, it is necessary to develop some techniques whereby we can operate phasor addition more readily. With more complicated networks, the summation of branch currents is made more difficult because they do not necessarily remain either in phase or in quadrature with one another. However we have seen in Chapter 2 that it is possible to resolve a current into components that are in phase or in quadrature with the supply voltage. The resolution of a current is shown in Fig. 8.

Consider Fig. 8, in which the current I is shown to lag (or to lead) the voltage **V** by a phase angle ϕ. The current can then be resolved

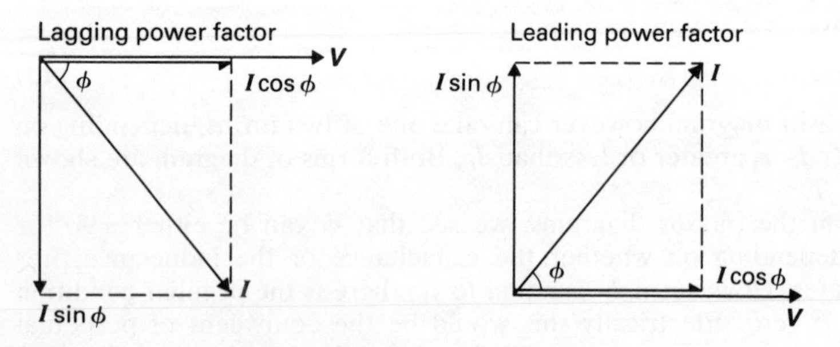

Fig. 8 Components of a current

into two components at right angles to one another:

(*a*) $\boldsymbol{I}$ **cos** $\boldsymbol{\phi}$, which is in phase with the voltage and is termed the active or power component.

(*b*) $\boldsymbol{I}$ **sin** $\boldsymbol{\phi}$, which is in quadrature with the voltage and is termed the quadrature or reactive component.

By the geometry of the diagram,

$$I^2 = (I \cos \phi)^2 + (I \sin \phi)^2$$

The term $I \cos \phi$ has already been met in the relation $P = VI \cos \phi$ and it is seen that the power dissipated by the load can be attributed to the current component $I \cos \phi$, hence the reason for it being termed the power component.

The reactive component either lags or leads the voltage by 90° depending on whether the current $\boldsymbol{I}$ lags or leads the voltage $\mathbf{V}$.

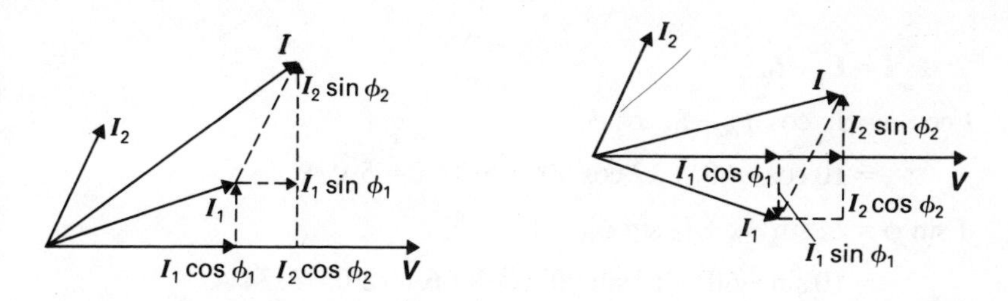

Fig. 9 Addition of current phasors

Consider the addition of the currents $\boldsymbol{I}_1$ and $\boldsymbol{I}_2$ as shown in Fig. 10, i.e.

$$\boldsymbol{I} = \boldsymbol{I}_1 + \boldsymbol{I}_2$$

The value of $\boldsymbol{I}$ can be determined by drawing a phasor diagram to scale, but it can also be calculated if the currents are resolved into components so that

$$I \cos \phi = I_1 \cos \phi_1 + I_2 \cos \phi_2$$

and $$I \sin \phi = I_1 \sin \phi_1 + I_2 \sin \phi_2$$

but $$I^2 = (I \cos \phi)^2 + (I \sin \phi)^2$$

hence $$I^2 = (I_1 \cos \phi_1 + I_2 \cos \phi_2)^2 + (I_1 \sin \phi_1 + I_2 \sin \phi_2)^2 \qquad (14)$$

Finally $$\cos \phi = \frac{I_1 \cos \phi_1 + I_2 \cos \phi_2}{I} \qquad (15)$$

Example 4 A parallel circuit consists of two branches A and B. $\boldsymbol{I}_A = 10\angle{-60°}$ A and $\boldsymbol{I}_B = 12\angle 90°$ A, all phase angles being relative to the supply voltage. Determine the supply current and the phase angle.

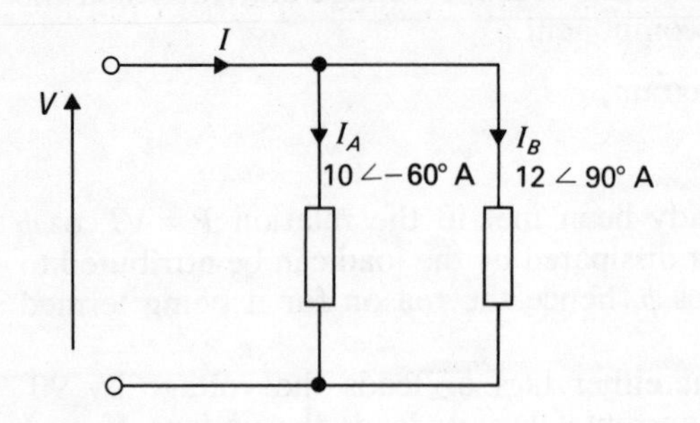

Fig. 10

$$\boldsymbol{I} = \boldsymbol{I}_A + \boldsymbol{I}_B$$

$$I \cos\phi = I_A \cos\phi_A + I_B \cos\phi_B$$
$$= 10 \cos -60° + 12 \cos 90° = 5{\cdot}0 + 0 = 5{\cdot}0 \text{ A}$$

$$I \sin\phi = I_A \sin\phi_A + I_B \sin\phi_B$$
$$= 10 \sin -60° + 12 \sin 90° = -8{\cdot}66 + 12{\cdot}0 = 3{\cdot}34 \text{ A}$$

Since $I \sin\phi$ is positive, the reactive current component is leading so the overall phase angle will also be leading.

$$I^2 = (I\cos\phi)^2 + (I\sin\phi)^2$$
$$I = (5{\cdot}0^2 + 3{\cdot}34^2)^{1/2} = 6{\cdot}0 \text{ A}$$

$$\cos\phi = \frac{I\cos\phi}{I} = \frac{5{\cdot}0}{6{\cdot}0} = 0{\cdot}83$$

$$\phi = 33{\cdot}5°$$

Thus $\boldsymbol{I} = \underline{6\angle 33{\cdot}5° \text{ A}}$

Consider the network shown in Fig. 11, in which two circuits are connected in parallel. One circuit is a capacitor which may be taken as pure, whilst the other circuit comprises a coil with resistance and inductance. The latter branch is therefore a series circuit. To analyse the arrangement, the phasor diagrams for each branch have been drawn separately.

A phasor diagram for a series circuit usually takes the current as reference, thus this is the basis for the branch phasor diagrams. When it comes to the drawing of the phasor diagram for the entire network, it is necessary to take the supply voltage as reference; hence the branch phasor diagrams are separately rotated and superimposed on one another to give the phasor diagram for the network. The current phasors may then be added to give the total current in correct phase relation to the voltage. The analysis of the diagram is carried out by resolution into components or by drawing to scale.

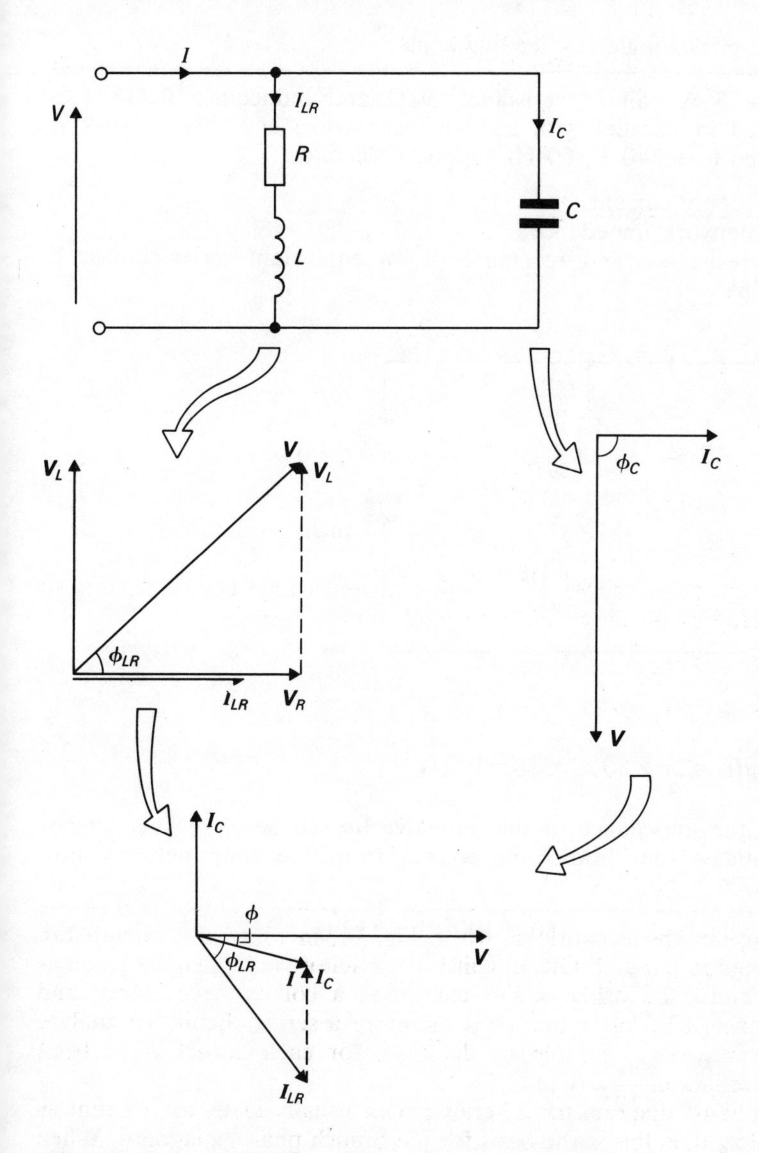

Fig. 11 *R–L* circuit in parallel with a capacitor

The phase angle of the network shown in Fig. 11 is a lagging angle since

$I_{LR} \sin \phi_{LR} > I_C$.

If

$I_C > I_{LR} \sin \phi_{LR}$

then the phase angle is a leading angle.

Example 5 A coil of resistance $50\,\Omega$ and inductance $0{\cdot}318$ H is connected in parallel with a 40-μF capacitor, and this network is connected to a 240-V, 50-Hz supply. Calculate:

(*a*) the supply current;
(*b*) the network impedance;
(*c*) the resistance and reactance of an equivalent series-connected circuit.

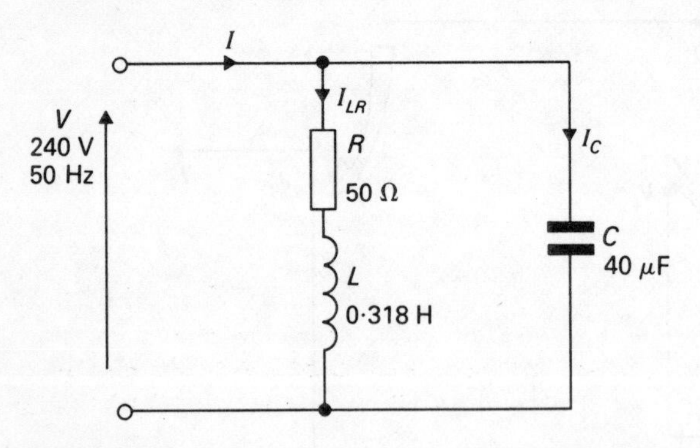

Fig. 12

$$X_L = 2\pi fL = 2\pi \times 50 \times 0{\cdot}318 = 100\,\Omega$$

Let the impedance of the inductive branch be Z_{LR}. This symbol differentiates that branch impedance from the total network impedance Z.

$$Z_{LR} = (R^2 + X_L^2)^{1\,2} = (50^2 + 100^2)^{1/2} = 112\,\Omega$$

$$I_{LR} = \frac{V}{Z_{LR}} = \frac{240}{112} = 2{\cdot}15 \text{ A}$$

$$\cos \phi_{LR} = \frac{R}{Z_{LR}} = \frac{50}{112} = 0{\cdot}447$$

$$\phi_{LR} = 63{\cdot}5° \text{ lag}$$

$$\mathbf{I_{LR} = 2{\cdot}15\angle{-63{\cdot}5°} \text{ A}}$$

$$X_C = \frac{1}{2\pi fC} = \frac{1}{2\pi \times 50 \times 40 \times 10^{-6}} = 80\ \Omega$$

$$I_C = \frac{V}{X_C} = \frac{240}{80} = 3{\cdot}0\ \text{A}$$

$$\boldsymbol{I}_C = 3{\cdot}0\angle 90^\circ\ \text{A}$$

$$\boldsymbol{I} = \boldsymbol{I}_{LR} + \boldsymbol{I}_C$$

$$I \cos \phi = I_{LR} \cos \phi_{LR} + I_C \cos \phi_C$$

$$= 2{\cdot}15 \cos -63{\cdot}5^\circ + 3{\cdot}0 \cos 90^\circ = 0{\cdot}96\ \text{A}$$

$$I \sin \phi = I_{LR} \sin \phi_{LR} + I_C \sin \phi_C$$

$$= 2{\cdot}15 \sin -63{\cdot}5^\circ + 3{\cdot}0 \sin 90^\circ = 1{\cdot}08\ \text{A}$$

$$I = ((I \cos \phi)^2 + (I \sin \phi)^2)^{1/2}$$

$$= (0{\cdot}96^2 + 1{\cdot}08^2)^{1/2} = \underline{1{\cdot}44\ \text{A}}$$

$$Z = \frac{V}{I} = \frac{240}{1{\cdot}44} = \underline{167\ \Omega}$$

$$R_{eq} = Z \cos \phi = Z \cdot \frac{I \cos \phi}{I} = 167 \times \frac{0{\cdot}96}{1{\cdot}44} = \underline{112\ \Omega}$$

$$X_{eq} = Z \sin \phi = Z \cdot \frac{I \sin \phi}{I} = 167 \times \frac{1{\cdot}08}{1{\cdot}44} = \underline{125\ \Omega}$$

In this example, the addition of the branch currents could also have been obtained from a phasor diagram drawn to scale as shown in Fig. 13. The diagram gives

$$\boldsymbol{I} = \boldsymbol{I}_{LR} + \boldsymbol{I}_C$$

$$= 1{\cdot}44\angle 48{\cdot}5^\circ\ \text{A}$$

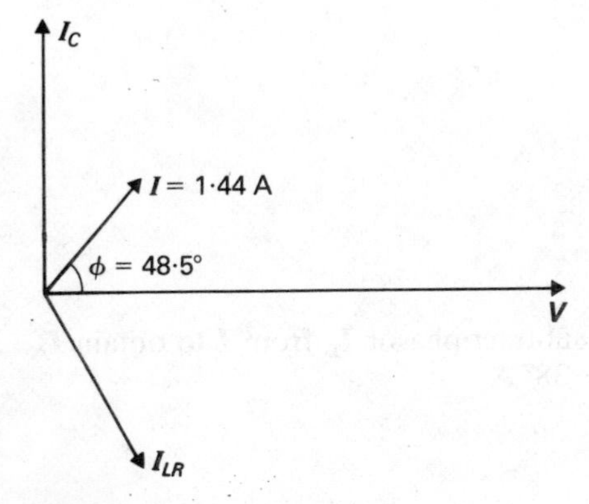

Fig. 13

Thus we obtain both the magnitude of the current and also the phase angle of the network from which the values of R_{eq} and X_{eq} may be derived.

Example 6 Two circuits, A and B, are connected in parallel to a 110-V, 850-Hz supply. The total taken by the network is 10·0 A at a leading phase angle of 30°. Circuit A consists of a resistive inductor and circuit B consists of a 20-μF capacitor. Determine, by means of a scale phasor diagram, the resistance and inductance of the coil.

$$X_C = \frac{1}{2\pi fC} = \frac{1}{2\pi \times 850 \times 20 \times 10^{-6}} = 9{\cdot}36\ \Omega$$

$$I_B = \frac{V}{X_C} = \frac{110}{9{\cdot}36} = 11{\cdot}8\ \text{A}$$

Taking the supply voltage as reference, the supply current $\boldsymbol{I}$ and the branch current $\boldsymbol{I}_B$ may be represented as shown in Fig. 14.

$$\boldsymbol{I} = \boldsymbol{I}_A + \boldsymbol{I}_B$$

$$\boldsymbol{I}_A = \boldsymbol{I} - \boldsymbol{I}_B$$

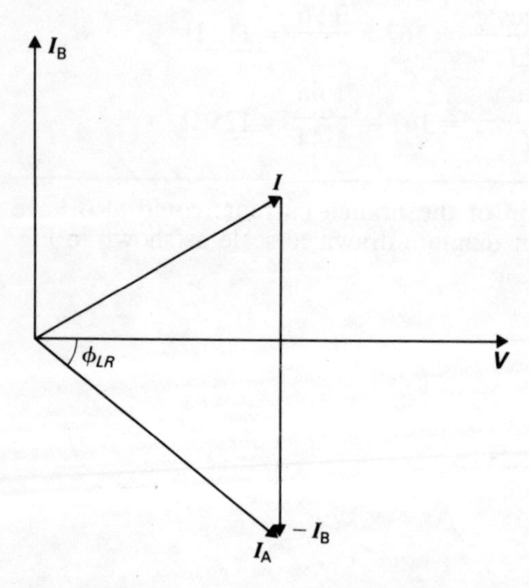

Fig. 14

From this relationship, we can subtract phasor $\boldsymbol{I}_B$ from $\boldsymbol{I}$ to obtain $\boldsymbol{I}_A$. From the diagram, $\boldsymbol{I}_A = 11{\cdot}0\angle{-38°}$ A

$$Z_{LR} = \frac{V}{I_A} = \frac{110}{11{\cdot}0} = 10{\cdot}0\ \Omega$$

$$R = Z_{LR} \cos \phi_{LR} = 10{\cdot}0 \cos 38° = \underline{7{\cdot}9\ \Omega}$$

$$X_L = Z_{LR} \sin \phi_{LR} = 10{\cdot}0 \sin 38° = 6{\cdot}2\ \Omega$$

$$L = \frac{X_L}{2\pi f} = \frac{6{\cdot}2}{2\pi \times 850} = 1{\cdot}16 \times 10^{-3}\ \text{H}$$

$$= \underline{1{\cdot}16\ \text{mH}}$$

If we finally consider a more complicated two-branch parallel network, we see that the emphasis lies in dealing with each branch separately as though it were a series circuit and then combining their effects by means of adding the branch currents. This may be observed from an example in which a resistive inductor is connected in parallel with a series circuit comprising capacitance and resistance.

Example 7 A coil of resistance 50 Ω and inductance 0·318 H is connected in parallel with a circuit comprising a 75-Ω resistor in series with a 159-μF capacitor. The resulting network is connected to a 240-V, 50-Hz a.c. supply. Calculate:

(*a*) the supply current;
(*b*) the network impedance;
(*c*) the equivalent series resistance and reactance.

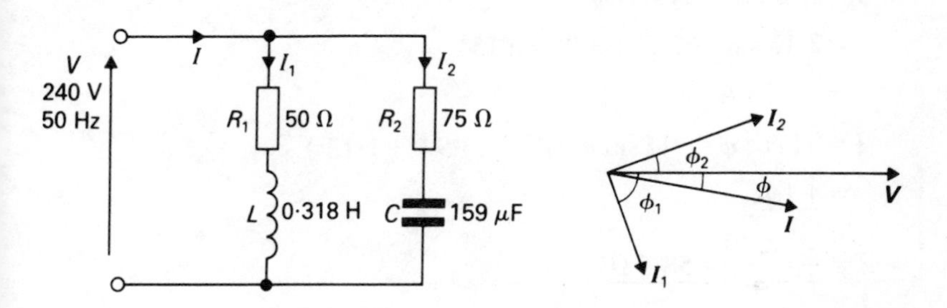

Fig. 15

$$X_L = 2\pi f L = 2\pi \times 50 \times 0{\cdot}318 = 100\ \Omega$$

$$Z_1 = (R_1^2 + X_L^2)^{1/2} = (50^2 + 100^2)^{1/2} = 112\ \Omega$$

$$I_1 = \frac{V}{Z_1} = \frac{240}{112} = 2{\cdot}15\ \text{A}$$

$$\cos \phi_1 = \frac{R_1}{Z_1} = \frac{50}{112} = 0{\cdot}447$$

$$\phi_1 = 63{\cdot}5°\ \text{lag}$$

$$\boldsymbol{I}_1 = 2{\cdot}15\angle{-63{\cdot}5°}\ \text{A}$$

$$X_C = \frac{1}{2\pi fC} = \frac{1}{2\pi \times 50 \times 159 \times 10^{-6}} = 20\ \Omega$$

$$Z_2 = (R_2^2 + X_C^2)^{1/2} = (75^2 + 20^2)^{1/2} = 77{\cdot}7\ \Omega$$

$$I_2 = \frac{V}{Z_2} = \frac{240}{77{\cdot}7} = 3{\cdot}09\ \text{A}$$

$$\tan \phi_2 = \frac{X_C}{R_2} = \frac{20}{75{\cdot}0} = 0{\cdot}267$$

$$\phi_2 = 15°\ \text{lead}$$

Note that the value of ϕ_2 was derived from the tangent because ϕ_2 is relatively small. With the accuracy available, it would not have been possible to obtain such a good estimate of the angle from $\cos \phi_2$.

$$\boldsymbol{I}_2 = 3{\cdot}09\angle 15°\ \text{A}$$

$$\boldsymbol{I} = \boldsymbol{I}_1 + \boldsymbol{I}_2$$

$$I \cos \phi = I_1 \cos \phi_1 + I_2 \cos \phi_2$$

$$= 2{\cdot}15 \cos -63{\cdot}5° + 3{\cdot}09 \cos 15°$$

$$= 3{\cdot}94\ \text{A}$$

$$I \sin \phi = I_1 \sin \phi_1 + I_2 \sin \phi_2$$

$$= 2{\cdot}15 \sin -63{\cdot}5° + 3{\cdot}09 \sin 15°$$

$$= -1{\cdot}13\ \text{A}$$

$$I = ((I \cos \phi)^2 + (I \sin \phi)^2)^{1/2} = (3{\cdot}94^2 + 1{\cdot}13^2)^{1/2}$$

$$= \underline{4{\cdot}1\ \text{A}}$$

$$Z = \frac{V}{I} = \frac{240}{4{\cdot}1} = \underline{58{\cdot}5\ \Omega}$$

$$R_{eq} = Z \cos \phi = Z \cdot \frac{I \cos \phi}{I} = 58{\cdot}5 \times \frac{3{\cdot}94}{4{\cdot}1} = \underline{56\ \Omega}$$

$$X_{eq} = Z \sin \phi = Z \cdot \frac{I \sin \phi}{I} = 58{\cdot}5 \times \frac{1{\cdot}13}{4{\cdot}1} = \underline{15\ \Omega}$$

Since $I \sin \phi$ is negative, the reactance must be inductive, hence the network is equivalent to a 56-Ω resistor in series with a 15-Ω inductor.

---

3 Polar impedances

In Example 7, the impedance was derived from the current and the voltage and this generally is the best and easiest method of calculation.

However, you may wonder why could not the parallel impedances have been handled in a similar manner to parallel resistors. Consider then two impedances connected in parallel as shown in Fig. 16. In the first branch,

$$\mathbf{V} = \mathbf{I}_1 \mathbf{Z}_1$$

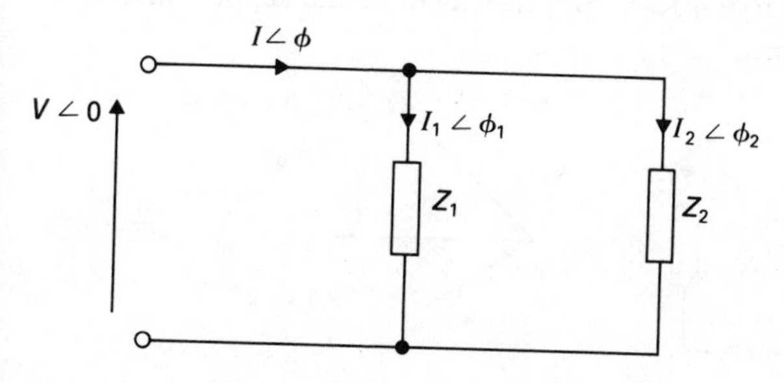

Fig. 16 Polar impedances in parallel

However, if consideration is given to the phase angles of $\boldsymbol{V}$ and $\boldsymbol{I}$, then to maintain balance, the impedance must also act like a complexor and have a phase angle, i.e.

$$V\angle\phi = I_1\angle\phi_1 \cdot Z_1\angle{-\phi_1}$$

The impedance phase angle is the conjugate of the circuit phase angle, i.e. if the circuit phase angle is ϕ, then the impedance angle is $-\phi$. This compares with the impedance triangles observed in Chapter 1. By applying Kirchhoff's First Law,

$$\boldsymbol{I} = \boldsymbol{I}_1 + \boldsymbol{I}_2$$

In polar notation,

$$I\angle\phi = I_1\angle\phi_1 + I_2\angle\phi_2$$

$$\frac{V\angle 0}{Z\angle{-\phi}} = \frac{V\angle 0}{Z_1\angle{-\phi_1}} + \frac{V\angle 0}{Z_2\angle{-\phi_2}}$$

$$\frac{1}{Z\angle{-\phi}} = \frac{1}{Z_1\angle{-\phi_1}} + \frac{1}{Z_2\angle{-\phi_2}} \qquad (16)$$

This relation compares with that for parallel resistors, but it has the complication of having to consider the phase angles. Because of this, it is not considered advisable, at this introductory stage, to use the polar notation approach to the analysis of parallel networks. The method used in Example 7 is much more suitable and less prone to arithmetical error.

It is most important that the impedance phase angles are not ignored and it would be quite incorrect to use

$$\frac{1}{Z}=\frac{1}{Z_1}+\frac{1}{Z_2}$$

Example 8 Two impedances, $\mathbf{Z}_1 = 20\angle{-45°}\ \Omega$ and $\mathbf{Z}_2 = 30\angle 30°\ \Omega$ are connected in parallel to a 400-V supply. Calculate the supply current.

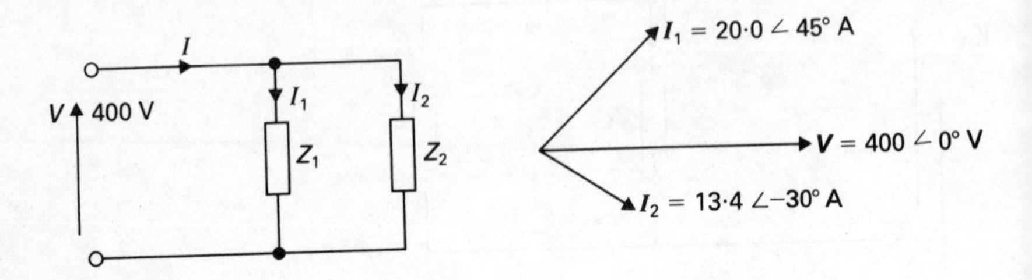

Fig. 17

$$I_1 = \frac{V}{Z_1} = \frac{400}{20} = 20{\cdot}0\ \text{A}$$

$$\phi_1 = 45°$$

$$I_2 = \frac{V}{Z_2} = \frac{400}{30} = 13{\cdot}4\ \text{A}$$

$$\phi_2 = -30°$$

$$I\cos\phi = I_1\cos\phi_1 + I_2\cos\phi_2$$

$$= 20{\cdot}0\cos 45° + 13{\cdot}4\cos -30° = 25{\cdot}7\ \text{A}$$

$$I\sin\phi = I_1\sin\phi_1 + I_2\sin\phi_2$$

$$= 20{\cdot}0\sin 45° + 13{\cdot}4\sin -30° = 7{\cdot}5\ \text{A}$$

$$I = ((I\cos\phi)^2 + (I\sin\phi)^2)^{1/2}$$

$$= (25{\cdot}7^2 + 7{\cdot}5^2)^{1/2} = \underline{26{\cdot}8\ \text{A}}$$

---

4 Power and power factor in parallel networks

We have already observed that the current in any branch of a network may be resolved into an active component and a reactive component.

The active components of the various branches can be added to give the total active component of the supply current and this permits the determination of the total active power.

Consider a two-branch network as shown in Fig. 18. The power in branch 1 is P_1 and the power in branch 2 is P_2, where

$$P_1 = VI_1 \cos \phi_1$$

and $$P_2 = VI_2 \cos \phi_2$$

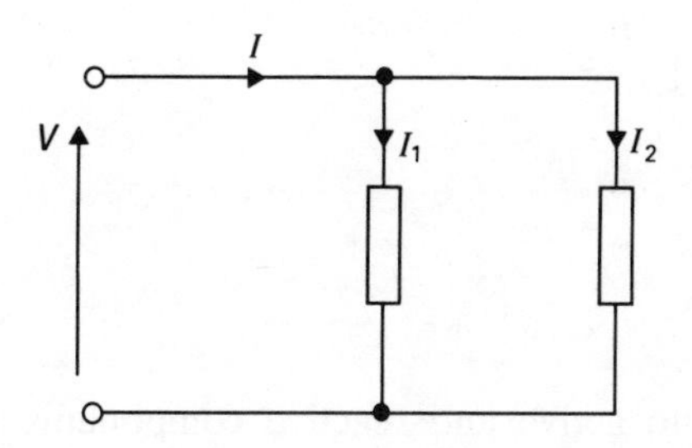

Fig. 18 Power in a parallel network

The active power is a mean value; thus the branch powers can be added directly, and the total active power P is given by

$$\begin{aligned} P &= P_1 + P_2 \\ &= VI_1 \cos \phi_1 + VI_2 \cos \phi_2 \\ &= V(I_1 \cos \phi_1 + I_2 \cos \phi_2) \\ &= VI \cos \phi \end{aligned} \tag{17}$$

Thus the active component of the supply current determines the total active power of the network.

In a similar manner, it can be shown that the reactive powers can be treated in the same way. In this case, however, the reactive powers may be added because they occur at the same instants, being peak instantaneous values. Thus

$$Q = VI \sin \phi \tag{18}$$

The power factor may be determined as in Chapter 1, being again defined as the ratio of active power to apparent power:

$$\text{Power factor} = \frac{P}{S} \tag{19}$$

$$= \frac{P}{(P^2 + Q^2)^{1/2}} \tag{20}$$

Example 9 A network comprises a capacitor in parallel with a resistor. The supply is rated at 50 V, 60 Hz and the supply current is 0·5 A at a power factor 0·8 leading. Determine the resistance and capacitance of the network components.

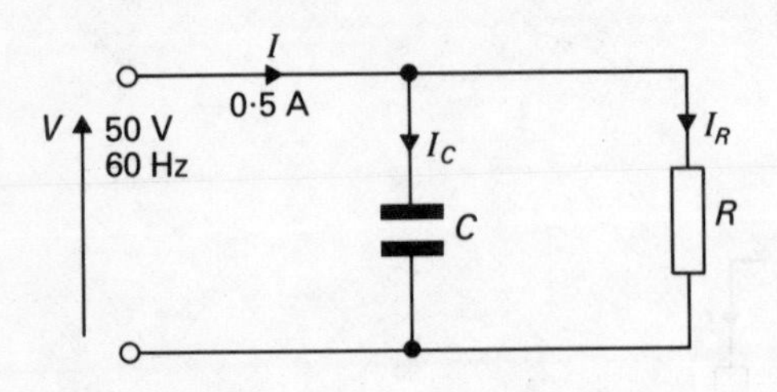

Fig. 19

Resolve the supply current into active and reactive components, thus

$$I \cos\phi = 0{\cdot}5 \times 0{\cdot}8 = 0{\cdot}4 \text{ A}$$

Also $\phi = 37°$

$$I \sin\phi = 0{\cdot}5 \sin 37° = 0{\cdot}3 \text{ A}$$

Only the resistor can dissipate energy, thus the active or power component must be entirely derived from the resistive branch. Similarly only the capacitor can store energy, thus the reactive component must be entirely derived from the capacitive branch.

Hence $I_R = I\cos\phi = 0{\cdot}4$ A

and $R = \dfrac{V}{I_R} = \dfrac{50}{0{\cdot}4} = \underline{125\ \Omega}$

Also $I_C = I \sin\phi = 0{\cdot}3$ A

$$X_C = \frac{V}{I_C} = \frac{50}{0{\cdot}3} = 167\ \Omega$$

$$C = \frac{1}{2\pi f X_C} = \frac{1}{2\pi \times 60 \times 167} = 15{\cdot}9 \times 10^{-6} \text{ F}$$

$$= \underline{15{\cdot}9\ \mu\text{F}}$$

Example 10 Two circuits A and B are connected in parallel to a 110-V, 850-Hz supply, and the supply current is 10·0 A. Circuit A consists of a coil of resistance 10 Ω and inductance 4·0 mH, and circuit B consists of a capacitor. By means of a phasor diagram, determine the current in the capacitor and hence find:

(*a*) the capacitance of the capacitor;
(*b*) the power and power factor of the network.

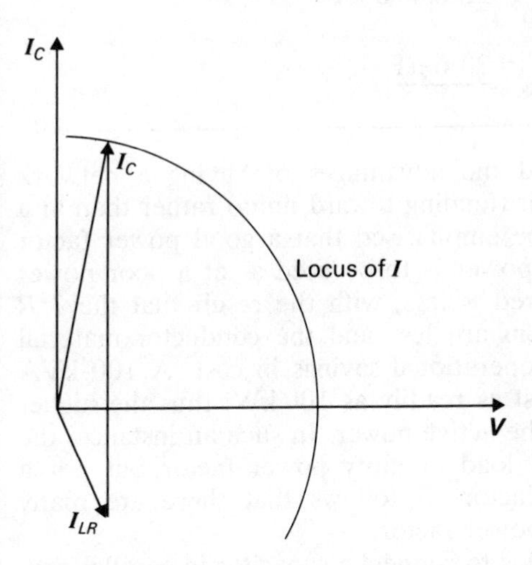

Fig. 20

$$X_L = 2\pi fL = 2\pi \times 850 \times 4 \times 10^{-3} = 21{\cdot}4\ \Omega$$

$$Z_{LR} = (R^2 + X_L^2)^{1/2} = (10^2 + 21{\cdot}4^2)^{1/2} = 23{\cdot}6\ \Omega$$

$$I_{LR} = \frac{V}{Z_{LR}} = \frac{110}{23{\cdot}6} = 4{\cdot}67\ \text{A}$$

$$\cos\phi_{LR} = \frac{R}{Z_{LR}} = \frac{10}{23{\cdot}6} = 0{\cdot}424$$

$$\phi_{LR} = 65°$$

The phasor diagram can now be drawn as shown in Fig. 20. The voltage is taken as reference and the phasor $\boldsymbol{I}_{LR}$ is shown lagging by 65°. The total current must lie on the circle, which has a radius equivalent to 10·0 A, and $\boldsymbol{I}_C$ must lead $\boldsymbol{V}$ by 90°. Adding $\boldsymbol{I}_C$ to $\boldsymbol{I}_{LR}$ must give a phasor lying on the circle, and this happens for only one point, thus defining the phase angle of the supply and also the magnitude of $\boldsymbol{I}_C$.

Note that there is a second point of intersection but this requires $\boldsymbol{I}_C$ to lag $\boldsymbol{V}$ by 90°, which is not possible for a capacitor. From the diagram,

$I_C = \underline{14{\cdot}0\ \text{A}}$

$$X_C = \frac{V}{I_C} = \frac{110}{14{\cdot}0} = 7{\cdot}86\ \Omega$$

$$C = \frac{1}{2\pi f X_C} = \frac{1}{2\pi \times 850 \times 7{\cdot}86} = 23{\cdot}8 \times 10^{-6}\ \text{F}$$

$$= \underline{23{\cdot}6\ \mu\text{F}}$$

In Chapter 2, we studied the advantages of having a network operate at a high power factor (tending toward unity) rather than at a poor power factor. It must be emphasised that a good power factor implies that the same active power is to be used as at a poor power factor, but the current required is less, with the result that the I^2R losses in the conductor system are less and the conductor material required is less, both giving operational savings in cost. A 100-kVA cable will supply 100 kvar just as readily as 100 kW; thus the higher the power factor, the better the active power. In such an instance, the cable could supply a 90-kW load at unity power factor but not a 90-kW load at 0·8 power factor. It follows that there are many attractions in improving the power factor.

The most common method is to connect a capacitor in parallel with the load. It happens that most loads operate at a lagging power factor and therefore a typical arrangement for power factor improvement can be represented by an *R–L* load shunted by a capacitor – which is the network that we have been studying.

The effect of the capacitor is to change the reactive power of the system without changing the active power; a capacitor does not dissipate power and therefore cannot contribute to the active power. It follows that the current in a capacitive branch is reactive and changes only the reactive component of the supply current. An example will help to illustrate this statement.

Example 11 A motor operates from a 240-V, 50-Hz supply with an output of 4·0 kW and an efficiency 0·85. The supply power factor is 0·65 lagging. Determine:

(*a*) the supply apparent power;
(*b*) the active and reactive components of the supply current.

It is required to improve the power factor to 0·9 lagging by connecting a capacitor in parallel with the motor. Determine the capacitance of the capacitor.

$$P_i = \frac{P_o}{\eta} = \frac{4000}{0{\cdot}85} = 4700\ \text{W}$$

$$S=\frac{P_i}{\cos\phi}=\frac{4700}{0{\cdot}65}=\underline{7230\text{ VA}}$$

$$I=\frac{S}{V}=\frac{7230}{240}=30{\cdot}12\text{ A}$$

$$I\cos\phi=30{\cdot}12\times0{\cdot}65=\underline{19{\cdot}6\text{ A}}$$

$$I\sin\phi=30{\cdot}12\times0{\cdot}76=\underline{22{\cdot}9\text{ A}}$$

If the power factor of the system is to be improved to 0·9, the reactive component of the supply must be reduced. However the active component remains at 19·6 A. The new supply current I' is there obtained from the active component of the current and the new power factor, thus

$$I'=\frac{19{\cdot}6}{0{\cdot}9}=21{\cdot}8\text{ A}$$

Also $\cos\phi'=0{\cdot}9$

hence $\sin\phi'=0{\cdot}436$

The reactive component of the improved supply current is given by

$$I'\sin\phi'=21{\cdot}8\times0{\cdot}436=9{\cdot}5\text{ A}$$

The capacitive branch must therefore contribute the difference in reactive current as illustrated in Fig. 21.

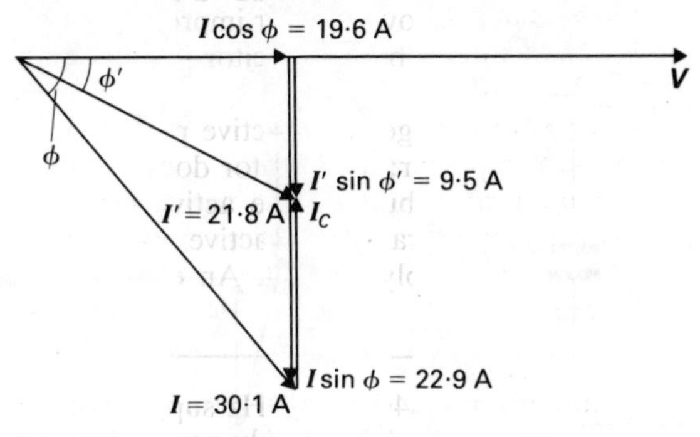

Fig. 21

$$I_C=I\sin\phi-I'\sin\phi'=22{\cdot}9-9{\cdot}5=13{\cdot}4\text{ A}$$

$$X_C=\frac{V}{I_C}=\frac{240}{13{\cdot}4}=17{\cdot}9\ \Omega$$

$$C=\frac{1}{2\pi fX_C}=\frac{1}{2\pi\times50\times17{\cdot}9}=178\times10^{-6}\text{ F}$$

$$=\underline{178\ \mu\text{F}}$$

In this example, you have seen that the effect of adding the capacitor to the system is that the supply current is reduced from 30·1 A to 21·8 A, which is a considerable saving. Nevertheless the motor continues to give out 4 kW as before. Finally it should be noted that the cost of the capacitor offsets the saving in operating costs, but an improvement to a power factor such as that instanced normally gives a nett saving.

5 Resonance in a simple parallel network

In Fig. 11, we investigated the action of a coil, with integral resistance, in parallel with a capacitor. The phasor diagrams were those appropriate for $I_{LR} \sin \phi_{LR}$ greater than I_C, so that the supply phase angle and the power factor were lagging ones. Also we investigated the case of I_C greater than $I_{LR} \sin \phi_{LR}$, for which case the supply phase angle and the power factor were leading ones. There remains the case in which

$$I_C = I_{LR} \sin \phi_{LR} \tag{21}$$

The network and phasor diagram corresponding to this condition are shown in Fig. 22. The condition may correspond to resonance in a similar manner to that of resonance in a series circuit.

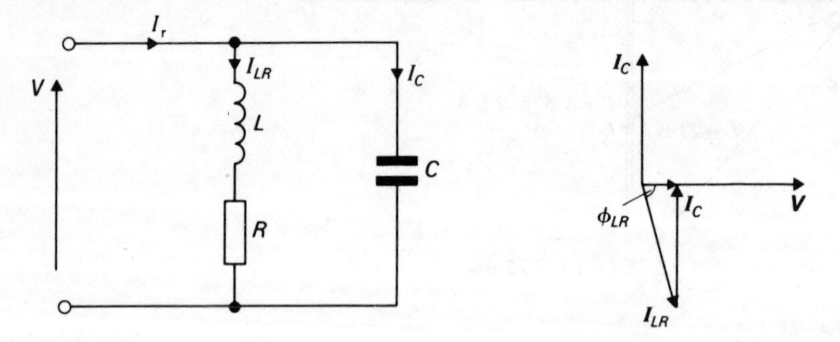

Fig. 22 Parallel resonant network

From the phasor diagram, we can see that I_C is greater than the supply current I. The network therefore magnifies the current (just as the series circuit magnified the voltage). If this magnification is greater than 10, the network is effectively in resonance. If the magnification is less, then the network is more appropriately considered to be operating at unity power factor.

Given a network with relatively small resistance, resonance can be arranged either by adjustment of the components or by adjustment of the frequency.

In the case of variation of frequency, when the frequency is zero (corresponding to a d.c. condition), the coil reactance is $X_L = 2\pi fL = 0$, so the resistance R limits the current in the coil. In the capacitive branch, $X_C = \frac{1}{2}\pi fC = \infty$, hence there is no current in the capacitor. Increase of frequency increases the reactance of the coil and consequently the coil impedance increases. The coil current I_{LR} therefore decreases and lags the voltage V by a progressively greater angle. The capacitive branch current, on the other hand, increases, and always leads the voltage by 90°. At some frequency, $I_C = I_{LR} \sin \phi_{LR}$ and resonance occurs. The variation of current with frequency is shown in Fig. 23, which bears a marked resemblance to Fig. 26 in Chapter 2 even though the latter illustrates the variation of impedance with frequency.

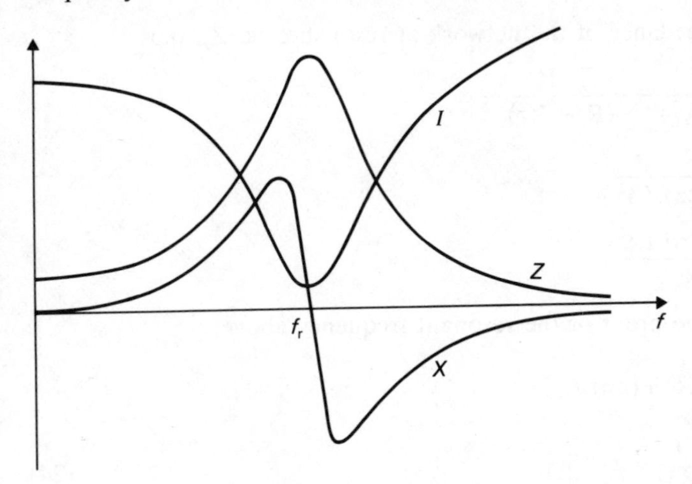

Fig. 23 Variation of current and reactance with frequency

From Fig. 22, we may observe that

$$I_C = I_{LR} \sin \phi_{LR}$$

$$\frac{V}{X_C} = \frac{V}{(R^2 + X_L^2)^{1/2}} \cdot \frac{X_L}{(R^2 + X_L^2)^{1/2}}$$

$$2\pi f_r C = \frac{2\pi f_r L}{R^2 + (2\pi f_r L)^2}$$

$$R^2 + (2\pi f_r L)^2 = \frac{L}{C}$$

$$(2\pi f_r L)^2 = \frac{L}{C} - R^2$$

$$f_r = \frac{1}{2\pi L}\left(\frac{L}{C} - R^2\right)^{1/2}$$

$$= \frac{1}{2\pi}\left(\frac{1}{LC} - \frac{R^2}{L^2}\right)^{1/2} \qquad (22)$$

This is the resonant frequency for the general parallel network. If R is small then

$$f_r = \frac{1}{2\pi(LC)^{1/2}} \qquad (23)$$

which is the same relation as in the general series circuit. Also

$I_r = I_{LR} \cos \phi_{LR}$

Let the impedance of the network at resonance be Z_r, thus

$$\frac{V}{Z_r} = \frac{V}{(R^2 + X_L^2)^{1/2}} \cdot \frac{R}{(R^2 + X_L^2)^{1/2}}$$

$$\frac{1}{Z_r} = \frac{R}{R^2 + (2\pi f_r L)^2}$$

$$Z_r = \frac{R^2 + (2\pi f_r L)^2}{R}$$

But from the proof of the resonant frequency above,

$$\frac{L}{C} = R^2 + (2\pi f_r L)^2$$

$$\text{thus} \quad Z_r = \frac{L}{CR} \qquad (24)$$

It is again of interest to note how the network responds to variation of frequency for fixed values of inductance and capacitance. At resonance, V and I_r are in phase and the impedance is a maximum, being equal to an effective resistance R_r. It follows that the minimum current is

$$I_r = \frac{VCR}{L} \qquad (25)$$

If R were zero in the ideal case, the current would be zero. Even so, there would be a current flowing around the circuit made by the coil and the capacitor. This is due to energy being transferred from the inductor to the capacitor and back again.

Resistance, however, must be present in the circuit and this prevents the continuous energy transfer process taking place without loss. Some current must therefore be drawn from the supply to make good the loss. The current needed from the supply is small and the branch currents are much greater than the supply current. At any other frequency, this situation does not arise and the supply current is greater. Also the impedance and the power factor both decrease, and these variations are illustrated in Fig. 24. Because the current rejects current at resonance, it is called a rejector network.

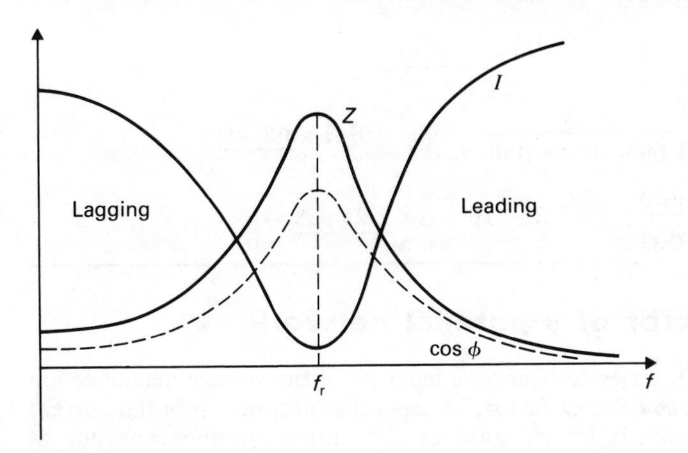

Fig. 24 Parallel resonance characteristics

Example 12 A coil of 1 kΩ resistance and 0·15 H inductance is connected in parallel with a variable capacitor across a 2·0-V, 10-kHz a.c. supply. Calculate:

(*a*) the capacitance of the capacitor when the supply current is a minimum;
(*b*) the effective impedance of the network;
(*c*) the supply current.

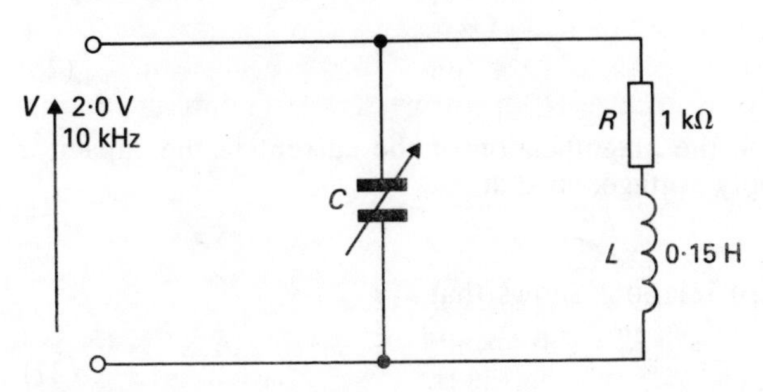

Fig. 25

 The network supply current is a minimum when the network is at resonance.

$$f_r = \frac{1}{2\pi}\left(\frac{1}{LC} - \frac{R^2}{L^2}\right)^{1/2}$$

$$\frac{1}{LC} = 4\pi^2 f_r^2 + \frac{R^2}{L^2} = 4\pi^2 \times 10^2 \times 10^6 + \frac{1000^2}{0{\cdot}15^2}$$

$$= 3960 \times 10^6$$

$$C = \frac{10^6}{3960L} = 1{\cdot}69 \times 10^{-9}\ \text{F} = \underline{1{\cdot}69\ \text{nF}}$$

$$Z_r = \frac{L}{CR} = \frac{0{\cdot}15}{1{\cdot}69 \times 10^{-9} \times 1000} = 890 \times 10^3\ \Omega = \underline{890\ \text{k}\Omega}$$

$$I_r = \frac{V}{Z_r} = \frac{2{\cdot}0}{890 \times 10^3} = 2{\cdot}25 \times 10^{-6}\ \text{A} = \underline{2{\cdot}25\ \mu\text{A}}$$

---

6 Q factor of a parallel network

In the case of series resonance, a measure of the voltage magnification was afforded by the Q factor. In a parallel network, it is the current that is magnified; let us consider the current in the capacitor at resonance in terms of the supply voltage V. At resonance,

$$V = I_r Z_r$$

$$= I_r \cdot \frac{L}{CR}$$

$$I_C = \frac{V}{X_C}$$

$$= \frac{I_r L}{CR} \cdot 2\pi f_r C$$

$$= I_r \cdot \frac{\omega_r L}{R} \qquad (26)$$

Let the Q factor be the magnification of the current in the capacitor relative to the supply voltage, so that

$$I_C = I_r Q$$

Comparison of these relations shows that

$$Q = \frac{\omega_r L}{R} \qquad (27)$$

This is the same expression as that for the series circuit.

As with the series circuit, it is more difficult to consider the Q factor in terms of the current associated with the coil, but it can be shown that the Q factor is also the ratio of the reactive component of the coil current to the supply current, as compared with the series circuit, in which it is the ratio of the reactive component of the coil voltage to the supply voltage. Thus

$$I_{LR} \sin \phi_{LR} = QI_r \tag{28}$$

and

$$I_C = QI_r \tag{29}$$

Q may be very large, especially if R is only the resistance of the inductor coil, and values of Q can be many hundreds. In such cases, the current in the capacitor will be many hundreds of times that of the supply current. The current in the inductor coil will also be many hundreds of times that of the supply current, but generally we cannot observe $I_{LR} \sin \phi_{LR}$.

The maximum value of network impedance is also given in terms of Q as follows:

$$Z_r = \frac{L}{CR}$$

$$= \frac{\omega_r L}{\omega_r CR}$$

$$= Q \cdot \frac{1}{\omega_r C} \tag{30}$$

When the capacitance is adjusted to make the network impedance maximum, the impedance, which is resistive, is sometimes termed the dynamic resistance of the network.

Example 13 A resonant arrangement is to be made from a coil of resistance 400 Ω and inductance 0·1 H and a 5000 pF capacitor. Find:

(*a*) the series and parallel resonant frequencies;
(*b*) the dynamic resistance;
(*c*) the Q factor, for parallel resonance.

For series resonance,

$$f_r = \frac{1}{2\pi(LC)^{1/2}} = \frac{1}{2\pi(0{\cdot}1 \times 5000 \times 10^{-12})^{1/2}} = \underline{7116 \text{ Hz}}$$

For parallel resonance,

$$f_r = \frac{1}{2\pi}\left(\frac{1}{LC} - \frac{R^2}{L^2}\right)^{1/2} = \frac{1}{2\pi}\left(\frac{1}{0{\cdot}1 \times 5000 \times 10^{-12}} - \frac{400^2}{0{\cdot}1^2}\right)^{1/2}$$

$$= \underline{7088 \text{ Hz}}$$

$$Z_r = \frac{L}{CR} = \frac{0 \cdot 1}{5000 \times 10^{-12} \times 400} = 50\,000\ \Omega = \underline{50\ \text{k}\Omega}$$

$$Q = \frac{\omega_r L}{R} = \frac{2\pi \times 7088 \times 0 \cdot 1}{400} = \underline{11 \cdot 1}$$

---

7 Parallel resonance in practice

In Example 13, we have seen that there can be a small difference between the resonant frequency of a series circuit and a parallel network. The higher the Q factor, the less is the effect of the resistance and the closer together are the frequencies. It is possible to obtain a relation between the frequencies in terms of the Q factor, but first there are comments that require to be made about the resonant frequency of the parallel network.

The resonant frequency of a coil in parallel with a capacitor is given by relation (22), yet around the circuit comprising the coil and the capacitor, the energy would naturally wish to resonate at a frequency appropriate to a series circuit.

The frequency appropriate to series resonance around the coil–capacitor loop is termed the natural frequency f_n and the frequency of resonance observed from the network terminals f_r is sometimes termed the forced resonant frequency. The latter is that appropriate for the condition that the supply voltage and current are in phase and also that the network impedance is a maximum.

From the coil–capacitor loop,

$$f_n = \frac{1}{2\pi(LC)^{1/2}} \tag{31}$$

This is related to the forced resonant frequency by

$$f_r = f_n\left(1 - \frac{1}{Q^2}\right)^{1/2} \tag{32}$$

Thus again we see that the forced resonant frequency is less than the natural frequency, but with even quite small values of Q, the difference tends to be negligible.

There are two forms of action in which the parallel resonant network plays an important role. If it is supplied from a source of constant voltage and low internal resistance, then the voltage across the network remains constant with variation of frequency. In this way, the network cannot be selective as in the case of the series circuit. However, when offered a variety of supply frequencies, it provides high impedance and permits little current at those frequencies close to resonance. At other frequencies, it offers little impedance and therefore tends to almost short circuit these other signals, which is the opposite action to that of the series resonant circuit. When the two arrangements are used together, they can very effectively select one signal at a given frequency and reject all others.

This action may be better understood by considering the network shown in Fig. 26. The input signal consists of a mixture of frequencies

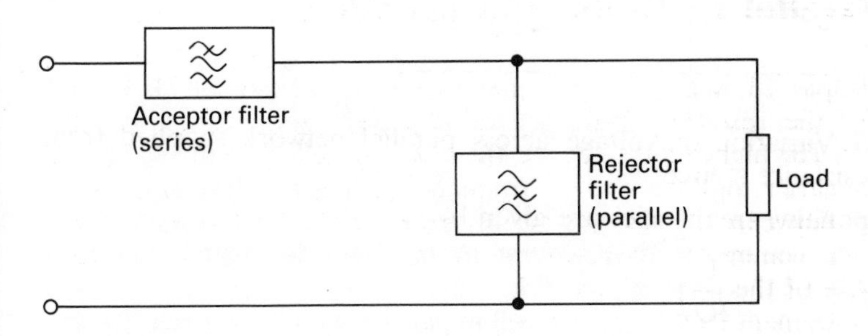

Fig. 26 Simple filter arrangement

and we wish to select only one, f_r, to be applied to the load. The input signal first meets the acceptor filter, which is a series circuit adjusted to resonate at f_r. The filter therefore offers little opposition to the signal at f_r, which it accepts, rejecting all other frequencies. However the rejection is not complete and although diminished, signals at other frequencies continue through to the rejection filter which is a parallel network again adjusted to resonate at f_r. This filter rejects the signal at f_r, which therefore must pass through the load as required. The signals at other frequencies are effectively short circuited by the rejector filter and therefore bypass the load.

This arrangement of filters is widely used to tune communications and other systems in order that they may select their frequency of operation. In practice, however, other varieties of filter construction may be used in order to obtain even better selection, but this is a development of the basic operation that we have observed.

If the source of supply to a parallel resonant network has high internal resistance, the operation compares with that of a constant current source. In this case, the voltage across the parallel network is proportional to the impedance of the parallel network and is a maximum when the impedance is a maximum. The network therefore becomes selective only when supplied from a source of high internal resistance. The appropriate characteristic is shown in Fig. 27.

The characteristic shown in Fig. 27 also represents the variation of network impedance with frequency, to a different scale. The half-power limits occur when the impedance falls to $(1/\sqrt{2})Z_r$, and the

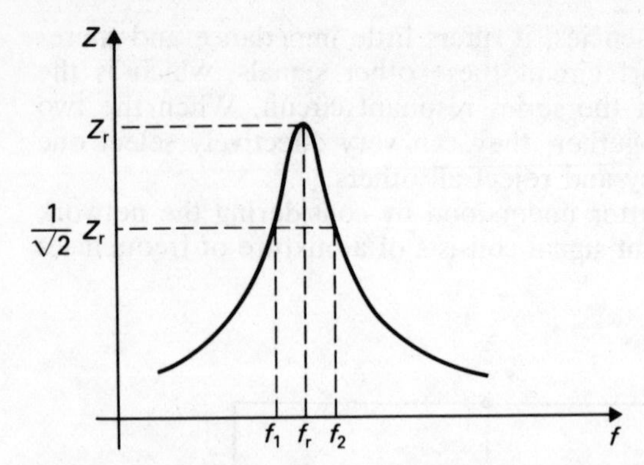

Fig. 27 Variation of voltage across parallel network supplied from high-resistance source

corresponding frequencies are given by

$$f_1 = f_r\left(1 - \frac{1}{2Q}\right)$$

and $$f_2 = f_r\left(1 + \frac{1}{2Q}\right)$$

The bandwidth of the network is therefore given by

$$\Delta f = f_2 - f_1 = \frac{f_r}{Q}$$

Hence $$Q = \frac{f_r}{\Delta f} \tag{33}$$

This is the same relation as that for series resonance.

Problems

1. A resistor of 10 Ω is connected in parallel with a capacitor of 318 μF, and a 100-V, 50-Hz a.c. supply is applied to this network. Find the supply current and the power factor.
2. A resistor of 10 Ω is connected in parallel with a 31·8-mH inductor. A 200-V, 50-Hz a.c. supply is applied to the network. Find the supply current and the power factor.
3. A 50-mH inductor coil of negligible resistance is connected in parallel with a 30-Ω resistor to a 240-V, 50-Hz a.c. supply. Calculate:
   (*a*) the supply current;
   (*b*) the power factor;
   (*c*) the supply power.
4. An inductor coil of negligible resistance is connected in parallel

with a resistor to a 240-V, 50-Hz a.c. supply. The supply power is 1440 W and the supply current is 10 A. Find:
(*a*) the current in the coil;
(*b*) the inductance of the coil;
(*c*) the supply power factor.

5. A resistor and a capacitor are connected in parallel to a 110-V, 850-Hz supply and the supply current of 5·0 A leads the voltage by 60°. Determine:
(*a*) the current in the resistor;
(*b*) the resistance of the resistor;
(*c*) the current in the capacitor;
(*d*) the capacitance of the capacitor;
(*e*) the supply power.
6. An inductor of inductance 96 mH and resistance 13 Ω is connected in parallel with a capacitor of capacitance 120 μF. The network is connected to a 50-V, 60-Hz supply. Determine the supply current and the power factor.
7. For the network shown in Fig. 28, the current I is
(*a*) 5 A with a leading power factor;
(*b*) 5 A with a lagging power factor;
(*c*) 7 A with a leading power factor;
(*d*) 7 A with a lagging power factor?

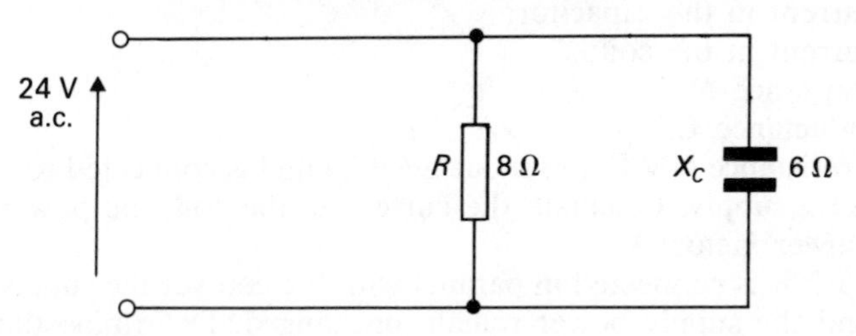

Fig. 28

8. A sinusoidal alternating current of 10 mA flows in the inductive coil shown in Fig. 29. The coil has inductance 1·0 H and resistance

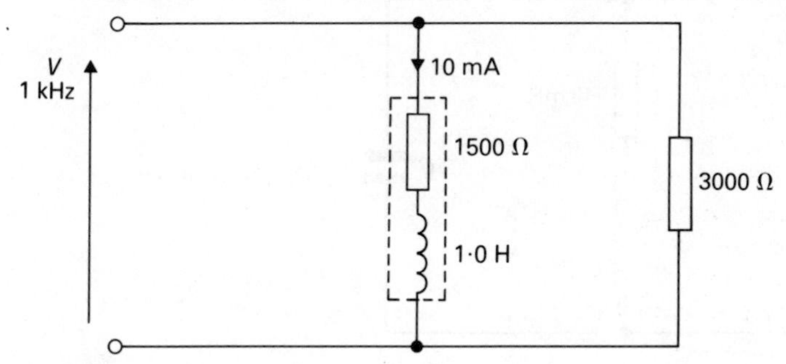

Fig. 29

1500 Ω. Calculate:
(*a*) the supply voltage;
(*b*) the current in the 3000-Ω resistor;
(*c*) the supply current;
(*d*) the supply phase angle.

9. A coil of inductance 35mH and resistance 8·0 Ω is connected in parallel with a capacitor to a 240-V, 50-Hz supply. The supply power factor is unity. Determine:
(*a*) the current in the coil;
(*b*) the supply current;
(*c*) the current in the capacitor;
(*d*) the capacitance of the capacitor.

10. A coil connected to a 100-V, 50-Hz supply takes a current 20·0 A at a power factor 0·5 lagging. A capacitor is then connected in parallel with the coil and the supply power factor becomes 0·9 lagging. Determine:
(*a*) the supply current;
(*b*) the current in the capacitor;
(*c*) the capacitance of the capacitor.

11. A coil of resistance R ohms and inductance L henrys is connected in parallel with a 110-μF capacitor to a 150-V, 50-Hz supply. The supply current is 8·0 A at unity power factor. Calculate:
(*a*) the current in the capacitor;
(*b*) the current in the coil;
(*c*) the resistance R;
(*d*) the inductance L.

12. A coil of resistance 600 Ω and inductance 24 mH is connected to a 10-V, 5-kHz supply. Calculate the current in the coil, the power and the power factor.

A capacitor is connected in parallel with the coil yet the supply current and the supply power remain unchanged. Determine the capacitance of the capacitor.

13. The network shown in Fig. 30 has an overall power factor of unity.

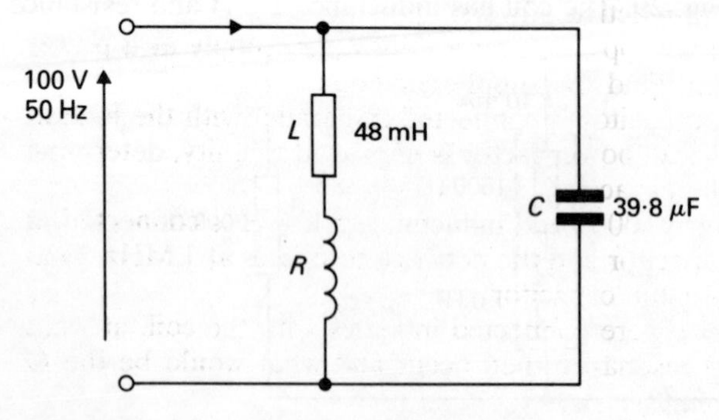

Fig. 30

Find:
(a) the value of R;
(b) the current in the inductive branch;
(c) the supply current.

14. A coil of resistance 12 Ω and inductance 0·12 H is connected in parallel with a 60-μF capacitor to a 100-V, variable-frequency supply. Determine:
    (a) the frequency at which the network will behave as a non-reactive resistor;
    (b) the dynamic impedance of the network.
15. A network consists of a coil of resistance 100 Ω and inductance 0·5 H in parallel with a capacitor of capacitance 20 μF. The network operates at unity power factor when connected to a 100-V, variable-frequency sinusoidal supply. Calculate:
    (a) the supply frequency;
    (b) the supply current;
    (c) the coil current.

    Given that the frequency is varied so that resonance occurs when the coil and the capacitor are connected in series across the supply, calculate:
    (d) the change in frequency;
    (e) the new supply current.
16. A 240-V, 50-Hz, 1-ph motor is loaded to have an output power 10 kW. Its operating efficiency is 85% and the power factor 0·75 lagging. Determine the capacitance of a capacitor connected in parallel with the motor which will change the supply power factor to unity. Determine also the supply current when the capacitor is connected into the circuit.
17. A 240-V, 1-ph supply feeds two loads connected in parallel. One load comprises incandescent lamps requiring a current of 12·0 A. The other load is a motor taking a current of 10·0 A at a power factor of 0·8 lagging. Determine:
    (a) the supply current;
    (b) the overall power factor;
    (c) the active and reactive powers.
18. A 20-kVA load is supplied from a 240-V a.c. supply at a power factor 0·7 lagging. Find the supply current.

    Given that a capacitor is connected in parallel with the load in order that the overall power factor is improved to unity, determine the current in the capacitor.
19. A coil of resistance 100 Ω and inductance 500 μH is connected in parallel with a capacitor and the network resonates at 1 MHz. Find the capacitance of the capacitor.

    If the capacitor were connected in series with the coil, at what frequency would resonance then occur and what would be the Q factor of the circuit?

    A 30-pF capacitor is connected in parallel with the first

capacitor, which is still connected in series with the coil. At what frequency does resonance now occur?

Answers

1. 14·1 A; 0·707 lead
2. 28·3 A; 0·707 lag
3. 17·3 A; 0·46 lag; 1920 W
4. 8·0 A; 95·5 mH; 0·6 lag
5. 2·5 A; 44 Ω, 4·33 A; 7·4 μF; 275 W
6. 1·13 A; 0·39 lead
7. (*a*)
8. 64·6 V; 21·5 mA; 25·7 mA; 22° lag
9. 17·6 A; 10·4 A; 14·2 A; 188 μF
10. 11·1 A; 12·5 A; 400 μF
11. 5·2 A; 9·5 A; 13·3 Ω; 27·2 mH
12. 10·4 mA; 64·6 mW; 0·62 lag; 0·052 μF
13. 31·3 Ω; 2·9 A; 2·6 A
14. 57·1 Hz; 167 Ω
15. 39 Hz; 0·4 A; 0·63 A; 11·3 Hz; 1·0 A
16. 574 μF; 49 A
17. 20·9 A; 0·96 lag; 4800 W; 1440 var
18. 83·3 A; 59·5 A
19. 50·7 pF; 1 MHz; 31·4; 792 kHz

Chapter 4

Three-phase supply

Most electrical energy is generated in large electricity generating stations and is then transmitted to the many consumers throughout the country. The transmission of the energy depends heavily on the use of three-phase (3-ph) alternating current systems and, apart from battery operated devices, almost every electrical piece of apparatus depends on a 3-ph system whereby it derives its power. It is therefore important that we should know something of these systems in order that we may better understand our supply of electricity.

Before describing the practice of 3-ph systems, we must study the theory on which it is based and from which its terminology is derived.

1 Polyphase systems

There are two polyphase systems of importance, namely 2-ph and 3-ph systems. The 2-ph system was quite common at one time but it is less flexible and less economical than the 3-ph system which has consequently been implemented as the standard national system. The 2-ph system has therefore disappeared almost completely as a system for power transmission – there are still a number of rural consumers supplied by this system – but the 2-ph system is growing in importance for certain forms of servomechanisms used in control systems. For the majority of power applications, the 3-ph system is that used and therefore most of the following theory will concern this system.

When a coil is rotated with uniform angular velocity in a uniform magnetic field, an e.m.f. is induced in it and this e.m.f. varies sinusoidally. This simple form of generation is shown in Fig. 1 along with the resulting e.m.f. wave.

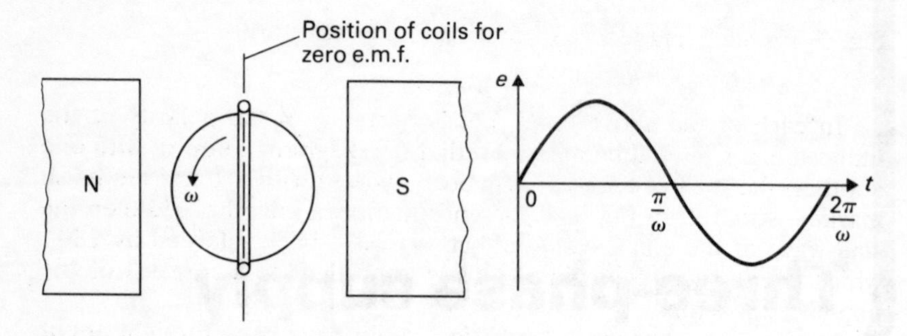

Fig. 1 Simple 1-ph generator and its e.m.f. waveform

If two coils are mounted at right angles to one another, as shown in Fig. 2, each has a sinusoidally varying e.m.f. induced in it, the waves being displaced by 90° or $\pi/2\omega$ seconds.

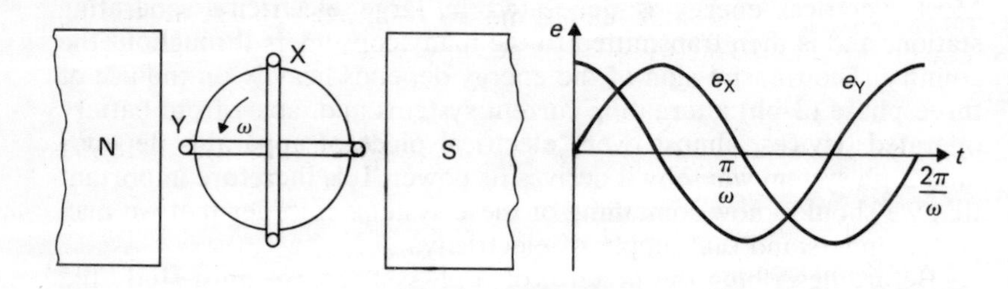

Fig. 2 Simple 2-ph generator and its e.m.f. waveforms

If three coils are mounted, each displaced by $120° = 2\pi/3$ radians from each other, as shown in Fig. 3, each coil again has a sinusoidally varying e.m.f. induced in it, the waves being displaced by 120° or $2\pi/3\omega$ seconds.

The coils of the 3-ph generator are designated red, yellow and blue (R, Y, B). For the arrangement shown, first the red coil, then the yellow coil and finally the blue coil reach the position at which each experiences the maximum positive e.m.f. induced in it. This is termed the RYB sequence and is now the national standard sequence.

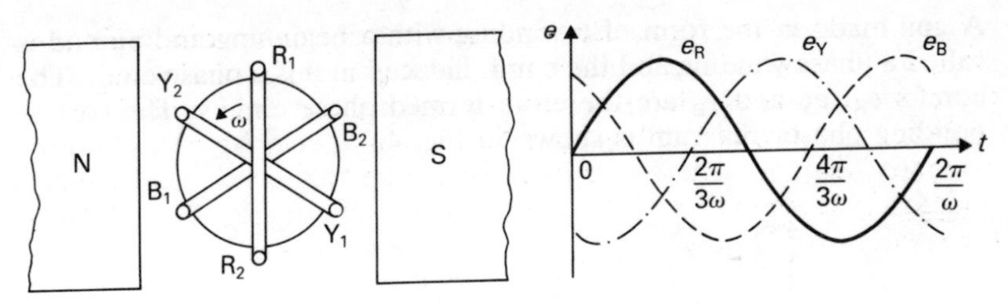

Fig. 3 Simple 3-ph generator and its e.m.f. waveforms

In each of the above cases, the importance of the polarity of the induced e.m.f. should be noted. So that the diagrams conform with one another, the start of each coil has been marked with a 1 and the finish marked with a 2. In Fig. 1, if the ends had been interchanged then the sine wave of the e.m.f. would effectively have been retarded by 180°, which is π/ω seconds in time. This is equivalent to a reversal of the phase of the supply.

In Fig. 2, a more important effect could have been brought about by one coil having its connections reversed whereby e_X would lead e_Y instead of the standard sequence shown.

In Fig. 3, the reversal of any one-phase coil would destroy the symmetry of the resulting waveform system; the symmetry is the outstanding feature of the 3-ph system. It is the normal case to generate a completely symmetrical system in which each e.m.f. has the same effective value. This is achieved by using identical coils in each phase and hence the maximum e.m.f. in each phase is the same, E_m. Let us take the R-phase as reference and hence

$$e_R = E_m \sin \omega t \tag{1}$$

The Y-phase coil e.m.f. lags by $2\pi/3\omega$ seconds and hence the e.m.f. induced in it is given by

$$e_y = E_m \sin\left(\omega t - \frac{2\pi}{3}\right) \tag{2}$$

$$= E_m \sin(\omega t - 120°)$$

The *B*-phase coil e.m.f. lags by a further $2\pi/3\omega$ seconds, or alternatively can be said to lead the R-phase e.m.f. by $2\pi/3\omega$ seconds, hence:

$$e_B = E_m \sin\left(\omega t - \frac{4\pi}{3}\right) \tag{3}$$

$$= E_m \sin\left(\omega t + \frac{2\pi}{3}\right)$$

$$= E_m \sin(\omega t - 240°)$$

A coil made in the form of a winding with a beginning and an end is called a phase winding and the e.m.f. induced in it is a phase e.m.f. The e.m.f.s e_R, e_Y and e_B are therefore termed phase e.m.f.s. The corresponding phasor diagram is shown in Fig. 4.

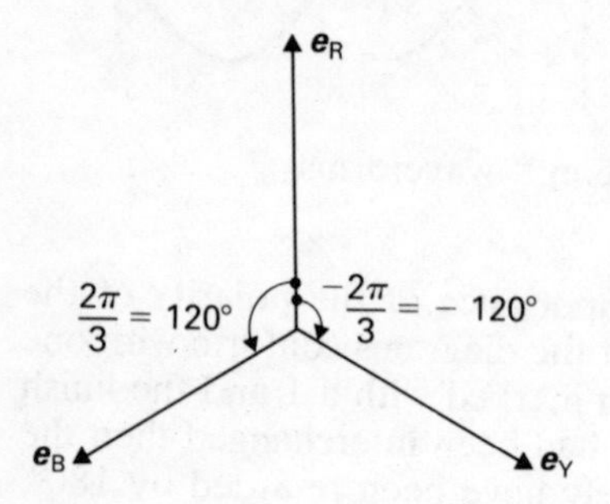

Fig. 4 Symmetrical 3-ph system phasor diagram

2 Symmetrical star-connected systems

The construction of the simple 3-ph generator is so arranged that the e.m.f.s can be tapped through slip-rings. If a load is connected by this means across each coil, then the arrangement becomes that shown in Fig. 5.

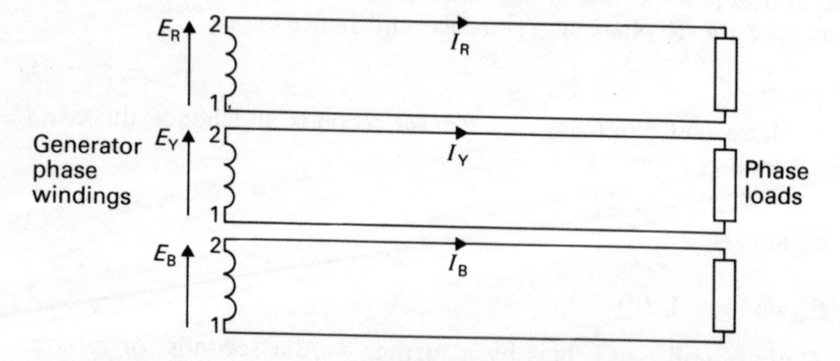

Fig. 5 Three loads supplied separately from three phases

The total transmission system therefore incorporates six conductors which could be a relatively costly installation; it is equivalent to three 1-ph systems. If each of the generator coil ends marked 1 is connected together and brought out from the generator through one common slip ring, the resulting network takes the form shown in Fig. 6.

The network diagram shown in Fig. 6 does not necessarily indicate

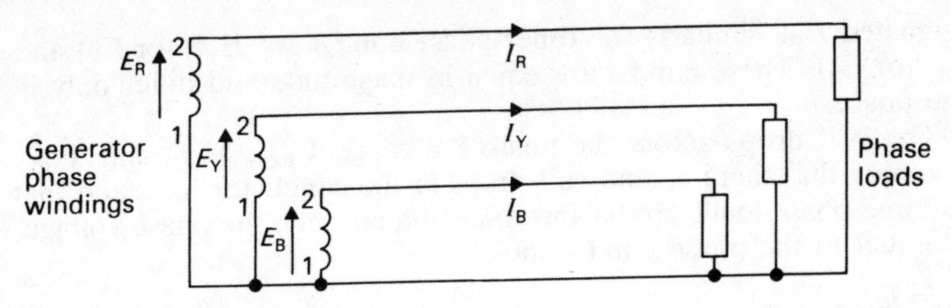

Fig. 6 Network made from circuits shown in Fig. 5 with one conductor common

that the system is a 3-ph one and this problem is overcome by rearranging the diagram as shown in Fig. 7. The network in this form is termed a 4-wire, star-connected system. The diagram is drawn to simulate the phasor diagram, hence the term 'star'. The fourth wire

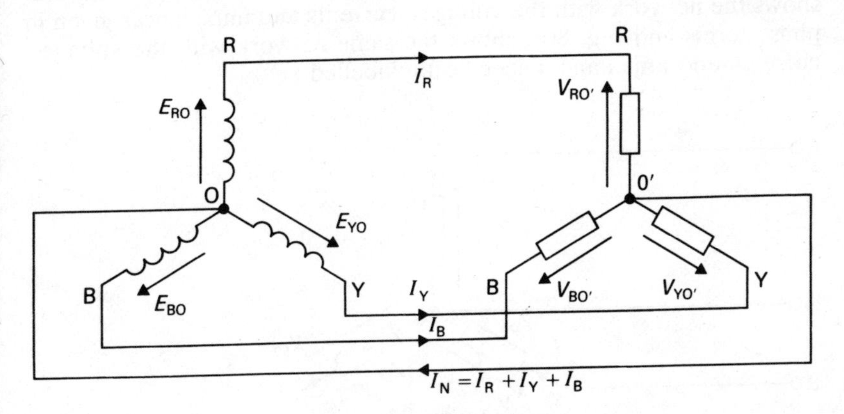

Fig. 7 3-ph, 4-wire star-connected load supplied from a star-connected generator

acts as a common 'return' wire for the currents in the three phases and is called the neutral wire. This part of the network is also referred to as the star or neutral point when discussing the e.m.f.s and the potential differences within the network.

It is reasonable to wonder about the suitability of reducing six wires to four wires in this way. For instance, might we not require a conductor of greater cross-sectional area to replace the 'return' wire since it is carrying the 'return' currents of the three phase circuits? To answer this question, we must analyse the operation of the network.

The e.m.f. between any line (R, Y or B) and the star point O is termed the phase e.m.f. E_{ph}. The R-phase e.m.f. is designated E_{RO} but reference to the star point may be omitted, in which case it is designated E_R. Similarly the other phase e.m.f.s are E_{YO} (or E_Y) and E_{BO} (or E_B). These e.m.f.s are equal in magnitude and differ only in time phase.

The volt drops across the phase loads are $V_{RO'}$, $V_{YO'}$ and $V_{BO'}$. Assuming that there are no volt drops in the conductor lines and that the three phase loads are balanced, i.e., equal, then the phase voltages are equal to the phase e.m.f.s and

$$V_{ph} = E_{ph} \tag{4}$$

Because we have assumed that there is no volt drop in the conductors, then for the R-phase,

$$V_{RO'} = V_{RO}$$

and as before it is permissible to omit the reference to the star point, in which case the volt drop across the R-phase is designated V_R. These observations may therefore be illustrated as shown in Fig. 8. Fig. 8(a) shows the network with the voltages, currents and impedances given in phase terms and Fig. 8(b) shows the same network with the voltages, currents and impedances specifically labelled.

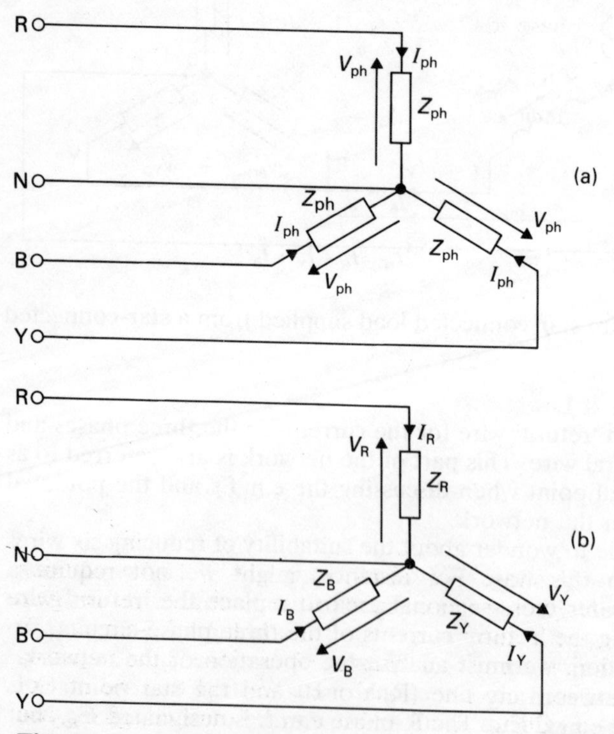

Fig. 8 Terminology of 3-ph, 4-wire star-connected networks

The current in the R-phase is given by

$$I_R = \frac{V_R}{Z_R} \tag{5}$$

The currents in the other phases are given by

$$I_Y = \frac{V_Y}{Z_Y}$$

and $\quad I_B = \dfrac{V_B}{Z_B}$

In more general terms, we may also observe that the phase current I_{ph} is given in each case by

$$I_{ph} = \frac{V_{ph}}{Z_{ph}} \tag{6}$$

It must be emphasised that this final step depends on the system's being balanced and that the three phase circuits are identical in every way except that their operations are displaced in time by a mutual 120°.

Consider if each phase load were a resistor *R*. It follows that

$$V_R = V_Y = V_B = V_{ph}$$

hence $\quad v_R = V_{ph_m} \sin \omega t$

and $\quad i_R = \dfrac{V_{ph_m}}{R} \sin \omega t = I_{ph_m} \sin \omega t$

Similarly $\quad i_Y = I_{ph_m} \sin\left(\omega t - \dfrac{2\pi}{3}\right) = I_{ph_m} \sin(\omega t - 120°)$

and $\quad i_B = I_{ph_m} \sin\left(\omega t - \dfrac{4\pi}{3}\right) = I_{ph_m} \sin(\omega t - 240°)$

By Kirchhoff's First Law:

$$\begin{aligned} i_N &= i_R + i_Y + i_B \\ &= I_{ph_m}(\sin \omega t + \sin(\omega t - 120°) + \sin(\omega t - 240°)) \\ &= 0 \end{aligned} \tag{7}$$

This result may also be shown graphically as in Fig. 9. The summation of the phase currents, being the current in the neutral wire, is again zero. Thus the current in the neutral wire of a balanced 3-ph, 4-wire star-connected system is zero at all instants. Thus rather than requiring a conductor of increased cross-sectional area, we find that we do not require the neutral conductor at all and it may be omitted, giving rise to a further economy.

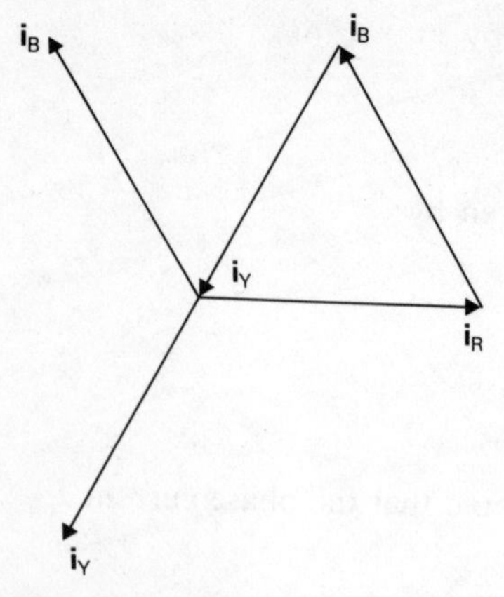

Fig. 9 Summation of phase currents in neutral wire

Although the case taken was that for equal load resistors, the same result would have been obtained for any system of star-connected similar impedances. The system may therefore be reduced to the 3-ph, 3-wire system shown in Fig. 10. Although the fourth wire has been omitted, O and O′ remain at the same potential so long as the system is balanced and hence for the system shown, E_{ph} is still equal to V_{ph}.

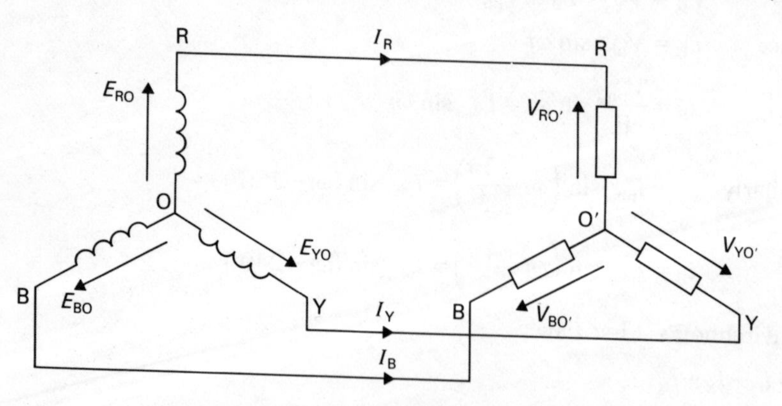

Fig. 10 3-ph, 3-wire star-connected network

With the omission of the fourth wire, it may no longer be possible to measure either E_{ph} or V_{ph} because we may be unable to obtain access to the star points. The line conductors remain and we can therefore measure the e.m.f.s between lines. These are termed the line e.m.f.s E_l and for the given system, they are E_{RY}, E_{YB} and E_{BR}.

The line e.m.f.s and the corresponding line voltages are the values given when describing a 3-ph system. The most common example is that of a 415-V, 3-ph supply which is one in which the line voltage is 415 V. Similarly the National Grid is rated at 275 kV, indicating that there is 275 000 V between lines.

The directions of current and potential differences shown in Fig. 11 are conventional. It can also be seen from the diagram that the phase current is also the current in each line, i.e. the line current I_l, hence for a star-connected system

$$I_l = I_{ph} \tag{8}$$

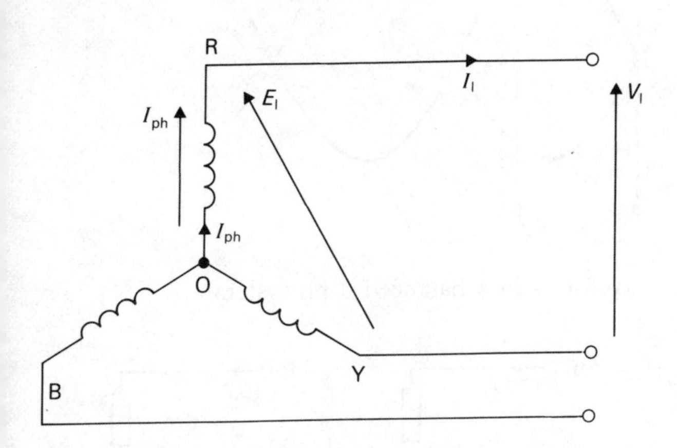

Fig. 11 3-ph, 3-wire star-connected generator supplying a balanced load

Now that there are only three wires, we should remember that at any instant there is a current flowing towards the load in one or two conductors and the return current is through the remaining two or one conductor. It is easy to overlook this point when diagrams such as Fig. 10 seem to indicate that all the currents flow toward the load. Such a diagram, however, indicate the conventional direction of r.m.s. flow but these currents vary in magnitude and direction from instant to instant. Let us look at the current waveforms as shown in Fig. 12. Three instants in the course of one cycle have been indicated as a, b and c. At instant a, i_R has attained its maximum value of I_{ph_m}. Let this be 100 A. At the same instant, i_Y is −50 A and i_B is also −50 A. These values have been indicated in Fig. 13(*a*) and it is now readily observed that the sum of the currents is zero.

Taking the more general point b, we find that i_R is 20 A, i_Y is 72 A and i_B is −92 A. Again these values are illustrated in Fig. 13(*b*) and we can observe that the sum of the instantaneous currents is zero.

Finally, there is the special case instanced at c, in which i_R is 0, i_Y is 87 A and i_B is −87 A as illustrated in Fig. 13(*c*). As we would anticipate, the sum of the currents is zero but we also observe that the

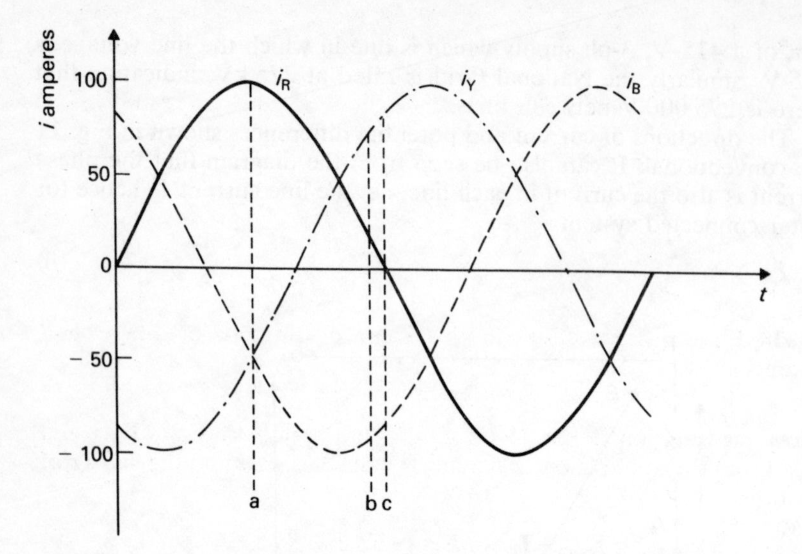

Fig. 12 Current waveforms in a balanced 3-ph system

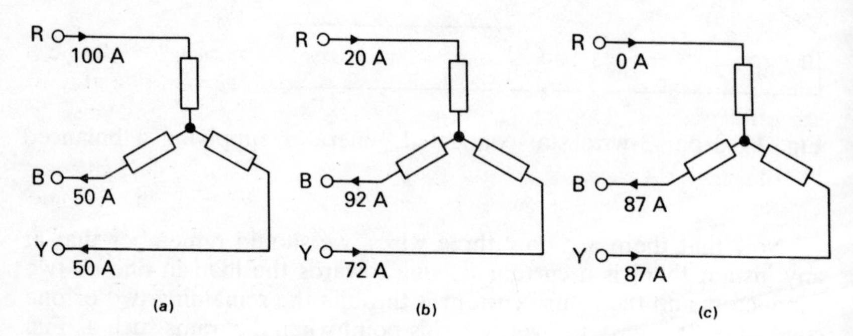

Fig. 13 Instantaneous current values

current in one of the phases is also zero and that therefore the network current at that instant is restricted to two conductors only.

This series of illustrations should therefore reinforce our understanding that for a balanced load, the sum of the currents in the conductors is zero and hence we have been able to omit the fourth wire. Also it should serve as a reminder to distinguish between the instantaneous values and the r.m.s. values with which we deal in the remainder of this chapter.

The phasor diagram for the phase currents and voltages of a star-connected system is shown in Fig. 14. To obtain the line voltages on a phasor diagram, we require to apply Kirchhoff's Second Law to the network shown in Fig. 11. In this case, you may find it easier to

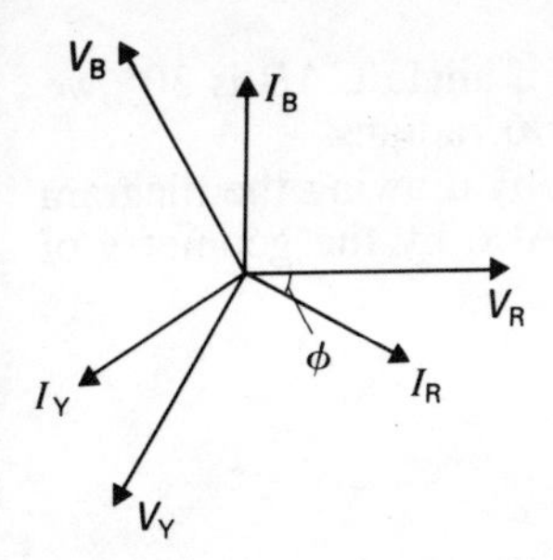

Fig. 14 Phasor diagram for a 3-ph system showing phase voltages and currents

follow the solution if the phase e.m.f.s are stated as E_{RO}, E_{YO} and E_{BO}. From the application of Kirchhoff's Second Law and the subscript notation,

$$\boldsymbol{V}_{RY} = \boldsymbol{E}_{RY} = \boldsymbol{E}_{RO} + \boldsymbol{E}_{OY} = \boldsymbol{E}_{RO} - \boldsymbol{E}_{YO} \tag{9}$$

Also $\quad \boldsymbol{Y}_{YB} = \boldsymbol{E}_{YB} = \boldsymbol{E}_{YO} - \boldsymbol{E}_{BO}$

and $\quad \boldsymbol{V}_{BR} = \boldsymbol{E}_{BR} = \boldsymbol{E}_{BO} - \boldsymbol{E}_{RO}$

The required subtraction is carried out in the normal phasor diagram manner by reversing the appropriate phasor and then adding this to the other phasor. This operation is illustrated in Fig. 15, and may be analysed by the following observations.

The angle between $\boldsymbol{E}_{RO}$ and $\boldsymbol{E}_{YO}$ is 120°, thus the angle between $\boldsymbol{E}_{RO}$ and $-\boldsymbol{E}_{YO}$, which is parallel to $\boldsymbol{E}_{YO}$, is also 120°. For the triangle designated *ABC*, the angle *ABC* is 120°. Also $E_{RO} = E_{YO} = E_{ph}$, thus $AB = BC$ and the triangle is isosceles. This implies that angle *CAB* is equal to angle *BCA* and, the angles of a triangle summing to 180°,

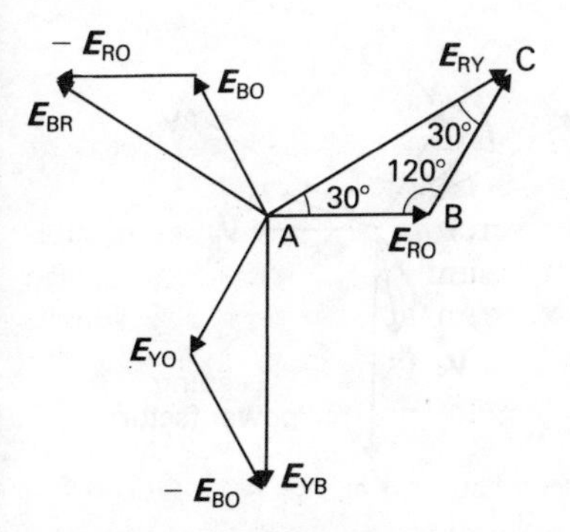

Fig. 15 Relationship between phase and line voltages in a 3-ph, star-connected system

these angles must therefore be each 30°. Thus if angle CAB is 30°, we may observe that E_{RY} leads E_{RO} by 30° or $\pi/6$ radians.

The same conclusion can also be obtained by drawing the diagram to scale and measuring the angle to be 30°. Also by the geometry of the diagram

$$V_{RY} = E_{RY}$$

$$= E_{RO} \cos 30° + E_{YO} \cos 30°$$

$$= 2E_{RO} \cos 30°$$

$$= 2E_{RO} \frac{\sqrt{3}}{2}$$

$$= \sqrt{3}E_{RO}$$

Since all phase voltages are equal and all line voltages are equal in a balanced system, it follows that

$$V_l = E_l = \sqrt{3}E_{ph} = \sqrt{3}V_{ph} \tag{10}$$

Example 1 Find the phase voltage of a 415-V, 3-ph supply.

$$V_l = 415\text{ V}$$
$$= \sqrt{3}V_{ph}$$

$$V_{ph} = \frac{415}{\sqrt{3}} = \underline{240\text{ V}}$$

The complete phasor diagram for a 3-wire, star-connected system takes the form shown in Fig. 16. You should note that the phase angle of the system is that between phase voltages and phase currents and

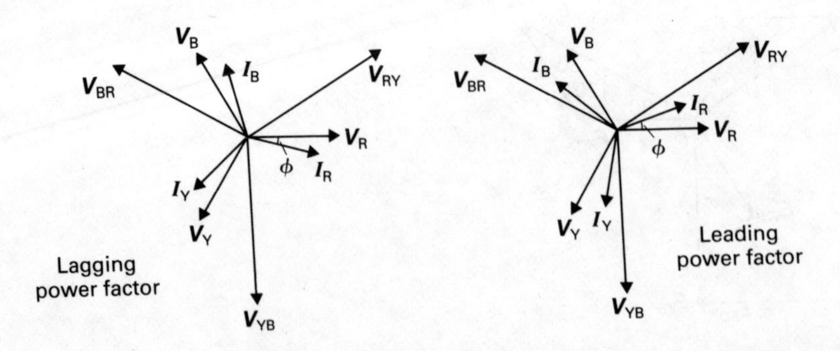

Fig. 16 Complete phasor diagram for 3-ph, 3-wire star-connected systems

the angle between the line voltages and the line currents is $(30° + \phi)$, which is not necessarily a larger angle for certain leading phase angles.

Finally in the case of the balanced load shown in Fig. 17, the star point O′ of the load is at the same potential as the star point O of the source, and hence $E_{ph} = V_{ph}$. It follows from relation (10) that

$$V_l = \sqrt{3}\, V_{ph}$$

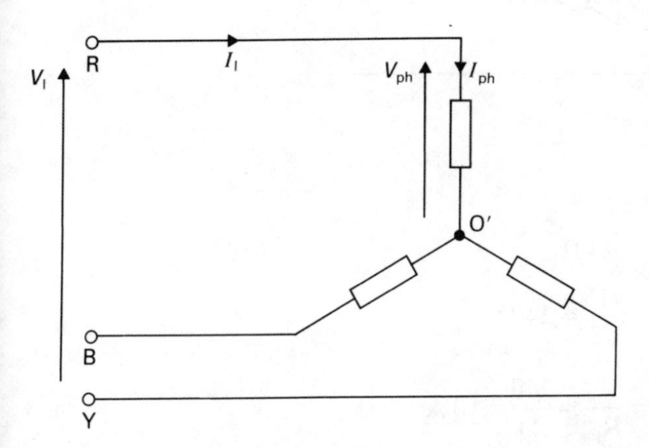

Fig. 17 3-ph, three-wire star-connected balanced load

In each phase, the line voltage leads the phase voltage by 30° ($= \pi/6$ radians) and as always there is a mutual displacement of 120° ($= 2\pi/3$ radians) between the line voltages and between the phase voltages.

The power developed in the system is three times that developed in each phase, hence

$$P = 3 V_{ph} I_{ph} \cos\phi \qquad (11)$$

$$= 3 E_{ph} I_{ph} \cos\phi$$

But $V_{ph} = V_l/\sqrt{3}$ and $I_{ph} = I_l$; hence

$$P = 3 \frac{V_l}{\sqrt{3}} I_l \cos\phi$$

$$= \sqrt{3}\, V_l I_l \cos\phi \qquad (12)$$

Example 2 Three 24-Ω resistors are connected in star to a 3-ph, 3-wire supply of line voltage 415 V. Calculate the line current and the total power dissipated by the resistors.

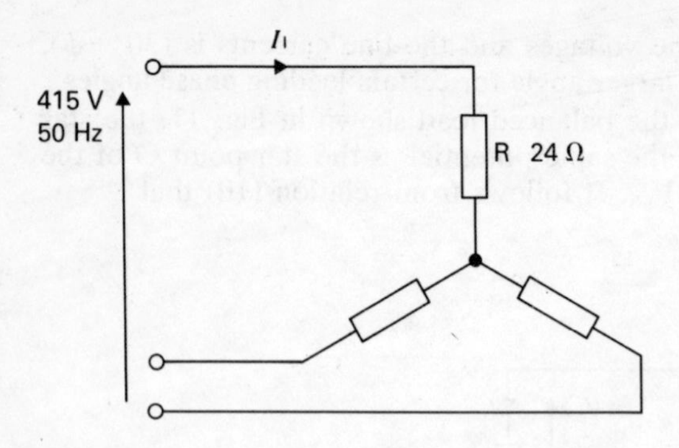

Fig. 18

$$Z_{ph} = R_{ph} = 24\ \Omega$$

$$V_{ph} = \frac{V_l}{\sqrt{3}} = \frac{415}{\sqrt{3}} = 240\ \text{V}$$

$$I_l = I_{ph} = \frac{V_{ph}}{R_{ph}} = \frac{240}{24} = \underline{10\ \text{A}}$$

$$P = 3I_{ph}^2 R_{ph} = 3 \times 10^2 \times 24 = \underline{7200\ \text{W}}$$

or $\quad P = \sqrt{3}V_l I_l \cos\phi = \sqrt{3} \times 415 \times 10 \times 1 = \underline{7200\ \text{W}}$ as before.

Example 3 Three similar coils, each of resistance 7·0 Ω and inductance 30 mH are connected in star to a 415-V, 3-ph, 50-Hz supply. Calculate the line current and the total power dissipated.

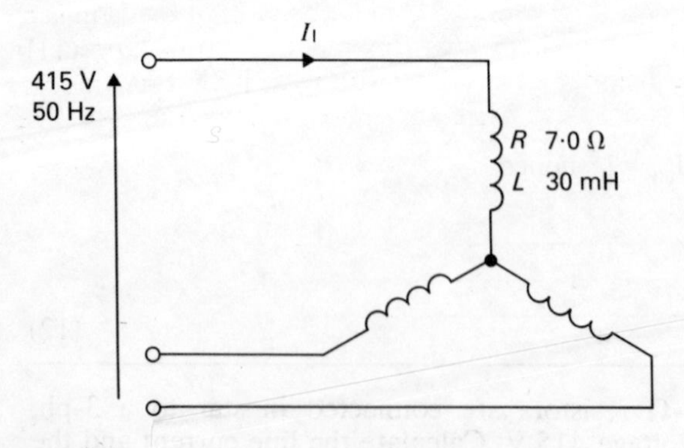

Fig. 19

$$X_L = 2\pi fL = 2\pi \times 50 \times 30 \times 10^{-3} = 9{\cdot}4\ \Omega$$

$$Z_{ph} = (R^2 + X_L^2)^{1/2} = (7{\cdot}0^2 + 9{\cdot}4^2)^{1/2} = 11{\cdot}7\ \Omega$$

$$V_{ph} = \frac{V_1}{\sqrt{3}} = \frac{415}{\sqrt{3}} = 240\ \text{V}$$

$$I_1 = I_{ph} = \frac{V_{ph}}{Z_{ph}} = \frac{240}{11{\cdot}7} = \underline{20{\cdot}5\ \text{A}}$$

$$P = 3I_{ph}^2 R_{ph} = 3 \times 20{\cdot}5^2 \times 7{\cdot}0 = \underline{8825\ \text{W}}$$

or $$\cos\phi = \frac{R_{ph}}{Z_{ph}} = \frac{7{\cdot}0}{11{\cdot}7} = 0{\cdot}6\ \text{lag}$$

$$P = \sqrt{3}\,V_1 I_1 \cos\phi = \sqrt{3} \times 415 \times 20{\cdot}5 \times 0{\cdot}6 = \underline{8825\ \text{W}}$$

---

3 Unbalanced star-connected loads

A 3-ph system serves to transmit energy from the generating stations at high values of power. Most loads however are 1-ph loads such as domestic appliances and lighting. The small 1-ph loads are great in number and create the combined effects of all the domestic consumers who take about half the energy provided by the supply authorities.

Also there is a limit to how small a load can be and remain economical operating from a 3-ph supply. Usually consumers requiring less than 15 kW from a supply authority are connected by means of a 1-ph system, and almost all consumers requiring more than 30 kW are connected by means of a 3-ph system. Thus we should consider how these various loads are connected to the 3-ph transmission system coming from the generating stations.

With 1-ph consumers, the problem is that each has a different load and therefore we cannot readily place the consumer loads in balanced groups of three. However if we were to connect any three loads in star, the resulting 3-ph load would, in all probability, be unbalanced. In order to maintain the balance of the load voltages, it becomes necessary to re-introduce the fourth wire, as illustrated in Fig. 20. The

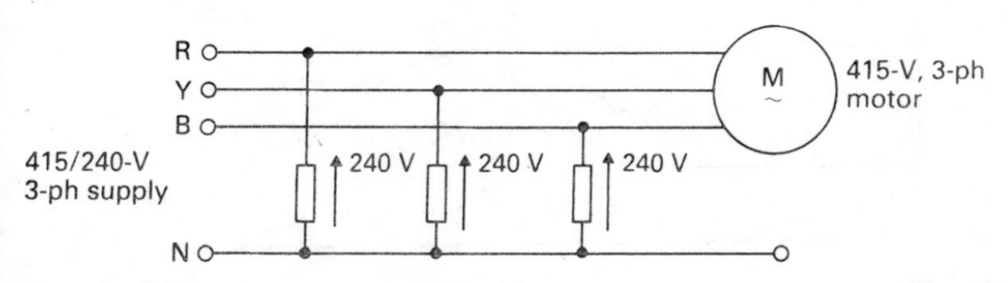

Fig. 20 Dissimilar 1-ph loads connected to a 3-ph, four-wire system

reintroduction of the fourth wire ensures that the neutral connection of each 1-ph load is at the same potential as that of the star point of the source, thus the voltage across each load remains the phase voltage V_{ph} given by $V_l/\sqrt{3}$.

Due to the dissimilarity of the loads, the currents in each load, and therefore in each line, are not necessarily equal to one another. This lack of balance results in a current flowing in the neutral conductor, as instanced in Fig. 22(*b*).

In most cases, a 3-ph, 4-wire system supplies a considerable number of consumers and it is found that the greater the number of 1-ph loads connected to each of the three phases, the better is the chance that their overall effect will balance. It is this tendency which ensures that any group of domestic consumers supplied from a 3-ph transmission system will give rise to a total load that is effectively balanced. Further, it may be noted that the larger the group, the greater is the probability of balance.

In cases involving larger commercial consumers, much of their demand arises from 3-ph motors that are already balanced. The remainder of their demand may stem from banks of lighting which can be connected in equal groups to give a balanced load. Such a consumer may therefore present a sufficiently balanced load to the supply system that a 3-wire supply can be used, but in such cases, a higher supply voltage such as 11 kV would be used. 415-V, 3-ph supplies usually incur the use of 4-wire connections, and an example of this form of supply is investigated in Example 4.

Example 4 A 415-V, 3-ph, 4-wire system supplies three resistive loads of 20 kW, 16 kW and 10 kW connected as shown in Fig. 21. Find the current in each of the conductors.

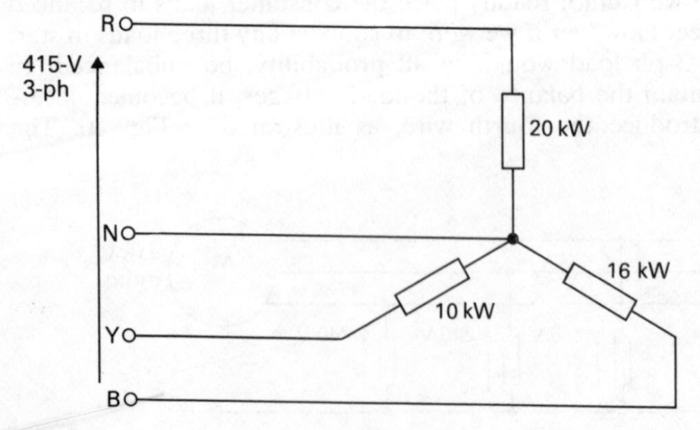

Fig. 21

$$V_{ph} = \frac{V_l}{\sqrt{3}} = \frac{415}{\sqrt{3}} = 240 \text{ V}$$

$$I_R = \frac{P_R}{V_{ph}} = \frac{20\,000}{240} = \underline{83 \text{ A}}$$

$$I_Y = \frac{P_Y}{V_{ph}} = \frac{16\,000}{240} = \underline{67 \text{ A}}$$

$$I_B = \frac{P_B}{V_{ph}} = \frac{10\,000}{240} = \underline{42 \text{ A}}$$

These three currents are shown in the phasor diagram of Fig. 22. Since each load is resistive, the currents are in phase with the phase voltages and are therefore mutually displaced by 120°. The current in the neutral conductor is given by

$$\boldsymbol{I}_N = \boldsymbol{I}_R + \boldsymbol{I}_Y + \boldsymbol{I}_B$$

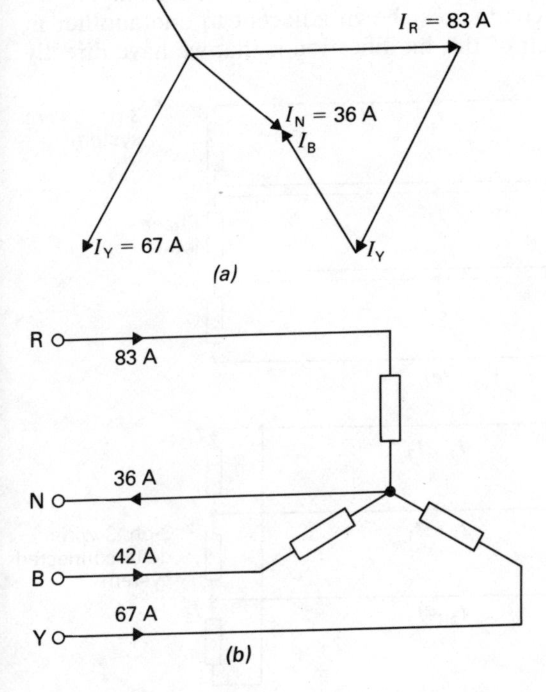

Fig. 22

and from the diagram drawn to scale, we find that $I_N = \underline{36\ A}$. The resulting distribution of currents is also illustrated in Fig. 22.

Finally it can be noted that the current in the neutral conductor cannot exceed the greatest current in any one line provided the various phase loads are reasonably similar in phase angle. The most common cases involve loads that are more or less resistive; hence it would be unusual to find a current in the neutral wire greater than in any one line. It is more likely that the largest neutral current would occur when only one phase load were connected and the neutral current would then be equal to the current in the line to which the load was connected. It follows that the cross-section of the neutral conductor should be at most equal to that of each line.

4 Symmetrical delta-connected systems

The three separate circuits shown in Fig. 23(*a*) requiring six conductors can be modified into a system alternative to that of star-connection. The second method is to connect the end of each coil to the beginning of the next as illustrated in Fig. 23(*b*). Effectively we have combined pairs of conductors shown adjacent to one another in the diagram and the result of this modification is that we have directly

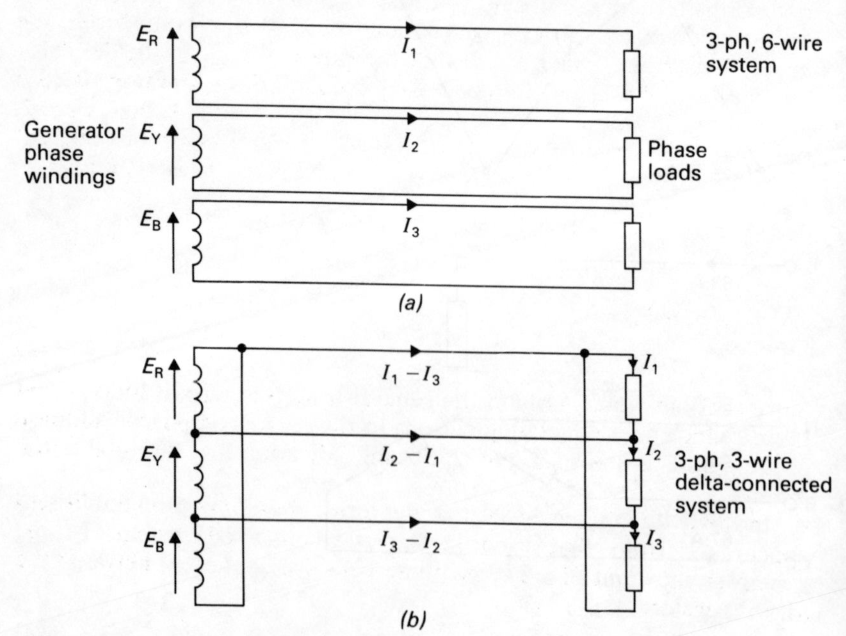

Fig. 23 Derivation of the delta-connected system

reduced the six wires to three. This affords a considerable económic advantage.

The layout of Fig. 23(*b*) is an awkward one and again gives little indication that we are dealing with a 3-ph system. The conventional method of drawing the network for this arrangement is shown in Fig. 24. Because of the layout, both the generator and the load are said to

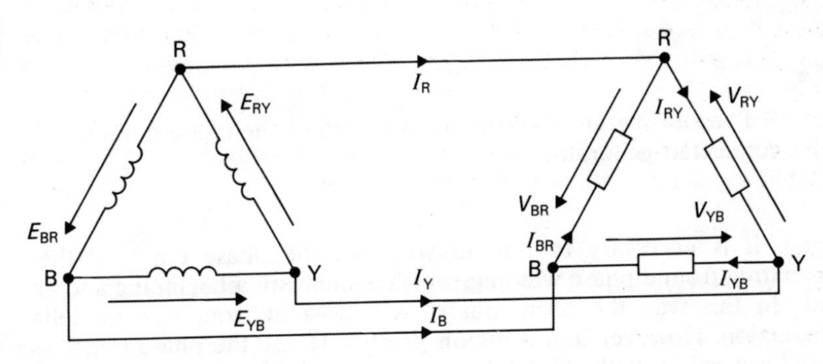

Fig. 24 Delta-connected generator supplying a delta-connected load

be delta-connected. An alternative term is mesh-connected, but this is not normally used; similarly, you may occasionally find the star connection described as wye connection or Y connection. From the network diagram, you will see that the phase e.m.f.s act in the same direction around the delta-connected coils of the generator. It also appears that there is no resulting volt drop around that circuit; therefore it is possible that there should be some enormous current around the generator loop. However if we add the phase e.m.f.s

$$\begin{aligned} e_R + e_Y + e_B &= E_m \sin \omega t + E_m \sin (\omega t - 120°) + E_m \sin (\omega t - 240°) \\ &= E_m (\sin \omega t + \sin (\omega t - 120°) + \sin (\omega t - 240°)) \\ &= E_m \times 0 \\ &= 0 \end{aligned} \tag{13}$$

Since the total e.m.f. around the generator loop is zero, it follows that there cannot be a circulating current in the loop. The phasor addition of the e.m.f.s is shown in Fig. 25, again indicating that the total e.m.f. is zero.

In the delta-connected system, the term 'phase' takes on a different appearance from that used in the star-connected system. In the generator shown in Fig. 24, the phase coils are connected between the lines; therefore

$$E_{ph} = E_l = V_l = V_{ph} \tag{14}$$

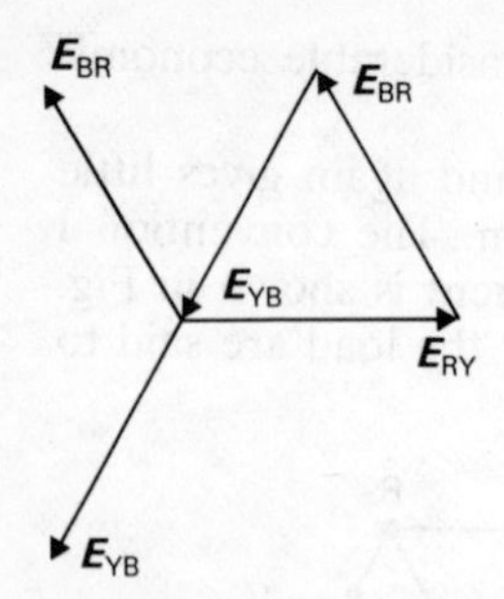

Fig. 25 Phasor diagram showing the addition of the phase e.m.f.s in a delta-connected generator

Again it is necessary to remind you that the phase e.m.f. is that emanating from a phase winding, which is one with a beginning and an end. In this way the term 'phase' is consistent from star to delta connection. However it is common practice to use the phase e.m.f. as $E_l/\sqrt{3}$ regardless of the form of connection and this misuse of the term can be confusing.

The phase current in a delta-connected system is taken as that in each phase winding. Two phase windings are connected to each line and, taking the conventional directions of current shown in Fig. 24, the line currents are obtained from the differences between the respective phase currents.

If the delta-connected loads are balanced, the line currents are found to be equal in magnitude. Let us apply Kirchhoff's First Law to each junction of the system, using the conventional notation indicated in Fig. 24:

$$\boldsymbol{I}_R = \boldsymbol{I}_{RY} - \boldsymbol{I}_{BR} \qquad (15)$$

$$\boldsymbol{I}_Y = \boldsymbol{I}_{YB} - \boldsymbol{I}_{RY}$$

$$\boldsymbol{I}_B = \boldsymbol{I}_{BR} - \boldsymbol{I}_{YB}$$

The phasor diagram corresponding to relation (15) is given in Fig. 26 and this shows the subtraction of one phase current from another to give the line current. This diagram bears a considerable resemblance to Fig. 15 and a similar form of analysis can be applied, with the result that we may observe that the phase angle between $\boldsymbol{I}_{RY}$ and $-\boldsymbol{I}_{BR}$ is 60°. By the symmetry of the phasor diagram, it follows that the phase current $\boldsymbol{I}_{RY}$ leads the line current $\boldsymbol{I}_R$, and by the geometry of the diagram,

$$I_R = I_{RY} \cos 30° + I_{BR} \cos 30°$$

$$= \sqrt{3} I_{RY}$$

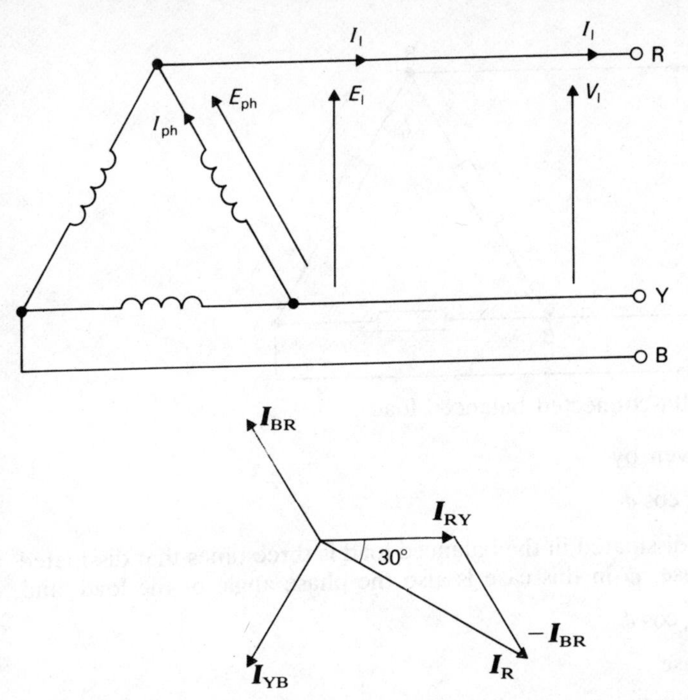

Fig. 26 Delta-connected generator supplying a balanced load

Given that the load is balanced, the line currents are equal in magnitude, as are the phase currents; thus

$$I_l = \sqrt{3} I_{ph} \tag{16}$$

This results applies to both the delta-connected generator and the delta-connected load.

In each phase, the line current lags the phase current by 30° ($=\pi/6$ radians) but, as always, there is a mutual displacement of 120° ($=2\pi/3$ radians) between the line currents and between the phase currents.

Turning our attention now to the delta-connected load, we find that it experiences similar voltages and currents to those on the generator. The phase voltage is also the line voltage, and this was the phase and line e.m.f. of the generator. The phase current in the load is equal to that in the generator in spite of there being a line current in between generator and load. The network diagram for a general load is shown in Fig. 27.

The power developed in the generator, when supplying a balanced load, is three times that developed in each phase, where again ϕ is the angle between the phase voltage and the phase current. Thus the total

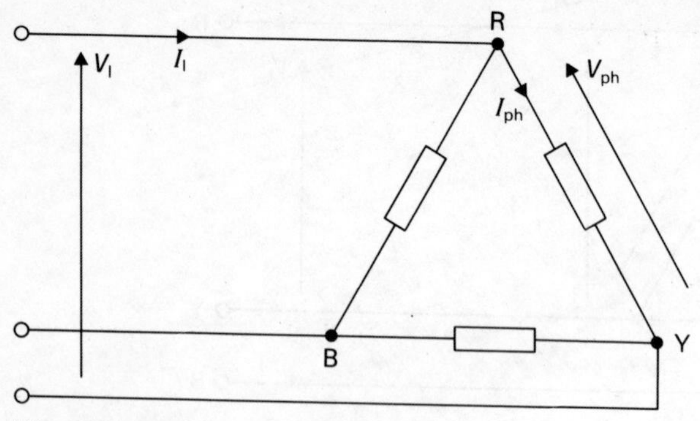

Fig. 27 Delta-connected balanced load

power is given by

$$P = 3E_{ph}I_{ph}\cos\phi$$

The power dissipated in the balanced load is three times that dissipated in each phase. ϕ in this case is also the phase angle of the load, and

$$P = 3V_{ph}I_{ph}\cos\phi$$

In either case

$$P = 3V_l \cdot \frac{I_l}{\sqrt{3}}\cos\phi$$

$$= \sqrt{3}V_lI_l\cos\phi \qquad (17)$$

This relation is the same as that for the power in a star-connected system and therefore it holds true for the power in any balanced 3-ph system.

Example 5 A 415-V, 3-ph induction motor is loaded to 10·4 kW output and operates at a power factor of 0·81 lagging with an efficiency 0·86. Calculate the line and phase currents if the motor is delta-connected.

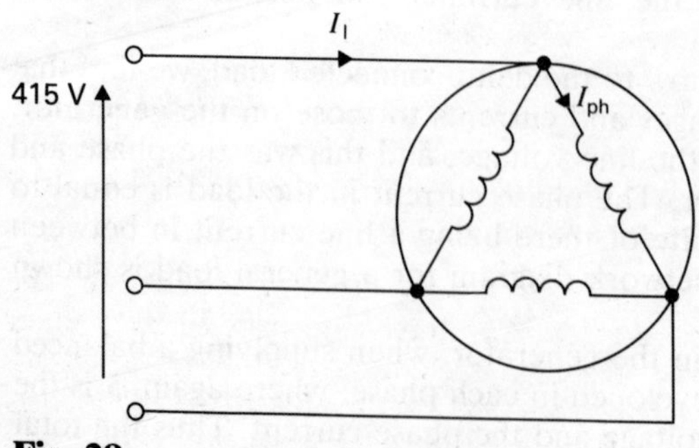

Fig. 28

$$P_i = \frac{P_o}{\eta} = \frac{10\,400}{0{\cdot}86} = 12\,100\ \text{W}$$

$$= \sqrt{3}\, V_l I_l \cos\phi$$

$$I_l = \frac{12\,100}{\sqrt{3} \times 415 \times 0{\cdot}81} = \underline{20{\cdot}8\ \text{A}}$$

$$I_{ph} = \frac{I_l}{\sqrt{3}} = \frac{20{\cdot}8}{\sqrt{3}} = \underline{12{\cdot}0\ \text{A}}$$

---

5 Comparison of star- and delta-connected systems

If you again look at the diagrams for the star- and delta-connected generators and loads (Figs. 11, 17, 26 and 27), you will see that the terminals for each arrangement are R, Y and B. Thus the output and input information in each case is the same, being the line voltages and the line currents. It follows that a star-connected generator can be coupled to a delta-connected load and also a delta-connected generator can be coupled to a star-connected load.

An extension of this observation is that a star-connected generator can operate in parallel with a delta-connected generator whilst a star-connected load can operate in parallel with a delta-connected load.

The case of the delta-connected generator is of little importance in practice for the following reasons:

1. For the same line voltage, the e.m.f. generated in each phase of the delta-connected generator is the full line voltage, whilst the e.m.f. generated per phase of the star-connected generator is only $1/\sqrt{3}$ of the line voltage. Thus for a given phase e.m.f., the star-connected generator gives a greater line voltage which is economically advantageous.
2. The star-connected generator gives a star point which is advantageous when supplying an unbalanced star-connected load. The neutral wire has been seen to ensure that each phase of the load receives the same applied voltage.
3. If the generator is not ideal and does not produce a pure sinusoidal e.m.f., it will generate excess heat losses.

It follows that most sources of 3-ph supply are star-connected. The loads may either be star-connected or delta-connected. In the former case, either the 3-wire or the 4-wire connection may be used. As previously stated, the 4-wire connection is useful when the load is not balanced. However, if the load is a balanced 3-ph one, only three wires need be used and the choice between star- and delta-connection depends on other factors.

Let us compare the power dissipated by three identical phase loads connected first in star and then in delta to the same supply. In star:

$$P = 3V_{ph}I_{ph}\cos\phi$$

$$= 3\frac{V_l}{\sqrt{3}}\frac{V_{ph}}{Z}\frac{R}{Z}$$

$$= 3\frac{V_l}{\sqrt{3}}\frac{V_l}{\sqrt{3}}\frac{R}{Z^2}$$

$$= \frac{V_l^2 R}{Z^2} \quad (18)$$

The loads are now reconnected in delta:

$$P = 3V_{ph}I_{ph}\cos\phi$$

$$= 3V_l\frac{V_{ph}}{Z}\cdot\frac{R}{Z}$$

$$= \frac{3V_l^2 R}{Z^2} \quad (19)$$

Hence it can be seen that the loads connected in delta dissipate three times more power than when connected in star.

Alternatively we may wish to dissipate the same power in each case. Let the power associated with the star-connected load be as before. If the power in the delta-connected system is to be the same, assuming the same power factor, i.e. the same ratio of R to Z, then the impedance must be three times that of the star-connected load. Thus, although the voltage across each phase is greater by a factor $\sqrt{3}$, the current is reduced by a factor of $\sqrt{3}/3 = 1/\sqrt{3}$.

For the same power, it follows that the loads connected in delta experience a greater voltage and a smaller current. To select the form of connection to use, it is therefore necessary to balance the greater insulation required in the delta-connection against the saving in conductor material due to the smaller current, and also to consider the lower costs due to insulation in a star-connected system which requires larger conductors to accept the greater currents. In many instances, the comparison indicates that the advantages of each system are equal, and other factors to determine the choice have to be sought; these are usually specific to the load in question.

Example 6 Three identical resistors of 30 Ω are connected in star to a 415-V, 3-ph, 50-Hz sinusoidal supply. Calculate the power dissipated in the load. Also calculate the power dissipated by the resistors if they are reconnected in delta to the same supply.

Given that one resistor is open-circuited in each case, draw network diagrams showing the currents in each line.

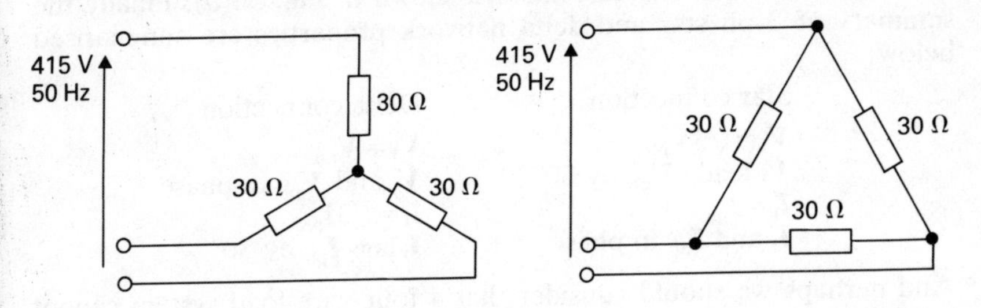

Fig. 29

In star: $V_{ph} = \dfrac{V_l}{\sqrt{3}} = \dfrac{415}{\sqrt{3}} = 240\ \text{V}$

$$I_{ph} = \frac{V_{ph}}{R} = \frac{240}{30} = 8{\cdot}0\ \text{A}$$

$$P = 3I_{ph}^2R = 3\times 8^2 \times 30 = \underline{5760\ \text{W}}$$

In delta: $P = 3 \times 5760 = \underline{17\,280\ \text{W}}$

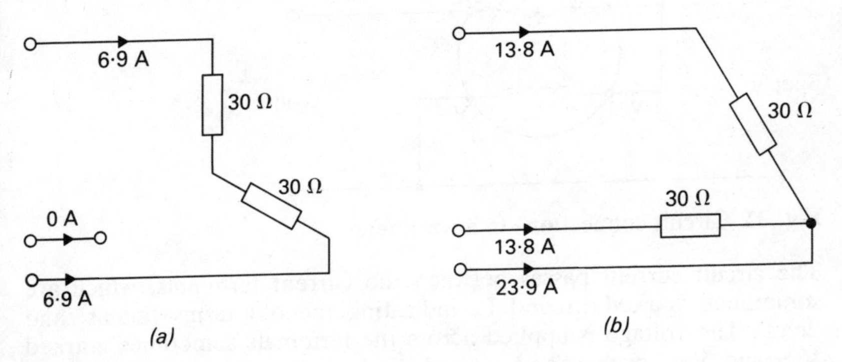

Fig. 30

In star: $I = \dfrac{V_l}{2R} = \dfrac{415}{2\times 30} = 6{\cdot}9\ \text{A}$

This is the current in two lines. The current in the remaining line is zero since the load to it has been disconnected. The currents are shown in Fig. 30(*a*).

In delta: $I_{ph} = \dfrac{V_l}{R} = \dfrac{415}{30} = 13{\cdot}8\ \text{A}$

This the current in two of the lines. In the remaining line, the current is given by the difference in the two remaining phase currents, i.e. $\sqrt{3} \times 13{\cdot}8 = 23{\cdot}9$ A. The currents are shown in Fig. 30(*b*). Finally the summary of 3-ph star and delta network properties are summarised below:

Star connection	Delta connection
$V_l = \sqrt{3} V_{ph}$	$V_l = V_{ph}$
$\mathbf{V}_l$ leads $\mathbf{V}_{ph}$ by 30°	$\mathbf{V}_l$ and $\mathbf{V}_{ph}$ in phase
$I_l = I_{ph}$	$I_l = \sqrt{3} I_{ph}$
$\boldsymbol{I}_l$ and $\boldsymbol{I}_{ph}$ in phase	$\boldsymbol{I}_l$ lags $\boldsymbol{I}_{ph}$ by 30°

And perhaps we should consider that a four-wire load system cannot be fully connected to a 3-wire supply system!

6 Three-phase power measurement

The power in an electrical circuit is measured in watts, and the device that measures the power is termed a wattmeter. Electrical power in a circuit is dependent on the voltage and on the current, thus the voltage and the current require to be supplied to a wattmeter in order to make it operate. Normally these are supplied separately, as shown in Fig. 31.

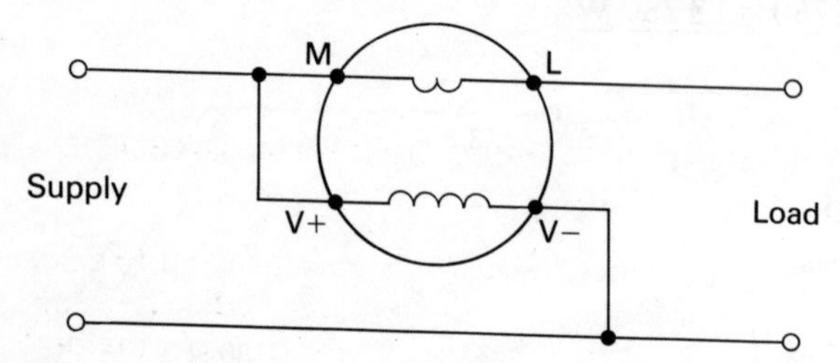

Fig. 31 Circuit connections to a wattmeter

The circuit current passes between the current terminals, which are sometimes marked M and L, indicating the old terms 'mains' and 'load'. The voltage is applied across the terminals sometimes marked V+ and V−, or V± and unmarked. In a.c. circuits, the wattmeter automatically takes into account the power factor, and thus indicates the active power. The operation of a wattmeter is a technique to be studied in its own right. For most applications, the form of connection indicated in Fig. 31 will permit the wattmeter to operate with minimal error, and it would certainly be the form of connection used in almost all 3-ph applications.

If the load is balanced, it is sufficient to measure the power in any one phase. The total active power in the 3-ph network is therefore given by multiplying the wattmeter reading by 3. Suitable methods of wattmeter connection for star- and delta-connected loads

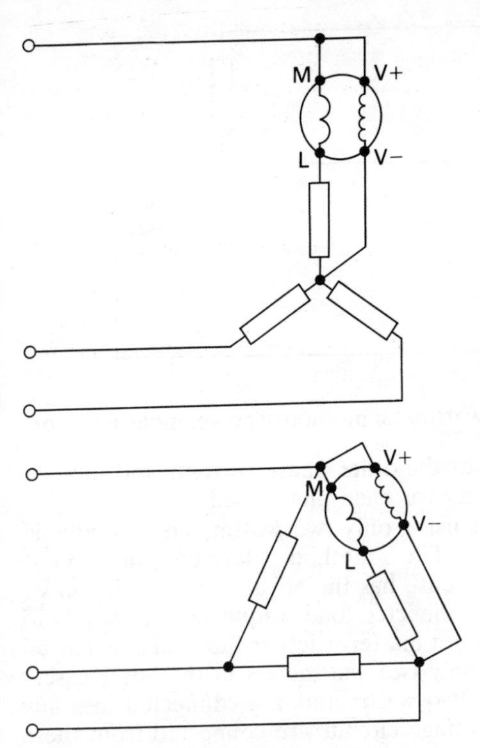

Fig. 32 Connections for one-wattmeter method of power measurement

are shown in Fig. 32. In each case, the wattmeter current circuit passes any phase current and the potential circuit is connected across the same phase voltage. This method is termed the 'one-wattmeter method of power measurement'. The principal disadvantage of this method is that it is not always possible to make the required connections in the load network. For instance, if the load were a 3-ph motor, only the line conductors would be available for connection.

If the load is not balanced, it is necessary to connect a wattmeter to each phase of the load. Since the load is unbalanced, the three wattmeters have differing indications, yet the total active power in the network is given by the sum of the three indications. The method of connection of the wattmeters is the same as that indicated in Fig. 32 except that there is a wattmeter connected to each phase of the load.

In practice, it is often not necessary to involve three wattmeters even in unbalanced systems. However, it is necessary to use three wattmeters when measuring the active power in a three-phase, four-wire system and the method of connection is shown in Fig. 33. This method is termed the 'three-wattmeter method of power measurement'. The presence of the fourth wire permits the current in each

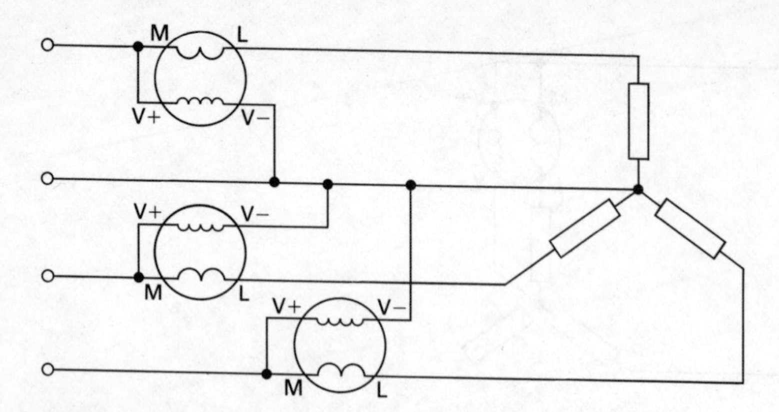

Fig. 33 Connections for three-wattmeter method of power measurement

phase load to be independent of the other phase currents and there is thus a need to have a wattmeter for each phase load.

A method of measurement using only two wattmeters is available for 3-ph, 3-wire systems. The method, termed the 'two-wattmeter method of measurement', has the advantages that it can be used with either star- or delta-connected loads requiring access only to the line conductors and it does not matter whether the load is balanced or unbalanced, still requiring only two wattmeters in the latter case.

The current circuits of the two wattmeters are connected into any two-line conductors and the voltage circuits are connected from these respective lines to the third-line conductor. The circuit diagram is shown in Fig. 34.

Although the load shown in Fig. 34 is a star-connected, the method of connecting the wattmeters would be the same for a delta-connected

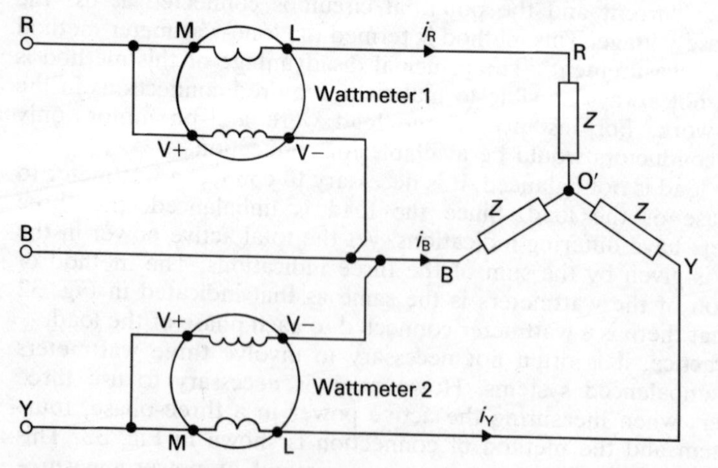

Fig. 34 Connections to two-wattmeter method of power measurement

load. In either case, the total active power dissipated by the load is given by the sum of the two wattmeter indications.

$$
\begin{aligned}
p_1+p_2 &= v_{RB}i_R+v_{YB}i_Y \\
&= (v_{RO'}-v_{BO'})i_R+(v_{YO'}-v_{BO'})u_Y \\
&= v_{RO'}i_R+v_{YO'}i_Y-v_{BO'}(i_R+i_Y) \\
&= v_{RO'}i_R+v_{YO'}i_Y+v_{BO'}i_B \\
&= p_R+p_Y+p_B
\end{aligned}
\qquad (20)
$$

This relationship has been derived for instantaneous values, thus at any instant the power experienced is equal to the power experienced by the load. It follows that the average power indicated by the wattmeters is equal to the average power dissipated by the load. You need to remember that the active power is the average rate of dissipation of energy in order to understand the above transfer from instantaneous to average values, but it is worth adding that a wattmeter indicates average values of power and therefore can indicate the active power supplied to the load.

The relation (20) applies to any 3-ph, 3-wire network, whether the load is balanced or unbalanced. In the case of the balanced load, there is a useful alternative proof that the sum of the wattmeter indications is equal to the active power supplied to the load. In the balanced case, the currents are equal and, in the typical phasor diagram shown in Fig. 35, they lag behind the respective phase voltages by the same angle ϕ.

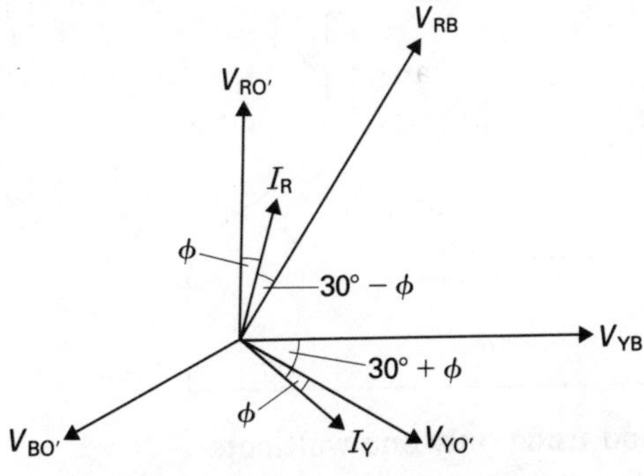

Fig. 35 Phasor diagram for two-wattmeter method of power measurement applied to a balanced load

$$
\begin{aligned}
P_1 &= V_{RB}I_R\cos(30°-\phi) = V_1I_1\cos(30°-\phi) \\
P_2 &= V_{YB}I_Y\cos(30°+\phi) = V_1I_1\cos(30°+\phi) \\
P_1+P_2 &= V_1I_1(\cos(30°-\phi)+\cos(30°+\phi)) \\
&= V_1I_1(2\cos 30°\cos\phi) \\
&= \sqrt{3}V_1I_1\cos\phi
\end{aligned}
$$

Thus again the wattmeter indications add to a value equal to the active power dissipated in the load. In the course of the above analysis, we note that one wattmeter indicates $V_1I_1 \cos(30° + \phi)$ and the other indicates $V_1I_1 \cos(30° - \phi)$. We should consequently take note of the following points:

1. If the load power factor is greater than 0·5, i.e. the phase angle is less than 60°, then both wattmeters indicate positively.
2. If the load power factor is 0·5, i.e. the phase angle is 60°, then one wattmeter indicates zero and the other indicates the total active power.
3. If the load power factor is less than 0·5, i.e. the phase angle is greater than 60°, then one wattmeter gives a negative indication.

In the last case, this indication must be recorded as a negative value. This requirement is easily brought to mind, since the wattmeter will be unable to give an indication until a switch is operated to operate the meter under negative-indication conditions. The negative value is added to the indicated power of the other wattmeter; thus the total power is a value somewhat less than that of the positively indicating wattmeter.

The two-wattmeter method may be carried out using only one wattmeter and a changeover switch. The circuit for this arrangement is shown in Fig. 36.

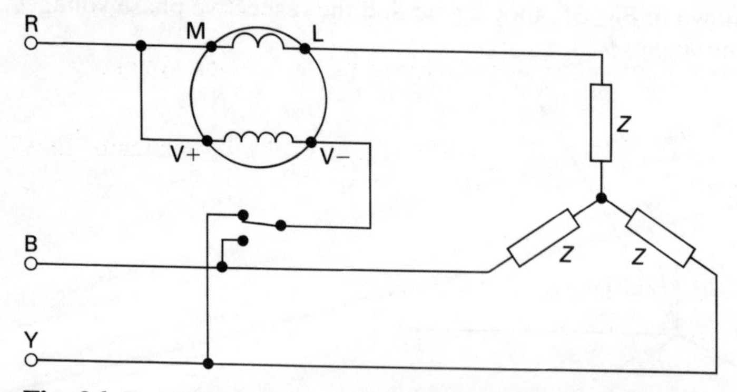

Fig. 36 Two-wattmeter method using only one wattmeter

For one position of the switch, the wattmeter gives an indication $V_1I_1 \cos(30° - \phi)$, and for the other position it indicates $V_1I_1 \cos(30° + \phi)$. This can be observed by comparing the connections with the phasor diagram given in Fig. 35, thus satisfying yourself that the indications are as stated. It has already been shown that the sum of these indications is equal to the active power dissipated in the load for the balanced condition. The limitation to the use of only one wattmeter is that the method of connection holds only for a balanced load and cannot be applied to an unbalanced load.

The load power factor for a balanced load can be determined either by the use of a voltmeter and an ammeter, or directly from the wattmeter indications. If a voltmeter and an ammeter are connected as shown in Fig. 37, the apparent power is

$$S = \sqrt{3} V_1 I_1$$

But $$P = \sqrt{3} V_1 I_1 \cos \phi$$

and $$\cos \phi = P/S$$

thus $$\text{power factor} = \frac{\text{sum of wattmeter indications}}{\sqrt{3} \times \text{voltmeter indication} \times \text{ammeter indication}}$$

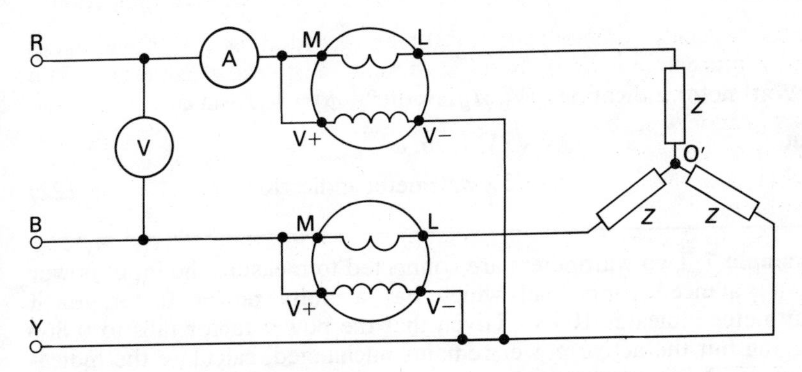

Fig. 37 Determination of power factor in balanced 3-ph system

If only the wattmeter indications are available, the power factor may be determined as follows. For the arrangement in Fig. 34,

$$\begin{aligned} P_1 - P_2 &= V_1 I_1 \cos (30° - \phi) - V_1 I_1 \cos (30° + \phi) \\ &= V_1 I_1 (\cos (30° - \phi) - \cos (30° + \phi)) \\ &= V_1 I_1 (2 \sin 30° \sin \phi) \\ &= V_1 I_1 \sin \phi \end{aligned}$$

$$\frac{P_1 - P_2}{P_1 + P_2} = \frac{V_1 I_1 \sin \phi}{\sqrt{3} V_1 I_1 \cos \phi} = \frac{\tan \phi}{\sqrt{3}}$$

$$\tan \phi = \frac{\sqrt{3}(P_1 - P_2)}{P_1 + P_2} \qquad (21)$$

From this relation, the phase angle ϕ and hence the power factor may be calculated.

Finally, a wattmeter can also be used to indicate the reactive power in a balanced 3-ph load. The method of connection is shown in Fig. 38.

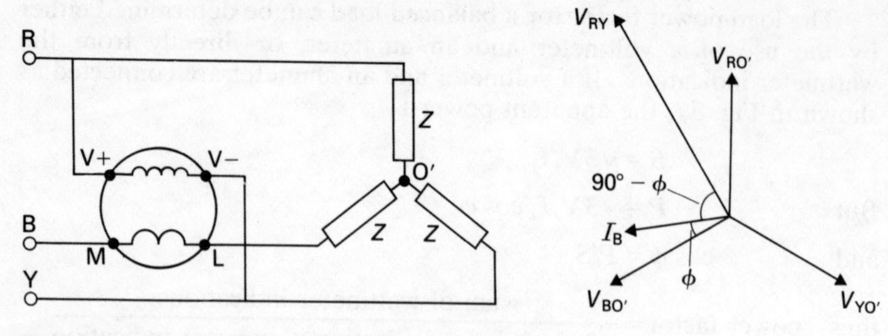

Fig. 38 Reactive power measurement in balanced load

$$\text{Wattmeter indication} = V_{RY} I_B \cos(90° - \phi) = V_1 I_1 \sin\phi$$

But

$$Q = \sqrt{3} V_1 I_1 \sin\phi$$
$$= \sqrt{3} \times \text{wattmeter indication} \qquad (22)$$

Example 7 Two wattmeters are connected to measure the input power to a balanced 3-ph load which has a unity power factor. Each wattmeter indicates 10 kW. Given that the power factor falls to 0·866 lagging but the active power remains unchanged, calculate the indications on the wattmeters.

$$P = P_1 + P_2 = 10 + 10 = 20 \text{ kW regardless of power factor}$$

$$\cos\phi = 0{\cdot}866 = \frac{\sqrt{3}}{2}$$

$$\tan\phi = \frac{1}{\sqrt{3}}$$

$$= \frac{\sqrt{3}(P_1 - P_2)}{P_1 + P_2} = \frac{\sqrt{3}(P_1 - P_2)}{20}$$

$$3P_1 - 3P_2 = 20$$

but $3P_1 + 3P_2 = 60$

thus $6P_1 = 80$

and $\underline{P_1 = 13{\cdot}3 \text{ kW}}$

also $P_2 = 20 - P_1 = 20 - 13{\cdot}3 = \underline{6{\cdot}7 \text{ kW}}$

Example 8 Two wattmeters are connected to measure the input power to a balanced 3-ph load. One wattmeter indicates 8·0 kW and the other −2·0 kW. Determine the load power factor.

$$\tan\phi = \frac{\sqrt{3}(P_1 - P_2)}{P_1 + P_2} = \frac{\sqrt{3}(8{\cdot}0 - (-2{\cdot}0))}{8{\cdot}0 + (-2{\cdot}0)}$$

$$= \frac{\sqrt{3} \times 10{\cdot}0}{6{\cdot}0} = 2{\cdot}89$$

$$\phi = 70{\cdot}9^\circ$$

$\cos\phi = \underline{0{\cdot}33 \text{ lead or lag}}$

---

7 The 3-ph supply system in practice

Most supplies of electrical energy are transmitted by 3-ph systems. Other forms of transmission system are both possible and practicable, yet the 3-ph system is that most commonly chosen by electrical engineers.

In order that we may better understand this preference, it is helpful to consider the development of electrical supply systems.

In the 1880s, the basic form of electricity supply came from a d.c. generating station operating at approximately 250 or 500 V. Such a station could supply consumers up to a distance of 1 or 2 km from the station but beyond that distance, the volt drop in the cables was so great that it was not practicable to supply consumers. It should be remembered that most electricity was supplied at that time for lighting which becomes appreciably poorer with a comparatively small drop in the supply voltage. Another use for electricity at that time was for tramcars, which required a supply for every 800 m of their route, again because longer distances involved volt drops which could be excessive.

There are two methods whereby we can decrease the volt drop in a d.c. system:

1. We can reduce the resistance of the transmission conductors. The length of cable between the generating station and the consumer is already fixed, not being a matter of choice, and it is reasonable to assume that the conductor material used is that with the optimum economical resistivity, and this is normally copper. It only remains to increase the cross-sectional area of the conductor in order that the resistance may be reduced. This is a costly process.

2. The second method is to use less current, but if we are to obtain the same power transmission, this can only be achieved provided that the voltage is increased to compensate. For instance, if we double the voltage, the current must be halved to transmit the same power. Half the current also produces only half the volt drop in a system now operating at twice the voltage. The overall result is effectively a considerable improvement in the relative volt drop.

Of the two methods, the second is the more attractive, yet a problem arises with d.c. systems because it is difficult to vary the system voltage. There is no difficulty in generating, say, 1000 V, but most consumers do not wish to be supplied at such a high voltage. To transmit the power at 1000 V is advantageous, but to change the voltage therafter from 1000 V to 500 V for the consumer requires complicated motor–generator equipment which is costly to operate. Thus for d.c. systems, the possibility of increasing the transmission voltage is largely limited by the voltage that the consumer is prepared to accept.

Some ingenious systems were devised whereby, with three wires, it was possible to transmit at 0 V to 250 V, and at −250 V to 0 V as indicated in Fig. 39. Provided the loads at the receiving end were

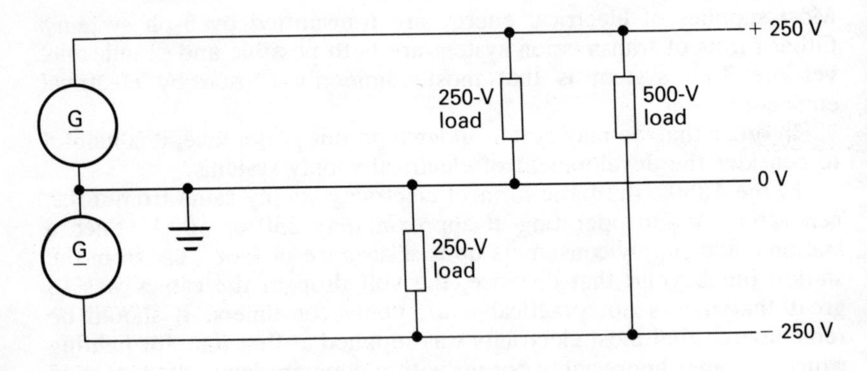

Fig. 39 Three-wire d.c. distribution system

approximately equal, then effectively they were in series and the current in the 0-V wire was approximately zero. The three wires were thus capable of dealing with two circuits, and the system had the added advantage that any consumer requiring a 500-V supply for a large machine could also be accommodated. The middle wire of the system was required to ensure that the lower voltage was maintained at 250 V regardless of the balance of the loads. Systems of this sort became very complicated with as many as six wires, but the d.c. system remained limited in its range of operation. However, it remained popular into the present century, mainly because of the advantage that the generating stations could be shut down at night and the supply maintained by large banks of lead–acid batteries!

About the end of the 1880s, 1-ph a.c. systems were introduced and these had the advantage that the transformer, which is a static device, takes power in at one voltage and gives it out at a different voltage of choice. In this way, the transmission voltage could be chosen to be as high as economically practicable whilst the consumer's voltage of utilisation could be varied to be that most suitable to him. Early systems transmitted the energy at 10 000 V and the consumer was supplied at 250 V.

A major advantage derived from this form of operation was that the generating station could be built in a remote area away from a town or city. This gave all the benefits of cheaper development and cheaper operation because the generating station could be placed wherever there was easy access to fuel and the necessary supply of water.

There are two disadvantages to the 1-ph a.c. systems. In a 1-ph circuit, the power pulsates with a frequency twice that of the supply frequency. The pulsating nature of the power makes it unsatisfactory for some purposes, particularly those involving machine outputs greater than 4 kW or those requiring speed control. By comparison, although the power supply to each phase of a 3-ph network is pulsating, the total 3-ph power to a balanced load is constant. The result of this is that a 3-ph machine has an output that is virtually steady and therefore free from pulsation.

Apart from improved machine performance, 3-ph machines are considerably smaller than the equivalent 1-ph machines.

However, we have observed that the number of conductors required for a 3-ph system is only half that required for three 1-ph circuits. This provides a major economical advantage which, added to those of being able to vary the operating voltage at will by means of transformers and of acceptable 3-ph machines, the 3-ph network is more adaptable than either the 1-ph a.c. system or the d.c. system. With all the advantages stated, it therefore becomes evident that electrical engineers are readily attracted to the 3-ph system in preference to the others.

Modern 3-ph generators operate at 22 kV or 33 kV, and the output is immediately transformed up to 275 kV or 132 kV for the National Grid, or often to 400 kV for the National Supergrid. The energy is therefore transmitted at a very high voltage and a comparatively low current. This is most suitable as the generating stations are often built in quite remote situations far from the areas of energy utilisation.

Once the energy has been transmitted to the area where it is to be used, the voltage is reduced for distribution throughout the area. This is generally done at 11 kV, although some very large consumers are supplied directly at 33 kV. Many commercial consumers are supplied at 11 kV by a 3-ph, 3-wire connection whilst the remaining smaller commercial and all the domestic consumers are supplied from 415/240-V 3-ph, 4-wire distribution lines in the manner described earlier in this chapter. These lines come from the many small substations that are a familiar part of our lives. Most people live within about 1 km from a substation because at this low voltage, the problem of volt drop in the cables arises. The volt drop must not exceed $6\frac{1}{2}$% of the nominal supply voltage. The volt drop may take the form of voltage magnification as in the series-resonant condition, and therefore the limit of voltage variation must be $\pm 6\frac{1}{2}\%$.

An illustration of a general form of transmission and distribution is given in Fig. 40.

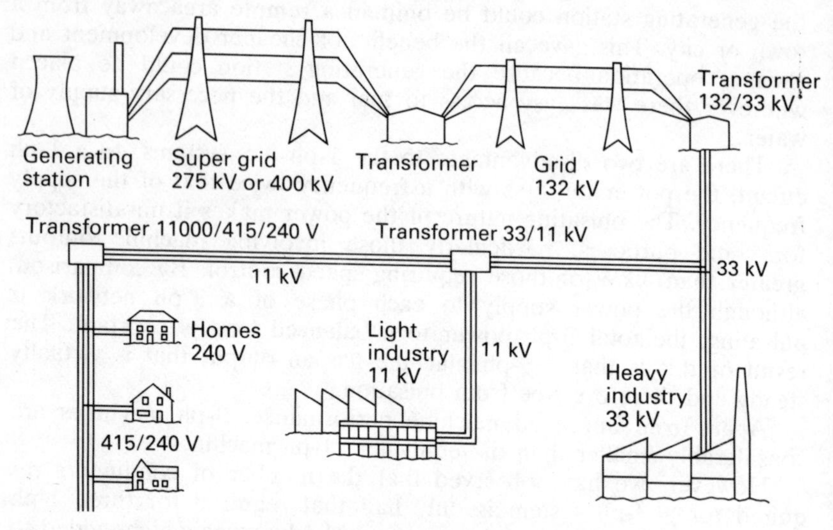

Fig. 40 Transmission and distribution of electrical energy

Almost all sources of 3-ph supply are star-connected. Apart from the possible advantage of being able to introduce a fourth wire if necessary, this also permits the star-point to be connected to earth. In this way, each line need only be insulated for a voltage of $1/\sqrt{3}$ that of the line voltage with respect to earth – it is still necessary to insulate for line voltage between lines.

Loads may be star- or delta-connected according to the choice already described. However to limit the starting current of 3-ph machines, which can be relatively high compared with the current required during normal running, many motors have their windings connected in star whilst gathering speed and are then reconnected in delta. The reason for this has already been observed; star-connected load takes only one-third the power that it does when connected in delta.

The input windings to 3-ph transformers are often delta-connected as this can be shown to help balance the loads on each phase.

Finally, although the 3-ph system has the advantage of high-voltage transmission, it suffers from a related disadvantage in that it is difficult to adjust the transmission voltage by small amounts. In recent years, it has become possible to do this with direct current, and systems operating at 200 kV d.c. are now being introduced. Such systems have the advantage that the variation of the operating voltage permits the regulation of the flow of current in different parts of the system. In these systems the energy is still generated by 3-ph a.c. generators and transformed to a suitably high voltage. The alternating current is then changed to direct current by a process termed rectification. At the receiving end of the transmission line, the direct current is inverted back to alternating current and the distribution remains the same as before.

One advantage of using direct current at high voltages (there are systems operating at 800 kV) is that the nominal voltage is that for which insulation needs to be provided. Further, a 200-kV system can operate at 100 kV above and below the zero potential of earth. By comparison, a 400-kV a.c. system requires to be insulated between lines for the maximum voltage which is $400\sqrt{2}$ kV, i.e. approximately 560 kV. However the pulsating nature of the voltage requires a further margin and such a system has to be insulated for a peak level of approximately 800 kV. Thus a.c. insulation is much more costly than that for a d.c. system and, added to the factor of flexibility in transmission control, there are strong arguments in favour of high-voltage d.c. transmission.

At the other end of the power scale, the 3-ph motor also provides a good source of power yet it also suffers from being difficult to control. With the development of thyristor techniques, the d.c. motor is becoming increasingly important, yet its supply remains a 3-ph system.

The 3-ph system remains the backbone of all distribution systems virtually throughout the world, and it is likely to continue without challenge for this purpose. Only at the extremely large transmission ratings and at the specialist motor control systems does the 3-ph system experience the challenge of direct current, and this in no way detracts from the importance of the 3-ph system.

Problems

1. The phase e.m.f. of a star-connected 3-ph generator is 150 V. Find the output line voltage.
2. The phase voltage of a star-connected system is 240 V r.m.s. Determine the maximum voltage between the lines.
3. The current in each phase of a star-connected source is 25 A. What is the current in each of the lines?
4. In the National Grid, the line voltage is 132 kV. If the load terminating such a line is star-connected, what is the voltage across each phase of the load?
5. A star-connected generator operates at 415 V and the current in each phase is 20 A at a power factor 0·8 lagging. Find
   (*a*) the phase voltage;
   (*b*) the power generated in each phase;
   (*c*) the total power generated.

6. A star-connected generator is rated at 120 MVA, 22 kV. Determine
   (*a*) the generator phase voltage;
   (*b*) the apparent power associated with each phase at full load;
   (*c*) the rated current in each phase.
7. A delta-connected 3-ph balanced load takes a line current 25 A. What is the current in each phase?
8. In a delta-connected 3-ph load, which is balanced, the phase current is 10 A. Determine the line current.
9. The phase e.m.f. of a 3000-V delta-connected 3-ph generator is
   (*a*) 1730 V;
   (*b*) 3000 V;
   (*c*) 5200 V;
   (*d*) 9000 V?
10. A 415-V, 3-ph motor requires a line current 16 A at a power factor 0·75 lagging. Determine
    (*a*) the power supplied by each phase;
    (*b*) the total power.
11. A 3-ph, delta-connected load has a phase voltage 415 V, a phase current 8 A and a power factor 0·85 lagging. Determine the power dissipated by the load, and the line current.
12. Three equal resistors connected in delta to a 3-ph supply dissipate 10 kW. If the supply voltage is 415 V, what is
    (*a*) the line current;
    (*b*) the phase current?
13. A 4-kW motor operating at full load has an efficiency 0·72. It is supplied from a 415-V, 3-ph supply at a power factor 0·8 lagging. Determine the line current.
14. A 3-ph generator has a line current 200 A at 6·6 kV. If the output power of the generator is 2·0 MW, what is the power factor?
15. Three 30-Ω resistors are connected in star to a 415-V, 3-ph supply. Determine
    (*a*) the phase current;
    (*b*) the line current;
    (*c*) the power dissipated.
16. Three 18-Ω resistors are connected in delta to a 415-V, 3-ph supply. Determine
    (*a*) the phase current;
    (*b*) the line current;
    (*c*) the power supplied.
17. Three resistors of resistance 6 Ω, 8 Ω and 10 Ω respectively are connected in star to a 415/240-V, 3-ph, four-wire supply. Determine the current in each of the four wires.
18. Three 10-Ω resistors are connected in delta to a 415-V, 3-ph supply. Calculate
    (*a*) the total power dissipated;
    (*b*) the power dissipated if they were connected in star to the same supply;

(*c*) the resistances of three equal resistors which would dissipate the same power as in (*a*) when connected in star.

19. Show that the power in a 3-ph balanced network can be measured by two wattmeters and deduce an expression giving the total active power in terms of the wattmeter indications. A 3-ph, 500-V motor operating at a power factor of 0·4 lagging takes 30 kW. Find the indication on each of two wattmeters correctly connected to measure the input power.
20. Two wattmeters are connected to measure the input power to a balanced 3-ph load which has a power factor of unity and which causes both wattmeters to indicate 20 kW. Given that the power factor falls to 0·8 lagging, the active power remaining the same, calculate the wattmeter indications.

    If the power factor were 0·8 leading, what would be the wattmeter indications?

    If one of the wattmeters were to indicate zero, what would be the power factor and what would be the indication on the other wattmeter?

Answers

1. 260 V
2. 588 V
3. 25 A
4. 76 kV
5. 240 V; 3840 W; 11 520 W
6. 12·7 kV; 40 MVA; 3150 A
7. 14·4 A
8. 17·3 A
9. (*b*)
10. 2·9 kW; 8·6 kW
11. 8·5 kW; 13·9 A
12. 13·9 A; 8·0 A
13. 9·7 A
14. 0·87 leading or lagging
15. 8·0 A; 8·0 A; 5760 W
16. 23·0 A; 39·8 A; 28·6 kW
17. 40 A; 30 A; 24 A; 14 A
18. 51·7 kW; 17·2 kW; 3·33 Ω
19. 34·9 kW; −4·9 kW
20. 28·7 kW, 11·3 kW; 11·3 kW, 28·7 kW; 0.5 lead or lag; 40 kW

Chapter 5

D.C. transients

Most of our electrical studies have concerned circuits in which the supply has been connected to the load, and the current in the circuit has been flowing for some time. With the passing of time, the circuit current settles down to a steady value and the circuit is then said to be in a steady-state condition. Immediately after the supply is connected to the load, however, there is a period during which the current is changing from zero to its final steady-state value. This change does not take place instantaneously but takes a short period of time. This change is said to be a transient.

In our previous studies, we have already come across three cases of transient conditions, these being the charging of a capacitor, the energising of an inductive coil and the starting of a d.c. motor, which we have only just considered. The cases of the capacitor and of the induction are of more general interest and require further study.

1 Charging of a capacitor

A capacitor may be charged from a battery by the circuit shown in Fig. 1. The galvanometer is included in the circuit to indicate the current, which is the rate of flow of charge into (or out from) the capacitor.

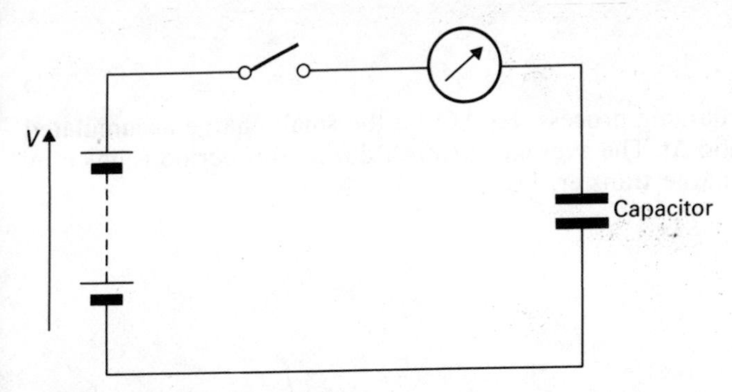

Fig. 1 Charging a capacitor

When the switch is closed, the galvanometer deflects for a short period, perhaps only momentarily, but this indicates that there is a current flow for that short period of time and that it has also ceased to flow. This burst of current is the flow of charge to or from the capacitor.

From the circuit diagram, you can deduce that the top plate of the capacitor finishes with a deficiency of electrons, these having been attracted to the positive plate of the battery. The transfer of these electrons is equivalent to a conventional current flowing into the top plate of the capacitor. The plates of the capacitor are insulated from one another by the dielectric and the circuit is incomplete, hence the current flow can only accumulate on the top plate making it positively charged. An equal but opposite action takes place at the lower plate, making it negatively charged.

The process of transferring charge cannot take place indefinitely because the accumulated charges on each plate repel further charge movements within the circuit. For instance, the quantity of charge to arrive first at a plate of the capacitor does so without opposition. However, the second quantity to arrive does so against the repelling electrostatic force of the first change. This difficulty increases with the arrival of each quantity of charge until it is impossible for any further quantity to be transferred.

This effect is better described in terms of the potential difference that appears between the plates due to the charges on the plates. The magnitude of the p.d. depends on the quantity of charge that has accumulated and is equal to the work done in moving a unit charge from one plate to the other.

The cause of the charge transfer is the battery e.m.f. which acts against the p.d. between the capacitor plates and, so long as it exceeds the capacitor p.d., charge continues to move round the circuit. As the difference between the e.m.f. and the p.d. decreases, the rate of charging also decreases until it ceases altogether. The capacitor is then charged such that the p.d. between the plates is equal to the e.m.f. of the battery. If the battery terminal voltage is V, then the charge is given by

$$Q = CV \tag{1}$$

During the charging process, let ΔQ be the small charge accumulated during a period Δt. The average current i during this period is given by the rate of charge transfer, i.e.

$$i = \frac{\Delta Q}{\Delta t}$$

$$= \frac{\Delta(CV)}{\Delta t}$$

It is usual to find that the capacitance C is constant unless there is movement between the plates of the capacitor, Assuming the capacitance to be constant,

$$i = C\frac{\Delta V}{\Delta t} \tag{2}$$

Although this relationship is derived on the basis of an accumulation of charge taking place in a particular time, you should remember that this can be either an increase or a decrease of charge.

Example 1 A direct potential 200 V is suddenly applied to a circuit comprising an uncharged capacitor in series with a 1000-Ω resistor. Calculate the initial rate of rise of voltage across the capacitor.

After the capacitor is completely charged, the source of e.m.f. is switched off and the terminals of the circuit are short-circuited. Calculate the initial rate of decrease of voltage across the capacitor.

Because the capacitor is initially uncharged, there is no initial p.d. between its plates. When the supply e.m.f. is applied, the corresponding voltage drop must appear across the 1000-Ω resistor. Let the initial current be i_0.

$$i_0 = \frac{V}{R} = \frac{200}{1000} = 0{\cdot}2\text{ A}$$

$$= C\frac{\Delta V}{\Delta t} = 100 \times 10^{-6} \times \frac{\Delta V}{\Delta t}$$

$$\frac{\Delta V}{\Delta t} = \frac{0{\cdot}2}{100 \times 10^{-6}} = 2000\text{ V/s}$$

$$= \underline{2{\cdot}0\text{ kV/s}}$$

When the capacitor is discharged, the full p.d. is applied to the resistor. This gives the same current initially as before, but the direction of current flow is reversed. It follows that the initial rate of voltage decrease is $\underline{-2{\cdot}0\text{ kV/s}}$.

For those familiar with the notation of calculus, the general case of the changing of a capacitor of capacitance C is given by increasing the applied voltage by $\mathrm{d}v$ in a time $\mathrm{d}t$ with a resulting increase in the charge $\mathrm{d}q$, hence

$$\mathrm{d}q = C\,\mathrm{d}v$$

If the charging current at that instant is i, then

$$\mathrm{d}q = i\,\mathrm{d}t$$

and $$i\,\mathrm{d}t = C\,\mathrm{d}v$$

hence $$i = C\frac{\mathrm{d}v}{\mathrm{d}t}$$

This is the simple relation (2) given in the notation of calculus.

It is possible to produce the charging current/time characteristic for the circuit shown in Fig. 1 by two methods. The simplest is achieved using a capacitor of quite large capacitance, say 8 μF, and a resistor also of high resistance, say 6·8 MΩ. Such a circuit has a transient lasting several minutes, and it is possible to note the current at, say 15-s intervals and to plot the characteristic as a result. A more complicated but useful method is to display the voltage drop across the circuit resistor on an oscilloscope and to photograph the display. The voltage across the resistor is proportional to the circuit current; thus the display is effectively that of the charging current/time characteristic.

No matter how the characteristic is obtained, it always takes the form shown in Fig. 2. As we have already noted, the current starts at a high value, the full supply voltage appearing across the circuit resistance. Even in circuits which comprise only a capacitor, the circuit resistance no matter how small it may be, initially limits the current. As the charge and hence the p.d. of the capacitor builds up, the current reduces until eventually it is zero.

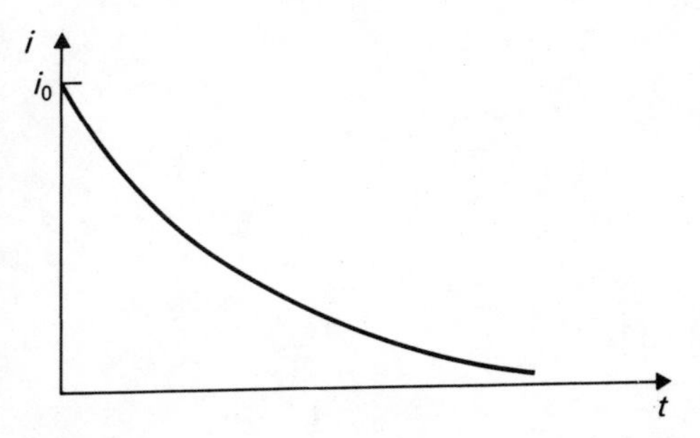

Fig. 2 Charging current/time characteristic of an R–C circuit

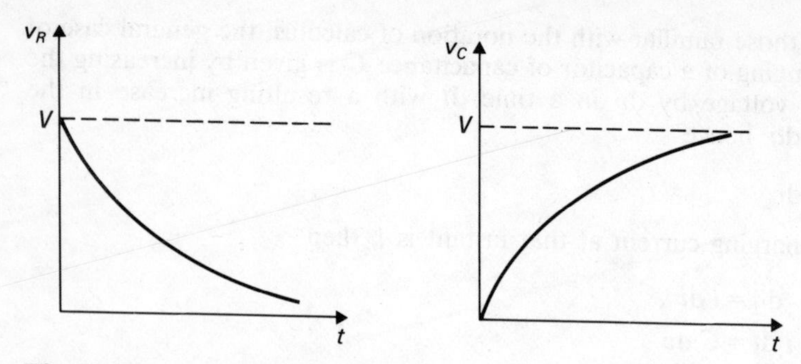

Fig. 3 Voltage/time characteristics of an *R–C* circuit being charged

The voltage drop across the circuit resistance is proportional to the circuit current, and therefore its variation with time has a characteristic of the form shown in Fig. 3(*a*). The difference between the voltage drop across the circuit resistance v_R and the supply voltage V is the voltage drop across the capacitor v_C, the time characteristic of which takes the form shown in Fig. 3(*b*). By Kirchhoff's Second Law,

$$V = v_C + v_R \tag{3}$$

There are two terms often used to describe the actions shown in Fig. 3. The voltage across the resistance is said to decay whilst the voltage across the capacitor is said to grow. The characteristics are therefore said to be decay and growth characteristics. Also when a capacitor is being charged, the current/time characteristic is also a decay characteristic.

If a capacitor is discharged through a resistor, the discharging current/time characteristic takes the form shown in Fig. 4. Again the current starts at a high value, the full voltage to which the capacitor had been charged appearing across the circuit resistance. As the charge of the capacitor is dissipated, the current reduces until eventually it is zero.

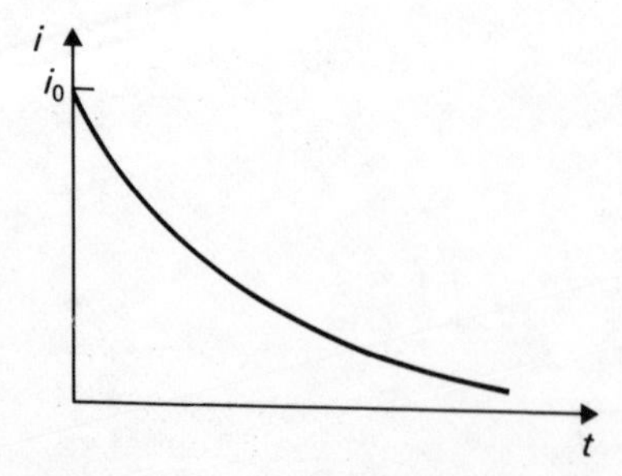

Fig. 4 Discharging current/time characteristic of a *R–C* circuit

If we compare the current/time characteristics of a circuit first being charged and then discharged, the characteristics are identical provided the circuit resistance and capacitance is the same in each case.

The voltage drop across the circuit resistance is proportional to the circuit current and therefore its variation with time has a characteristic of the form shown in Fig. 5(*a*). However, this is due to the voltage applied to the resistance by the capacitor; hence it is also the voltage drop across the capacitor, the time characteristic taking the form shown in Fig. 5(*b*). Both voltage/time characteristics during discharge are identical.

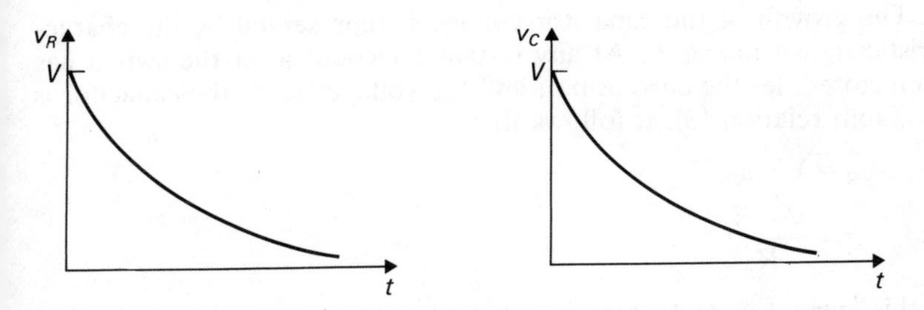

Fig. 5 Voltage/time characteristics of an *R–C* circuit being discharged

When a capacitor is being discharged, all the characteristics based on time show decay. Finally, whether a circuit capacitor is being charged or discharged, all the time-based characteristics describe the transient conditions and may therefore be termed the transient characteristics.

2 Graphical derivation of the transient characteristics of an *R–C* circuit

The transient characteristics of an *R–C* circuit can be derived graphically from the values of the resistance *R*, the capacitance *C* and the applied voltage *V*. For the circuit shown in Fig. 6, the circuit current i_0 at the instant of the supply being switched on is given by

$$i_0 = \frac{V}{R}$$

assuming that the capacitor is initially uncharged and that therefore the full supply voltage appears across the resistance. The voltage across the capacitor v_C builds up from zero until it is eventually equal to the supply voltage *V*.

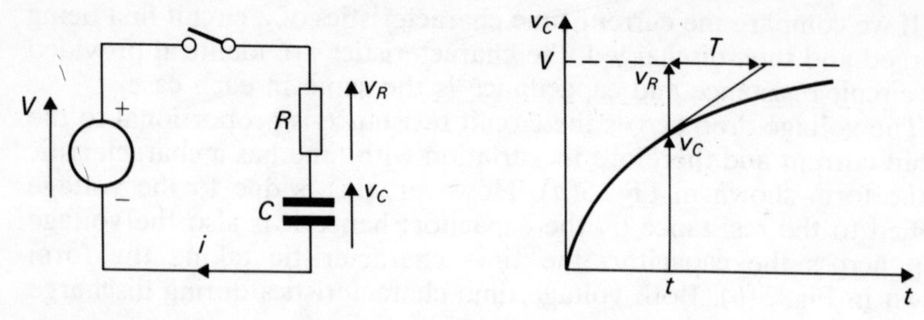

Fig. 6 Growth of voltage v_C across the capacitor in an R–C circuit

The growth of the capacitor voltage is represented by the characteristic shown in Fig. 6. At any instant t seconds after the switch has been closed, let the current be i and the voltage across the capacitor is v_C. From relation (3), it follows that

$$v_R = V - v_C$$

and $$i = \frac{V - v_C}{R}$$

If this current were to remain constant until the capacitor were fully charged, and if the time required for such an action were T seconds, then the charge supplied in that period would be

$$iT = \left(\frac{V - v}{R}C\right)T$$

During the same period, the voltage across the capacitor would increase uniformly from the value v_C to the supply voltage V. This would require it to continue to rise at the same rate throughout as at the instant chosen and the characteristic would then take the form of a straight line which would be a tangent to the curve as shown. The charge on the capacitor would rise from Cv_C to CV and the difference in charge would be $C(V - v_C)$.

Thus $$C(V - v_C) = \left(\frac{V - v_C}{R}\right)T$$

hence $$T = RC \tag{4}$$

This relation indicates that so long as the capacitance and the resistance of a circuit remain constant, the time taken to complete the charging of a capacitor from any given state of charge is the same and T is therefore termed the time constant of the circuit. From the above observations, it is possible to construct the characteristic shown in Fig. 7. Let the final capacitor voltage V be represented by OA and let AB be constructed to represent the time constant derived from relation (4). If we join O to B, OB represents the tangent to the curve at the point O.

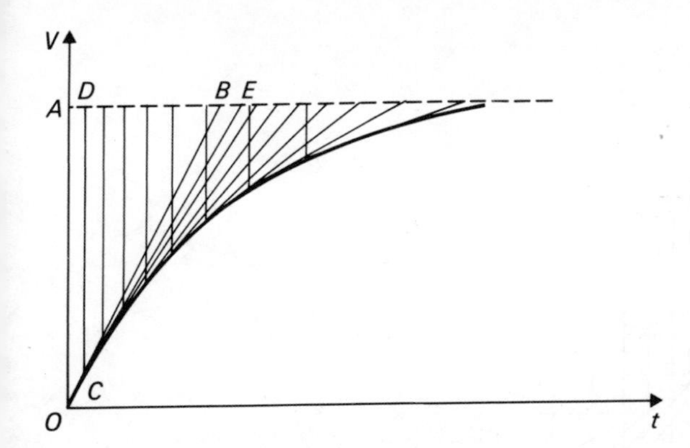

Fig. 7 Construction of the capacitor voltage growth characteristic in an R–C circuit

From a point C quite near to O and lying on line OB, drop the perpendicular CD and measure along the distance DE equal to AB (the time constant). Join CE, which is the tangent to the curve of C. By drawing a family of such tangents, the shape of the characteristic soon becomes apparent. Obviously the greater the number of constructions drawn, the better is the accuracy of this method of producing the characteristic.

It has already been noted that the current/time characteristic takes the same form for both charging and discharging. Also the voltage transient characteristics also take the same form, and, with suitable adaptation (illustrated in Example 2), this method of construction can be applied to all transient characteristics.

Example 2 A 10-μF capacitor is connected in series with a 20-kΩ resistor across a 100-V d.c. supply. Calculate the time constant of the circuit and the initial charging current, assuming that the capacitor is initially uncharged.

Draw the current/time characteristic of the circuit and estimate the current and the voltage across the capacitor at an instant at which the time constant has elapsed after closing the switch to apply the supply voltage to the circuit.

$$T = RC = 20 \times 10^3 \times 10 \times 10^{-6} = \underline{0{\cdot}2\ \text{s}}$$

$$i_0 = \frac{V}{R} = \frac{100}{20 \times 10^3} = 5{\cdot}0 \times 10^{-3}\ \text{A}$$

$$= \underline{5{\cdot}0\ \text{mA}}$$

The construction of the current/time characteristic is shown in Fig. 8. OA represents the initial current 5·0 mA and OB the time constant

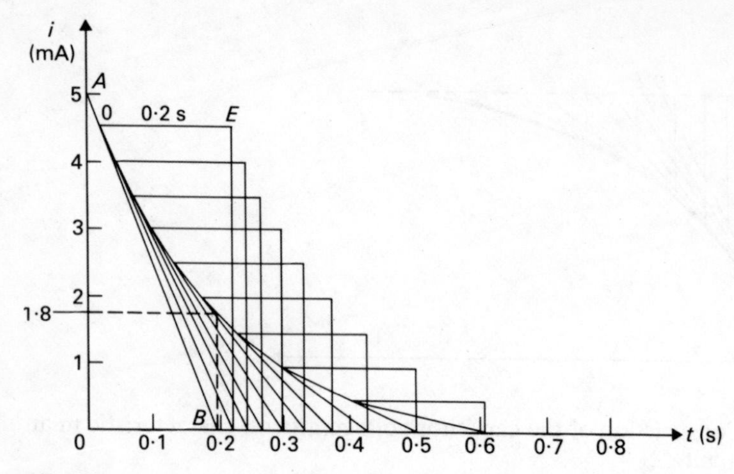

Fig. 8

0·2 s. The point D is taken 10 per cent along the line AB measured from A. This is equivalent to an instantaneous current 0·5 mA less than 5·0 mA, i.e. an instantaneous current 4·5 mA. The procedure is repeated at 0·5 mA intervals and the current/time characteristic constructed as shown.

From the characteristic, at an instant 0·2 s after switch on, the instantaneous value of the current is found to be 1·8 mA.

It follows that

$$v_R = iR = 1{\cdot}8 \times 10^{-3} \times 20 \times 10^3 = 36\ \text{V}$$

and $$v_C = V - v_R = 100 - 36 = 64\ \text{V}$$

---

3 Transient analysis of R–C circuit

Rather than attempt the graphical analysis of the transient behaviour of an R–C circuit, it may be that you would prefer either to use the exponential tables given in most books of mathematical data or to use the exponential function available on your electronic calculator.

For each of the charging characteristics shown in Fig. 9, it is possible to derive the mathematical relations associated with each. In each instance, there is a term $e^{-t/RC}$, and we have already observed that RC is the time constant measured in seconds. The index of e is therefore a pure number since the instant t is also measured in seconds.

In those characteristics which are decay characteristics, the expression takes the form of the initial value multiplied by the term $e^{-t/RC}$.

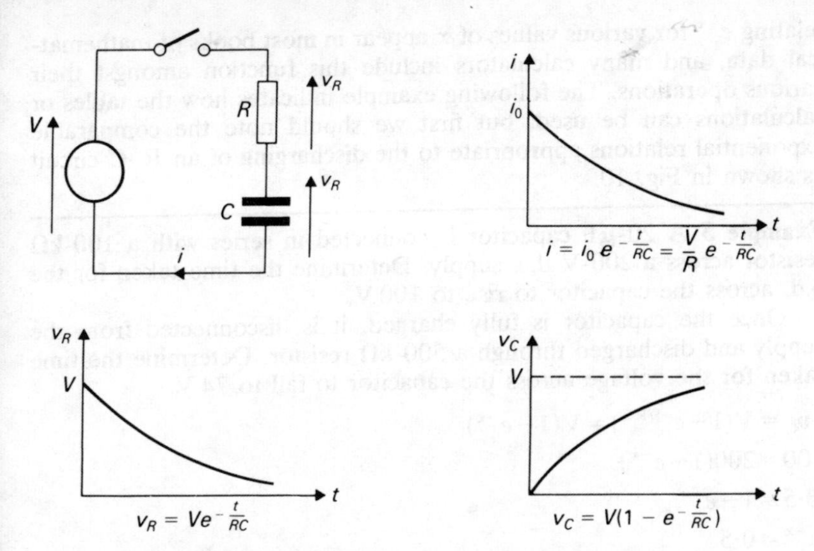

Fig. 9 Charging characteristics and relations of an *R–C* circuit

This is described as the exponential term and hence the curve is an exponential decay.

In those characteristics which take the form of exponential growth, the expression consists of the final value multiplied by the term $(1-e^{-t/RC})$.

In either case, e is the natural logarithmic base (=2·718). Tables

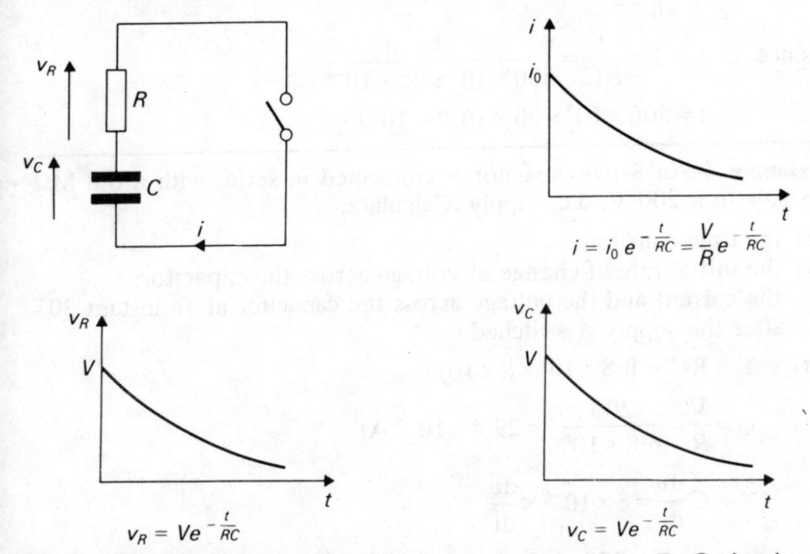

Fig. 10 Discharging characteristics and relations of an *R–C* circuit

relating e^{-x} for various values of x appear in most books of mathematical data, and many calculators include this function amongst their various operations. The following example indicates how the tables or calculations can be used, but first we should note the comparable exponential relations appropriate to the discharging of an R–C circuit as shown in Fig. 10.

Example 3 A 20-μF capacitor is connected in series with a 100-kΩ resistor across a 200-V d.c. supply. Determine the time taken for the p.d. across the capacitor to rise to 100 V.

Once the capacitor is fully charged, it is disconnected from the supply and discharged through a 500-kΩ resistor. Determine the time taken for the voltage across the capacitor to fall to 74 V.

$$v_C = V(1-e^{-t/RC}) = V(1-e^{-x})$$

$$100 = 200(1-e^{-x})$$

$$0{\cdot}5 = 1-e^{-x}$$

$$e^{-x} = 0{\cdot}5$$

From tables or from a calculator,

$$x = 0{\cdot}7 = \frac{t}{RC} = \frac{t}{100\times10^3\times20\times10^{-6}}$$

$$t = 0{\cdot}7\times100\times10^3\times20\times10^{-6} = \underline{1{\cdot}4\text{ s}}$$

$$v_C = Ve^{-t/RC} = Ve^{-x}$$

$$74 = 200e^{-x}$$

$$e^{-x} = 0{\cdot}37$$

Hence $$x = 1 = \frac{t}{RC} = \frac{t}{500\times10^3\times20\times10^{-6}}$$

$$t = 500\times10^3\times20\times10^{-6} = \underline{10{\cdot}0\text{ s}}$$

Example 4 An 8-μF capacitor is connected in series with a 6·8-MΩ resistor to a 200-V, d.c. supply. Calculate:

(*a*) the time constant;
(*b*) the initial rate of change of voltage across the capacitor;
(*c*) the current and the voltage across the capacitor at an instant 30 s after the supply is switched on.

(*a*) $$T = RC = 6{\cdot}8\times10^6\times8\times10^{-6}$$

(*b*) $$i_0 = \frac{V}{R} = \frac{200}{6{\cdot}8\times10^6}\ (=29{\cdot}4\times10^{-6}\text{ A})$$

$$= C\frac{dv}{dt} = 8\times10^{-6}\times\frac{dv}{dt}$$

$$\frac{dv}{dt} = \frac{200}{6{\cdot}8\times10^6\times8\times10^{-6}} = 3{\cdot}68\text{ V/s}$$

From this calculation, you can see that the initial rate of change of voltage is given by the supply voltage divided by the time constant.

(c) $i = i_0(e^{-t/RC}) = i_0(e^{-x})$

where $x = \frac{t}{RC} = \frac{30}{54{\cdot}4} = 0{\cdot}552$

hence $e^{-x} = 0{\cdot}576$

and $i = 29{\cdot}4 \times 10^{-6} \times 0{\cdot}576 = 16{\cdot}9 \times 10^{-6}\ \text{A}$

$= 16{\cdot}9\ \mu\text{A}$

$v_C = V(1 - e^{-t/RC}) = V(1 - e^{-x})$

$= 200(1 - 0{\cdot}576) = \underline{84{\cdot}8\ \text{V}}$

Before leaving the R–C circuit, it is worth recalling that the energy stored in the capacitor is given by

$W_f = \frac{1}{2}Cv_C^2$

In charging up a capacitor, the passage of the current through the circuit resistance dissipates energy, as does the discharge of a capacitor. It can be shown that the energy dissipated by the resistance is equal to the energy stored in a capacitor.

The circuit resistance is mainly derived from the conductors of the circuit but also includes a component caused by the dielectric loss in the capacitor, which is comparable to the hysteresis loss in ferromagnetic materials.

4 Growth and decay of current in an inductor

Having investigated the transient behaviour of the R–C circuit, let us carry out a similar investigation into the behaviour of an R–L circuit. A suitable network is shown in Fig. 11, but in this instance an additional assistance R_2 has been connected in parallel with the main R–L circuit. The function of this additional resistive circuit is to protect the switch and we shall discuss this later.

When the switch is closed, the galvanometer A_2 indicating the current in the short resistor R_2 increases instantly to its final value – any delay in indication is due to the inertia of the meter movement. By comparison, the indication of the galvanometer A_1 increases slowly to its final steady-state value, thus showing that the current has gradually increased to its final value.

The cause of this action is somewhat similar to the action of the capacitor, except that in this case the inductor builds up a flux in its core whereas the capacitor built up a charge (which of course gave rise

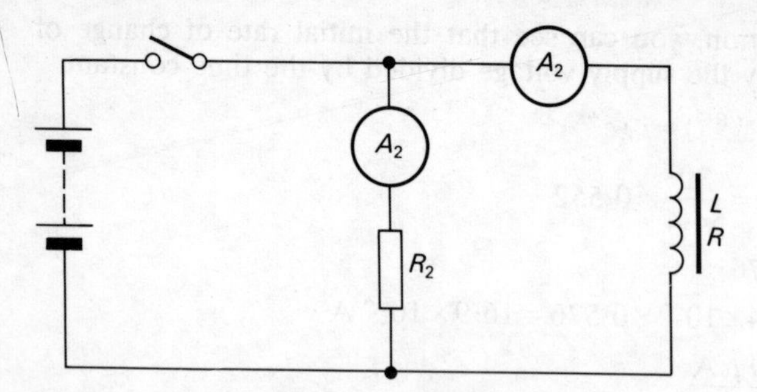

Fig. 11 Measurement of transient behaviour of an R–L circuit

to a flux). By Lenz's Law, we expect that any change of flux linkage will be opposed by an induced e.m.f., and in this case the e.m.f. opposes the applied voltage.

Immediately after the supply is switched on, the rate of change of flux linkage is sufficient to induce an e.m.f. equal and opposite to the applied voltage, with the result that the current does not change instantly but commences to grow.

At the same initial instant, because there is no current flowing in the inductor coil, it follows that there can be no voltage drop in the resistive component of that circuit, hence all the voltage must be balanced by the back e.m.f. induced in the coil.

As soon as current starts to flow in the circuit, a voltage drop appears across the resistive element of the circuit and therefore there is less voltage to appear across the coil. It follows that the e.m.f. to be induced in the coil is less and therefore the rate of change of flux linkage is reduced. As the induced e.m.f. can also be derived from

$$e = L\frac{\Delta I}{\Delta t} \qquad \left(\text{or } e = L\frac{\mathrm{d}i}{\mathrm{d}t}\right)$$

we may observe that the rate of change of current decays as the current builds up. Eventually the current reaches its final steady-state value and grows no further. The current/time growth characteristic is shown in Fig. 12.

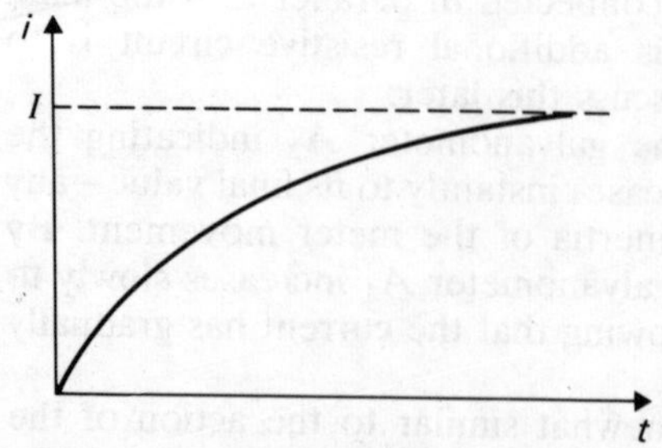

Fig. 12 Current/time growth characteristic of an R–L circuit

Example 5 A coil of inductance 5·0 H and resistance 100 Ω is suddenly connected to a 200-V d.c. supply. Determine the initial rate of increase of current in the coil.

At the instant of switch on, there is no current in the coil and

$$V = L\frac{di}{dt}$$

$$\frac{di}{dt} = \frac{V}{L} = \frac{200}{5 \cdot 0} = \underline{40\ \text{A/s}}$$

The determination of the current/time characteristic of an inductive coil is more difficult than the comparable experiment concerning a capacitor. This difficulty arises from the resistance, which is an integral part of the coil. Nevertheless if we use a coil of high inductance, say 10 H, and of low resistance, say 10 Ω, and connect it in series with a resistor of high resistance, say 90 Ω, it is possible to display the voltage across the resistor on an oscilloscope and to photograph that display. The voltage across the resistor is proportional to the circuit current and therefore the display is effectively that of the circuit current.

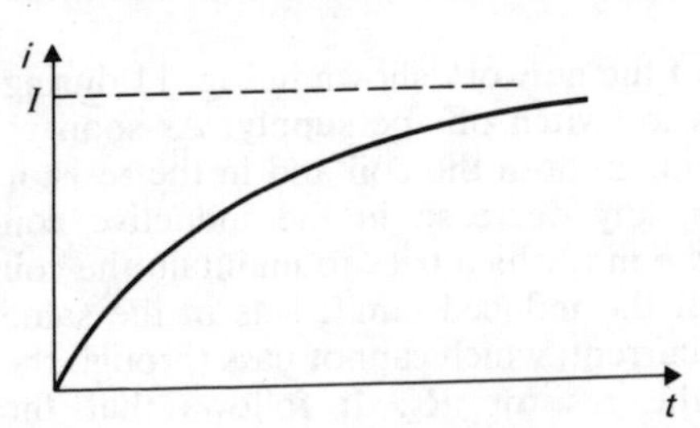

Fig. 13 Current growth/time characteristic of an *R–L* circuit

No matter how the characteristic relating the growth of current to time is obtained, it always takes the form shown in Fig. 13. The current starts at zero and at the same time the full supply voltage appears across the circuit inductance. Even in circuits with little inductance, the circuit inductance will oppose the increase of current by inducing a back e.m.f., but this is ineffectual unless there is a reasonable inductance available. As the current builds up, the back e.m.f. decreases until it is eventually zero, yet the voltage drop across the circuit resistance increases until it is equal to the supply voltage V.

The voltage drop across the circuit resistance is proportional to the circuit current, and therefore its variation with time has a characteristic of the form shown in Fig. 14(a). The difference between the voltage drop across the circuit resistance v_R and the supply voltage V is the back e.m.f. across the inductance v_L, the time characteristic of which

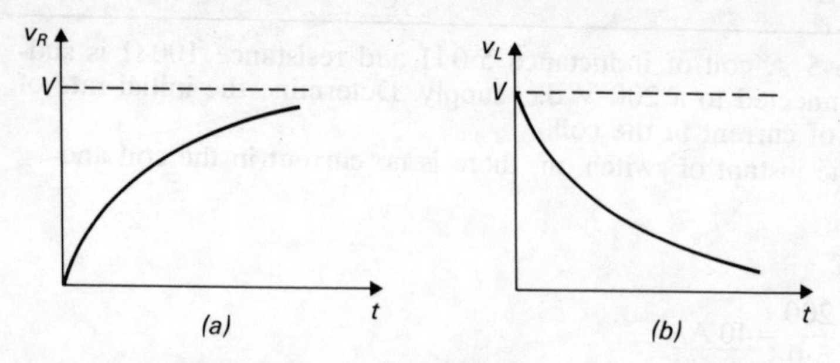

Fig. 14 Voltage/time characteristics of an R–L circuit being switched on to a d.c. supply

takes the form shown in Fig. 14(b). By Kirchhoff's Second Law,

$$V = v_R + v_L \tag{5}$$

In this instance the voltage across the resistance is said to grow whilst the voltage across the inductance decays. At the same time, the current/time characteristic is a growth characteristic. If you compare these with the characteristics of the R–C circuit, you can see that these characteristics are opposite.

Having examined the action of the network shown in Fig. 11 during the switch-on period, it remains to switch off the supply. As soon as the switch is operated, the currents in both the coil and in the resistor R_2 decrease to zero. However, any decrease in the inductive coil current is opposed by an induced e.m.f. which tries to maintain the coil current as before. It follows that the induced e.m.f. acts in the same direction as the current and this current, which cannot pass through the open switch, passes through the resistor R_2. It follows that the currents indicated by each of the galvanometers are the same; you will recall that the currents were not necessarily the same when they were switched on because each branch took a steady-state current appropriate to the branch resistance.

If the experiment is carried out without a shunt resistor R_2, the growth of the current in the coil is unaffected. However, when the switch is opened, an immediate difference becomes apparent in the form of arcing at the switch. This arcing could be quite severe and badly burn the switch. It is caused by the e.m.f. induced by the coil attempting to maintain the current flow, and the faster the switch is operated, the quicker does the current collapse and the greater is the induced e.m.f. The shunt resistor was therefore introduced into the experiment in order to minimise the effect of arcing at the switch.

If the supply voltage is interrupted and the coil current is permitted to decay through a resistor, the decay current/time characteristic takes the form shown in Fig. 15. The resistor dissipates the energy stored in the inductive coil and it is therefore said to discharge the coil. The

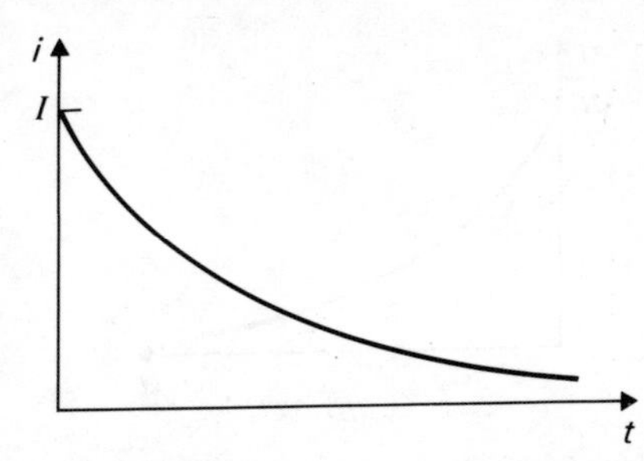

Fig. 15 Discharging current/time characteristic of an R–L circuit

resistor may therefore be called a discharge resistor. In this case, the current starts at a high value, the full induced e.m.f. appearing across the circuit resistance. As the energy of the inductive coil is dissipated, the e.m.f. and the current reduce until eventually they are zero.

If we compare the current/time characteristics of a circuit first being charged and then discharged, the characteristics are opposite and, since the circuit resistance is generally different in each case, the curves are not mirror images of one another. As an instance of the difference between the two curves, the charge and discharge current/time characteristics for the network shown in Fig. 11 is given in Fig. 16.

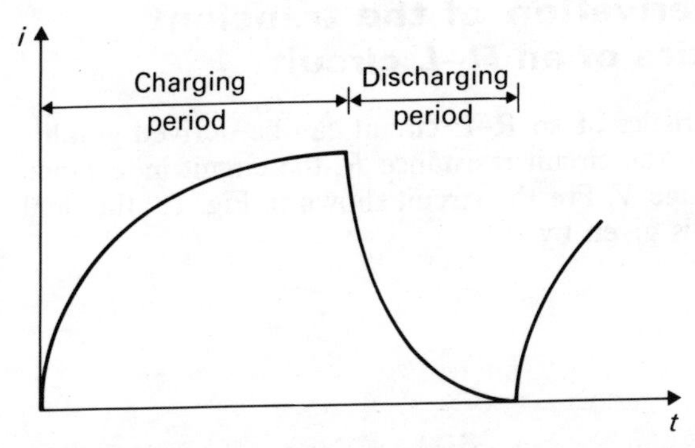

Fig. 16 Charge and discharge current/time characteristic of the experimental network shown in Fig. 11

The voltage drop across the circuit resistance during the discharge period is proportional to the discharge circuit current and therefore its variation with time has a characteristic of the form shown in Fig. 17(a). However, this is due to the voltage applied to the resistance by the inductive coil; hence it is also the voltage drop across the coil inductance, or which the time characteristic takes the form shown in Fig. 17(b). Both voltage/time characteristics during the discharge period

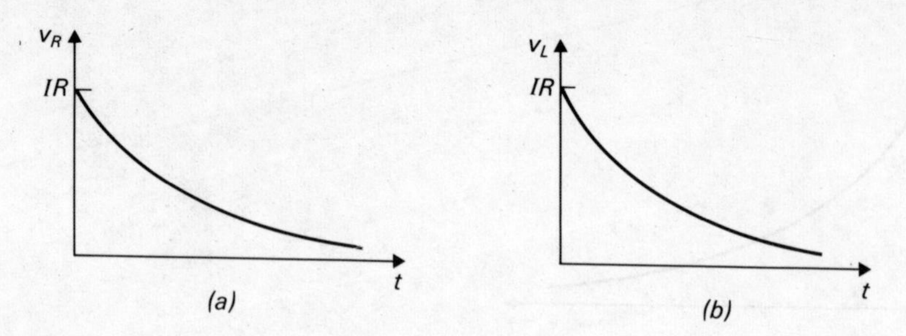

Fig. 17 Voltage/time characteristics of an R–L circuit being discharged

are identical, but note that the voltage across the coil inductance is not the terminal voltage, there being an internal voltage drop due to the internal resistance of the coil.

Like the capacitor arrangement, when a coil is being discharged all the characteristics based on time show decay. Thus all the transients are decay transients.

5 Graphical derivation of the transient characteristics of an *R–L* circuit

The transient characteristics of an R–L circuit can be derived graphically from the values of the circuit resistance R, the circuit inductance L and the applied voltage V. For the circuit shown in Fig. 18, the final steady-state current I is given by

$$I = \frac{V}{R}$$

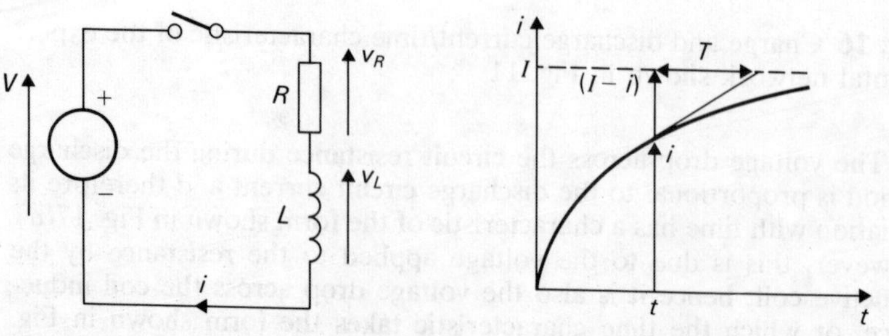

Fig. 18 Growth of current in an R–L circuit

During the transient period, the circuit current is i at any instant t seconds after the closure of the supply switch.

The growth of the current is represented by the characteristic shown in Fig. 18. The instantaneous current i is represented by the line AB on the diagram, and it follows that the current has yet to increase by $(I-i)$ in order to reach its final value. If the current were to continue to increase at the rate at which it is increasing at the instant t, let it take T seconds to reach the final steady-state value as indicated in Fig. 18. By continuing to increase at the same rate, the increase is linear and tangential to the curve.

The slope of the tangent is given by

$$\frac{I-i}{T}$$

hence the necessary e.m.f. to produce such a change is

$$e_L = L\frac{I-i}{T}$$

The total supply voltage V partly covers the back e.m.f. of the inductance and the remainder of it is accounted for by the voltage drop across the circuit resistance; thus

$$V = iR + L\frac{I-i}{T}$$

However, when the steady-state conditions are achieved, $V = IR$, where I is the final steady-state current and

$$IR = iR + L\frac{I-i}{T}$$

hence $$(I-i)R = L\frac{(I-i)}{T}$$

and $$T = \frac{L}{R} \tag{6}$$

This relation indicates that so long as the inductance and the resistance of a circuit remain constant, the time taken to complete the charging of an inductor from any given state of charge is the same, and T is therefore termed the time constant of the circuit.

From the above observations, it is possible to construct the characteristic shown in Fig. 19. Let the final inductor current I be represented by OA and let AB be constructed to represent the time constant derived from relation (6). If we join O to B, OB represents the tangent to the curve at the point O. From a point C quite near to O and lying on line OB, drop the perpendicular CD and measure along the distance DE equal to AB (the time constant). Join CE,

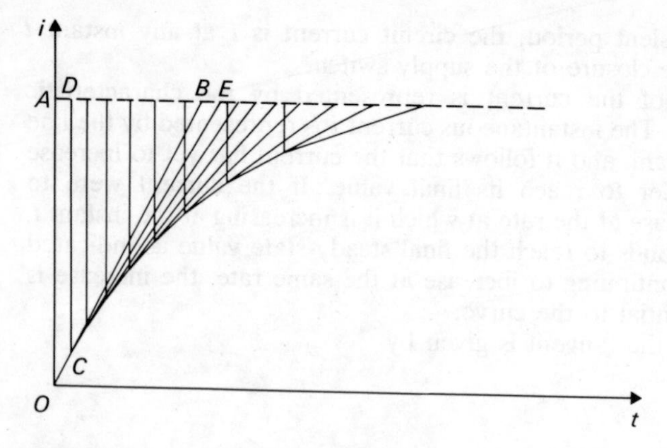

Fig. 19 Construction of the current growth characteristic of an *R–L* circuit

which is the tangent to the curve at *C*. If a family of such tangents is drawn, the shape of the characteristic soon becomes apparent. The greater the number of constructions drawn, the better is the accuracy of this method of producing the characteristic.

It has already been noted that the current/time characteristic takes the form indicated when charging and takes the inverse form when discharging. With suitable adaptation, the method of characteristic construction can be applied to either form of current transient. The method of construction can also be applied to both the voltage growth and decay characteristics.

Example 6 A 5·0-H inductor has an internal resistance 10 Ω and is connected across a 50-V, d.c. supply. Calculate the time constant of the circuit and the final steady-state current. Assuming that there is no current initially flowing in the inductor, draw the current/time characteristic of the circuit and estimate the current and the back e.m.f. at an instant 0·4 s after the voltage has been applied to the inductor circuit.

$$T=\frac{L}{R}=\frac{5\cdot0}{10}=\underline{0\cdot5\text{ s}}$$

$$I=\frac{V}{R}=\frac{50}{10}=\underline{5\cdot0\text{ A}}$$

The construction of the current/time characteristic is shown in Fig. 20. *OA* represents the final steady-state current and *OB* the time constant 0.5 s. The point *D* is taken 10 per cent along line *AB* measured from *A*. This is equivalent to an instantaneous current of 0.5 A. The procedure is repeated at 0.5 A intervals and the current/time characteristic constructed as shown.

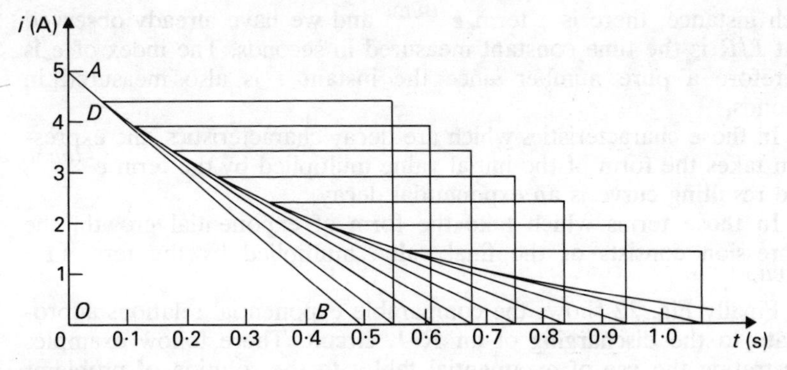

Fig. 20

From the characteristic, at an instant 0·4 s after switch on, the instantaneous value of the current is found to be <u>2·8 A.</u>

It follows that

$$v_R = iR = 2{\cdot}8 \times 10 = 28\text{ V}$$

and $\quad v_L = V - v_R = 50 - 28 = \underline{22\text{ V}}$

---

6 Transient analysis of *R–L* circuit

Rather than attempt the graphical analysis of the transient behaviour of the *R–L* circuit, it may again be that you would prefer either to use the exponential tables used in most books of mathematical data or to use the exponential function available on your electronic calculator.

For each of the charging characteristics shown in Fig. 21, it is possible to derive the mathematical relations associated with each. In

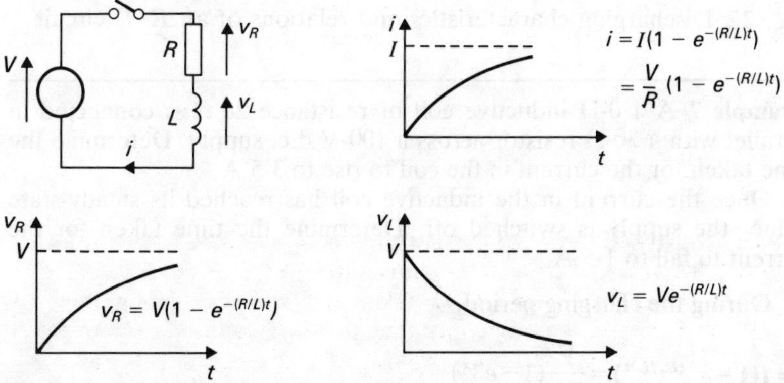

Fig. 21 Charging characteristics and relations of an *R–L* circuit

 each instance, there is a term $e^{-(R/L)t}$ and we have already observed that L/R is the time constant measured in seconds. The index of e is therefore a pure number since the instant t is also measured in seconds.

In those characteristics which are decay characteristics, the expression takes the form of the initial value multiplied by the term $e^{-(R/L)t}$. The resulting curve is an exponential decay.

In those terms which take the form of exponential growth, the expression consists of the final value multiplied by the term $(1-e^{-(R/L)t})$.

Finally Fig. 22 shows the comparable exponential relations appropriate to the discharging of an R–L circuit. There follow examples illustrating the use of exponential tables to the solution of problems concerning the charging and discharging of R–L circuits.

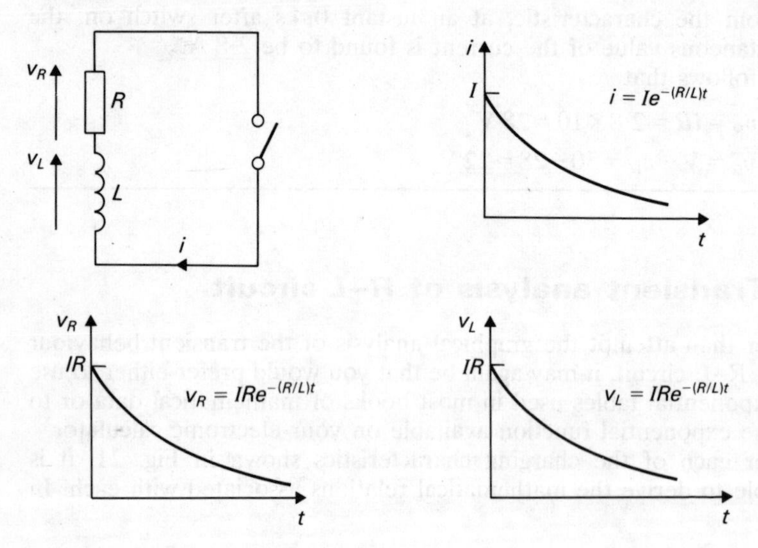

Fig. 22 Discharging characteristics and relations of an R–L circuit

---

Example 7 A 1·0-H inductive coil of resistance 20 Ω is connected in parallel with a 30-Ω resistor across a 100-V d.c. supply. Determine the time taken for the current in the coil to rise to 3·5 A.

Once the current in the inductive coil has reached its steady-state value, the supply is switched off. Determine the time taken for the current to fall to 1·5 A.

During the charging period,

$$i = I(1-e^{-(R_L/L)t}) = \frac{V}{R_L}(1-e^{-x})$$

$$3{\cdot}5=\frac{100}{20}(1-e^{-x})$$

$$0{\cdot}7=1-e^{-x}$$

$$e^{-x}=0{\cdot}3$$

From tables or from a calculator,

$$x=1{\cdot}2=\frac{R_L}{L}t=\frac{20}{1}\times t$$

$$t=\frac{1{\cdot}2}{20}=\underline{0{\cdot}06\text{ s}}$$

During the discharging period,

$$i=Ie^{-(R/L)t}=Ie^{-x}$$

In this instance, the circuit resistance consists of R_L and the shunt resistance R_s; hence

$$x=\frac{Rt}{L}=\frac{(20+30)t}{1}=50t$$

and $$I=\frac{V}{R_L}=\frac{100}{20}=5{\cdot}0\text{ A}$$

thus $$1{\cdot}5=5{\cdot}0e^{-x}$$

and $$e^{-x}=0{\cdot}30$$

hence $$x=1{\cdot}20=50\,t$$

and $$t=\frac{1{\cdot}20}{50}=\underline{0{\cdot}024\text{ s}=24\text{ ms}}$$

Example 8 A 50-mH inductor of resistance 50 Ω is connected to a 50-V d.c. source. Calculate:

(*a*) the time constant;
(*b*) the initial rate of rise of current in the inductor;
(*c*) the current and the e.m.f. at an instant 1·0 ms after the supply is switched on.

(*a*) $$T=\frac{L}{R}=\frac{50\times10^{-3}}{50}=1{\cdot}0\times10^{-3}\text{ s}=1{\cdot}0\text{ ms}$$

(*b*) $$V=v_R+v_L=iR+L\frac{di}{dt}$$

$$i=0$$

thus $$50=0+50\times10^{-3}\times\frac{di}{dt}$$

hence $\dfrac{di}{dt} = \dfrac{50}{50 \times 10^{-3}} = \underline{1000 \text{ A/s}}$

From this calculation you can see that the initial rate of increase of current is given by the supply voltage divided by the time constant.

(*c*) $i = I(1 - e^{-(R/L)t}) = \dfrac{V}{R}(1 - e^{-x})$

where $x = \dfrac{Rt}{L} = \dfrac{50 \times 1 \times 10^{-3}}{50 \times 10^{-3}} = 1$

hence $e^{-x} = 0{\cdot}37$

and $i = \dfrac{50}{50} \times (1 - 0{\cdot}37) = \underline{0{\cdot}63 \text{ A}}$

$v_R = iR = 0{\cdot}63 \times 50 = 31{\cdot}5 \text{ V}$

and $v_L = V - v_R = 50 - 31{\cdot}5 = \underline{18{\cdot}5 \text{ V}}$

---

7 Transient behaviour in practice

If you look again at the transient characteristics, you can see that although the curves tend towards some final steady-state value, they in fact never actually reach it. However, there comes a time when the difference between the transient value and the final steady-state value is so small that it is negligible. After a period equal to that of the time constant, 63 per cent of the change has taken place. After a period of twice the time constant, 86 per cent of the change has taken place, and after a period three times that of the time constant, 95 per cent of the change has taken place. For many purposes this represents a big enough part of the change, and to a first approximation it can be taken that the transient condition is completed after a period three times that of the time constant.

However, there are cases where you might wish the transient change to have died out to a greater extent than simply 95 per cent of the change. In such cases, it is sufficient to take a period 10 times that of the time constant, in which time the transient effectively ceases.

Most transient conditions are of relatively short duration, but they have an effect which cannot readily be ignored. For instance, if we are pulsing current through a circuit by applying a direct voltage for short periods of time, the current does not exactly correspond to the applied voltage as shown in Fig. 23. This difference in waveform may not be significant in many instances, yet if the transient becomes too prominent, it may prevent the current reaching its maximum value

Some circuits make use of this observation and thus delay the current from rising, so that switching can take place after the passage of a discrete period of time following the application of the voltage to

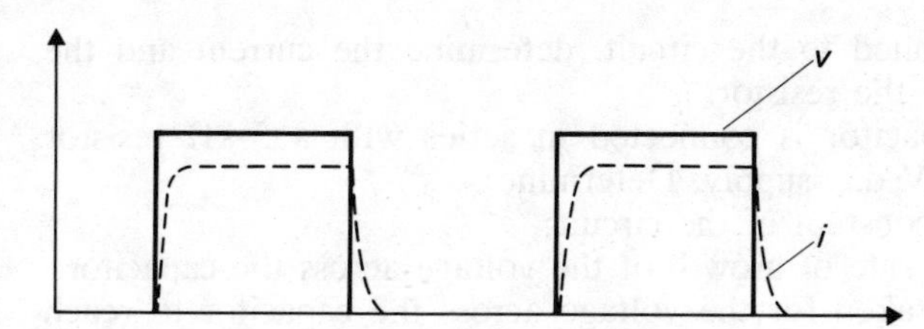

Fig. 23 Effect of transient behaviour on pulse system

the circuit. Thus we make use of transient behaviour for the benefit of certain control functions.

Finally, we need to distinguish clearly between the transient condition and the steady alternating condition. In the transient condition, we are dealing with instantaneous values, and at any instant, the sum of the volt drops across the circuit components is equal to the applied voltage. This is an arithmetic addition and should not be confused with the phasor addition of sinusoidal alternating volt drops, which involves the phase difference between the voltages. This is a common mistake which often becomes apparent when students try to determine the reactance of the capacitor or the inductor when determining the transient condition. Reactances only occur in alternating circuits and do not arise in d.c. circuits under transient conditions.

Nevertheless, we should also remember that in an a.c. circuit there is a similarity of operation. For instance, in an $R-C$ circuit, the charge rushes into and out from the capacitor operating under a.c. conditions, just as it rushes into or out from the capacitor under d.c. transient conditions. The difference is that the objective in the a.c. situation keeps on changing and no steady-state condition can be achieved, whereas in the d.c. situation, the steady-state condition is attainable. In either case, the charge moves into or out from the plates of the capacitor simulating a current, but in neither case does the current flow through the capacitor.

Problems

1. A 100-μF capacitor is connected in series with an 8·0-kΩ resistor. What is the time constant of the circuit?

   If the circuit is suddenly connected to a 100-V, d.c. supply, find:
   (*a*) the initial rate of rise of voltage across the capacitor;
   (*b*) the initial charging current;
   (*c*) the final charge on the capacitor.
2. A 4-μF capacitor is connected in series with a 1·0-MΩ resistor to a 100-V d.c. supply. Draw the current/time characteristic assuming the capacitor to be initially uncharged. At an instant 2·5 s after the

voltage is applied to the circuit, determine the current and the voltage across the resistor.

3. An 8-μF capacitor is connected in series with a 5-kΩ resistor across a 110-V d.c. supply. Determine:
   (*a*) the time constant of the circuit;
   (*b*) the initial rate of growth of the voltage across the capacitor;
   (*c*) the time taken for the voltage across the capacitor to reach 85 V.
4. A 50-μF capacitor is connected in series with a voltmeter of resistance 500 kΩ. The circuit is switched on to a 500-V d.c. supply. For intervals of 5 s after closing the circuit, determine:
   (*a*) the voltage across the capacitor;
   (*b*) the reading on the voltmeter.

   Find the charging current when the voltmeter reads 250 V and the time taken to reach this current.
5. An R–C series circuit is suddenly connected to a 500-V d.c. supply. The capacitor is initially uncharged and has a capacitance 0·1 μF; the resistance of the resistor is 100 kΩ. When the current has decayed to 80 per cent of its initial value, determine the voltage across the capacitor and the time that has elapsed since the supply was connected to the circuit.
6. A 5-μF capacitor is connected in series with a 10-kΩ resistor to a 100-V d.c. supply. Calculate the initial value of the charging current.

   Once the capacitor is fully charged, it is disconnected from the supply (and from the 10-kΩ resistor) and is discharged through a 5-kΩ resistor. Determine the initial value of the discharge current and the energy dissipated by the 5-kΩ resistor.
7. An 80-μF capacitor is charged to 100 V d.c. and is then discharged through a voltmeter of resistance 10 kΩ. What are the following after 0·4 s:
   (*a*) the reading of the voltmeter;
   (*b*) the voltage across the capacitor;
   (*c*) the charge remaining in the capacitor?

   How long does it take for the voltmeter reading to fall to 50 V?
8. A 20-μF capacitor is connected in series with a 2-kΩ resistor to a 100-V d.c. supply. Determine:
   (*a*) the initial charging current;
   (*b*) the current in the circuit at the end of a period equal to the time constant after the supply is applied.

   Once the capacitor is fully charged, the supply is disconnected and the circuit terminals short-circuited. Find:
   (*c*) the current in the circuit at the end of a period equal to the time constant after the short-circuit is applied;
   (*d*) the voltage across the resistor at that instant;
   (*e*) the total energy dissipated by the resistor during the complete charging and discharging processes.

9. A coil of resistance 50 Ω and inductance 0·5 H is connected to a 250-V d.c. supply. Find:
   (*a*) the final steady value of the current;
   (*b*) the initial rate of change of the current;
   (*c*) the time constant of the circuit;
   (*d*) the energy stored in the magnetic field once the current reaches its final steady value;
   (*e*) the value of the current after a time equal to the time constant.
10. A circuit of resistance 1·0 Ω and inductance 1·0 H is suddenly connected to a 10-V d.c. supply. Construct the current/time characteristic for the period 0–5 s after the supply is connected and hence determine the circuit current after 1·0 s.
11. A coil of resistance 50 Ω and inductance 1·0 H is connected to a 100-V d.c. supply. Find:
   (*a*) the time taken for the current to reach half its final steady-state value;
   (*b*) the time taken for the current to reach 87·5 per cent of its final steady-state value.
12. The field coil of a d.c. motor has inductance 4·0 H and resistance 80 Ω and is connected to a 220-V d.c. supply. Determine
   (*a*) the current 50 ms after the supply is connected;
   (*b*) the time taken for the current to rise to half its final steady-state value.
13. A buzzer of inductance 10 H and resistance 600 Ω operates when its current is in excess of 10 mA. When a voltage 10 V is applied to the buzzer, how long does it take until the buzzer sounds?
14. A coil of resistance 100 Ω and inductance 2·5 H is connected in parallel with a resistor of resistance 400 Ω, and the network is energised from a 200-V d.c. supply. Once the steady-state conditions in the network have been achieved, the supply is disconnected. Determine:
   (*a*) the current/time characteristic;
   (*b*) the current 4 ms after the supply is disconnected;
   (*c*) the voltage across the 400-Ω resistor.
15. An *R*–*L* circuit has a time constant 4·0 ms. The circuit is connected to a 100-V d.c. supply and after a period of 4·0 ms, the current is 0·5 A. Determine the final steady-state current and hence find the resistance and inductance of the circuit.

Answers

1. 0·8 s; 125 V/s; 12·5 mA; 10 mC
2. 54 μA; 54 V
3. 40 ms; 2·75 kV/s; 59 ms
4. 91 V, etc; 409 V, etc; 0·5 mA; 17·3 s
5. 100 V; 2·2 ms
6. 10 mA; 20 mA; 25 mJ

7. 61 V; 61 V; 4·9 mC; 0·56s
8. 50 mA; 18·5 mA; 18·5 mA; 37 V; 0·2 J
9. 5·0 A; 500 As; 10 ms; 6·25 J; 3·2 A
10. 6·3 A
11. 13·9 ms; 41·6 ms
12. 1·74 A; 35 ms
13. 15·3 ms
14. 0·90 A; 360 V
15. 0·8 A; 125 Ω; 0·5 H

Chapter 6

Single-phase transformer

A transformer is a machine that has no moving parts but is able to transform alternating voltages and currents from high to low values and vice versa. Transformers are used extensively in all branches of electrical and electronic engineering, from the large power transformer employed in the National Grid to the small signal transformer of an electronic amplifier.

1 Principle of operation

When an alternating voltage is applied to a concentrated coil wound on a ferromagnetic core, a back e.m.f. is induced in the coil due to the continual alternation of the self flux linkage. This is the principle of the simple induction shown in Fig. 1. In such an arrangement, the applied alternating voltage causes an alternating current to flow in the conductor coil. This current gives rise to an alternating m.m.f. which causes a core flux linking the coil. This alternating flux linkage induces an e.m.f. in the coil which, by Lenz's Law, opposes the change of current. If the resistance of the conductor is negligible, then balance is obtained when there is just enough current to produce just enough m.m.f. to create

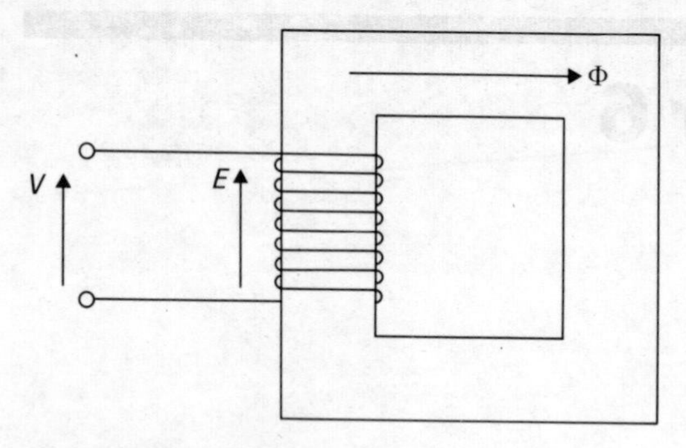

Fig. 1 Simple inductor

just enough flux to induce just enough e.m.f. to exactly oppose the applied voltage.

If the core of the inductor is made from a good ferromagnetic material which is easily magnetised, the inductance of the inductor is large and the current which flows in the coil is very small. At this stage, it is not unreasonable to assume that the current required to magnetise the core is so small that it is negligible, and our present studies will now be limited to inductive arrangements requiring negligible current to excite them.

Again the back e.m.f. induced in the coil is due to the change of flux linkage within the coil. If another coil were placed around the core, as shown in Fig. 2, this other coil also would have an e.m.f. induced in it. This e.m.f. is caused by the effect of mutual inductance whereby the e.m.f. in the second coil is induced as a result of the flux emanating from the first coil. It should be noted that the two coils are insulated from one another and are therefore electrically separate.

The first coil, connected to the supply, is termed the primary winding, and the other coil is termed the secondary winding. The e.m.f. E_1 induced in the primary winding is equal to the applied voltage V_1

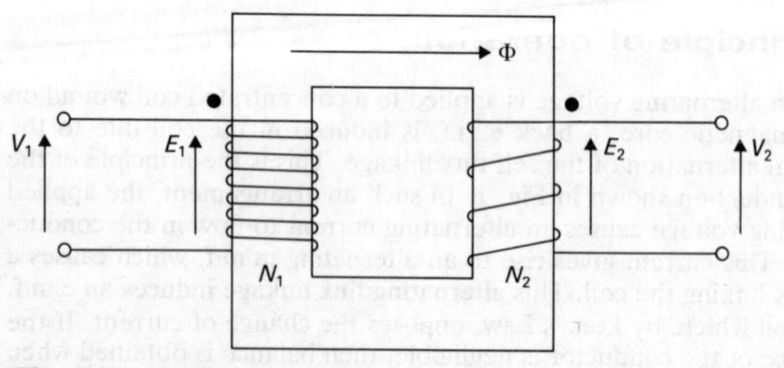

Fig. 2 Simple transformer

and E_1 is termed the primary e.m.f. whilst V_1 is the primary voltage.
The same flux links each turn of the primary winding and therefore each turn has the same amount of e.m.f. induced in it. If there are N_1 turns in the primary winding, the induced e.m.f. of each turn is E_1/N_1.

Each turn of the secondary winding is also linked by the same flux as is each turn of the primary winding; thus the induced e.m.f. in each turn is again E_1/N_1. This only holds because we have assumed that there is no loss of flux linkage due to leakage. If there are N_2 turns in the secondary winding, then the induced e.m.f. E_2 in the secondary winding is given by

$$E_2 = N_2 \frac{E_1}{N_1} \tag{1}$$

thus $$\frac{E_1}{E_2} = \frac{N_1}{N_2} \tag{2}$$

E_2 is termed the secondary e.m.f. and the terminal voltage V_2 of the secondary winding is termed the secondary voltage. The ratio of the e.m.f.s (and of the terminal voltages) of the two windings is therefore seen to be the same as the ratio of the turns; thus

$$V_1 = E_1$$

and $$V_2 = E_2$$

hence $$\frac{V_1}{V_2} = \frac{E_1}{E_2} = \frac{N_1}{N_2} \tag{3}$$

The simple device of Fig. 2 thus provides a basic means of transforming voltages.

Example 1 A transformer has a primary winding of 150 turns and a secondary winding of 45 turns. The primary voltage applied to the transformer is 240 V. Find the secondary voltage.

$$\frac{V_1}{V_2} = \frac{N_1}{N_2}$$

thus $$V_2 = V_1 \frac{N_2}{N_1} = 240 \times \frac{45}{150} = \underline{72\ \text{V}}$$

In general, we may assume that the applied alternating voltage is sinusoidal. In the ideal transformer, the resulting core flux is also sinusoidal; let this flux be represented by

$$\phi = \Phi_m \sin \omega t$$

For a coil of N turns, the e.m.f. induced in it by this flux ϕ is given by

$$e = \frac{\delta\psi}{\delta t}$$

$$= N\frac{\delta\phi}{\delta t}$$

$\delta\phi/\delta t$ is the rate of change of flux. Because the flux varies sinusoidally, it follows that the rate of change is cosinusoidal. It should therefore come as no surprise that the expression for the induced e.m.f. can be shown to be

$$e = \omega N\Phi_m \cos \omega t$$

and hence $$e = 2\pi fN\Phi_m \sin(\omega t + 90°) \qquad (4)$$

This represents an induced e.m.f. of maximum value $2\pi fN\Phi_m$. The induced e.m.f. leads the flux by 90°. The maximum instantaneous e.m.f. E_m is given when

$$\cos \omega t = \sin(\omega t + 90°) = 1,$$

thus

$$E_m = 2\pi fN\Phi_m$$

The e.m.f. is sinusoidal; thus the r.m.s. value of the e.m.f. E is given by

$$E = \frac{E_m}{\sqrt{2}}$$

$$= \frac{2\pi fN\Phi_m}{\sqrt{2}}$$

$$= 4{\cdot}44\, fN\Phi_m \qquad (5)$$

From this general expression, it follows that the primary e.m.f. in the simple transformer is given by

$$E_1 = 4{\cdot}44fN_1\Phi_m$$

and the secondary e.m.f. by

$$E_2 = 4{\cdot}44fN_2\Phi_m$$

hence $$\frac{E_1}{E_2} = \frac{4{\cdot}44fN_1\Phi_m}{4{\cdot}44fN_2\Phi_m}$$

$$= \frac{N_1}{N_2}$$

This is the relation (2) derived in an alternative manner.

In earlier studies of magnetism, we have observed that the maximum flux for operation is determined by the maximum magnetic flux density to which the core can be magnetised without reaching a value associated with the saturation of the core material. It is therefore convenient to express relation (5) in terms of the maximum flux density; thus, given that the cross-sectional area of the core is A,

$$\Phi_m = B_m A$$

and $$E = 4{\cdot}44 f N B_m A \tag{6}$$

In making a transformer, the number of turns in a winding are therefore chosen to ensure that the maximum operation flux density is not exceeded. Remember that you cannot have a fraction of a turn since any circuit must be complete if it is to operate electrically.

Example 2 A 1-ph, 415/22 000-V, 50-Hz transformer has a nett core area of 10 000 mm² and a maximum operating flux density 1·3 T. Estimate the number of turns in each winding.

$$E_1 = 4{\cdot}44 f N_1 B_m A$$

$$N_1 = \frac{E_1}{4{\cdot}44 f B_m A} = \frac{415}{4{\cdot}44 \times 50 \times 1{\cdot}3 \times 10\,000 \times 10^{-6}}$$

$$= 143{\cdot}6 \text{ turns}$$

The number of turns must be an integer, hence

$$N_1 = 144 \text{ turns}$$

$$N_2 = \frac{V_2}{V_1} N_1 = \frac{22\,000 \times 144}{415} = \underline{7634 \text{ turns}}$$

In a calculation of this type, the number of turns in the low-voltage winding is always estimated first. This is because any correction to the number of turns to give an integer is relatively more important to the smaller number. When adjusting the number of turns to give an integer, this is always made to the next largest integer, otherwise the maximum permitted flux density would be exceeded.

Having transformed the primary voltage, the consequence is almost certain to be that the secondary voltage is applied to a load. In such a case the secondary e.m.f. E_2 causes a current I_2 to flow through the load, as illustrated in Fig. 3. The secondary current I_2 in the secondary winding produces an m.m.f. F_2 in the core. This m.m.f. is in phase with the current and is given by

$$F_2 = I_2 N_2$$

Such an m.m.f. should create further flux in the core, thus altering the core flux from the value previously experienced. This core flux, however, must remain unchanged from the original condition because this was the flux which, when linking the primary winding and varying at the supply frequency, gave rise to the primary back e.m.f. equal to the

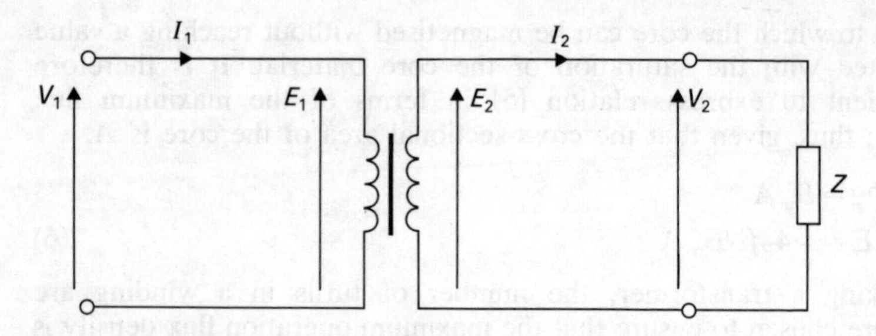

Fig. 3 Transformer on load

supply voltage. This equality must be maintained at all times as we know from the observations of Kirchhoff's Second Law.

If the m.m.f. F_2 were to vary the core flux, then the e.m.f. induced in the primary winding would not be equal to the applied voltage and the equality required would not result. In order to maintain the equality, an additional current must flow in the primary winding to give rise to an m.m.f. equal in magnitude but opposite in effect to the secondary m.m.f. F_2. Let this primary current be I_1; hence

$$F_1 = I_1 N_1$$

But $\quad F_1 = F_2$

thus $\quad I_1 N_1 = I_2 N_2$

$$\frac{N_1}{N_2} = \frac{I_2}{I_1} \tag{7}$$

In this explanation, it has been noted that we are dealing with additional current in the primary winding. However, we also earlier assumed that the current required to excite the core initially was negligible; thus effectively the primary current is made up from the additional current required for the m.m.f. balance. In practical terms, this approximation has very little effect on transformers operating at more than 25 per cent of the rated loading.

Finally the m.m.f. balance is sometimes described at the ampere-turn balance of the transformer, which is rather descriptive of the action taking place. The ampere-turn balance results in relation (7), which involves the ratio N_1/N_2. This ratio of the turns is termed the turns ratio and may be generally expressed as

$$\frac{N_1}{N_2} = \frac{E_1}{E_2} = \frac{V_1}{V_2} = \frac{I_2}{I_1} \tag{8}$$

Example 3 A 5-Ω resistor is connected across the secondary winding of a transformer and the secondary voltage is 200 V. The primary current is 10 A. Find the primary voltage.

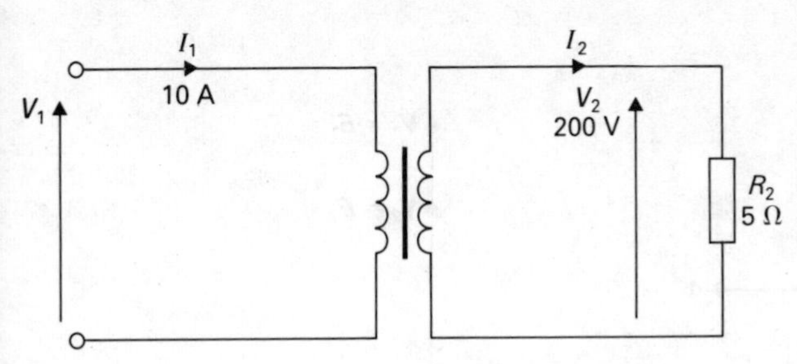

Fig. 4

$$I_2 = \frac{V_2}{R_2} = \frac{200}{5} = 40\text{ A}$$

$$\frac{I_2}{I_1} = \frac{V_1}{V_2}$$

$$V_1 = V_2\frac{I_2}{I_1} = \frac{200 \times 40}{10} = \underline{800\text{ V}}$$

2 Ideal transformer

An ideal transformer is one that has no losses either in the electric circuits or in the magnetic circuits. Added to this is the important property that it requires negligible current to magnetise the core.

When a transformer is loaded, an m.m.f. (ampere-turn) balance is set up between the primary and secondary circuits in order that the core flux remains unchanged. Little mention has yet been made as to how this initial core flux was created, other than the explanation of a negligible current setting up a flux to produce the back e.m.f. If the inductance of the primary winding is infinite, zero current is required to create the required core flux. In such a transformer under loaded conditions, all the primary current is therefore used to balance the secondary current in accordance with relation (7). It follows that relation (8) applies to such an ideal transformer and the corresponding phasor diagram is shown in Fig. 5.

In the case of an ideal transformer under no-load conditions, i.e. with the secondary winding open circuit, the phasor diagram need only show the winding e.m.f.s, the terminal voltages and the flux. In such phasor diagrams, the flux is taken as reference and the primary e.m.f. in taken to lead the flux by 90°. This relationship depends on the choice of terminals from which the e.m.f. is observed; thus if you were to consider the e.m.f. with the winding terminals reversed, then the

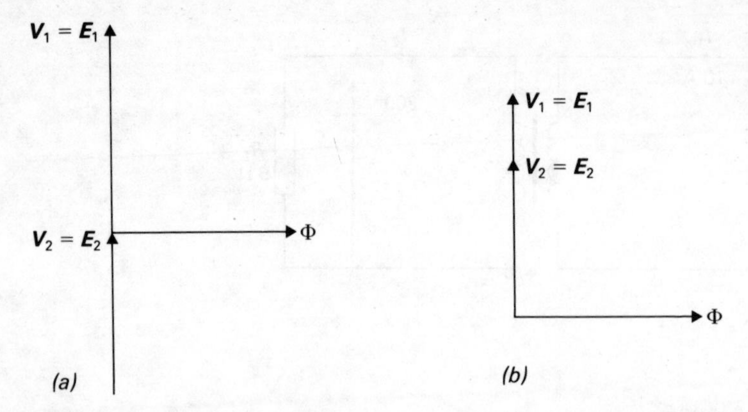

Fig. 5 Phasor diagram for an ideal transformer on no load: (*a*) preferred diagram; (*b*) confused diagram

e.m.f. would appear to lag the same flux by 90°. Again, therefore, we have a problem of choice of convention which ought not be confused with magnetic statements such as Lenz's Law.

The same problem arises with the choice of polarity of the secondary e.m.f. but it may be observed that the windings of most transformers are so mounted that the secondary e.m.f. is in phase with the primary e.m.f. Only transformers that are made for special purposes do not have the winding terminals mounted in this way, and these are quite rare.

That the e.m.f.s are in phase produces another problem because normally they should then lie in the same position on a phasor diagram with the result that the diagram becomes rather cluttered. To avoid this, we draw the secondary e.m.f. on the opposite side of the reference for clarity as shown in Fig. 5(*a*). Finally, the e.m.f.s are equal to the terminal voltages in an ideal transformer because we have noted that there are no electric or magnetic losses. And the current to magnetise the core is negligible, hence it cannot appear on the diagram.

If the transformer is loaded, you only have to add the currents to the no-load phasor diagram. The primary and secondary currents must be in phase for the purpose of the ampere-turn balance. The phase relationship between the secondary current and the secondary voltage is determined by the nature of the secondary load. Because the corresponding primary e.m.f. and current are in phase with their secondary counterparts, it follows that the phase relationship between the primary current and voltage is the same as that between the secondary current and voltage. Let this phase relationship be ϕ. ϕ can be chosen to produce leading, lagging and in phase conditions. The corresponding forms of phasor diagram are shown in Fig. 6.

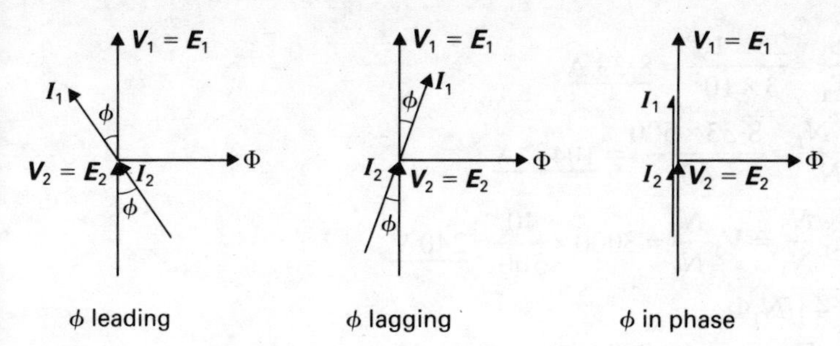

Fig. 6 Phasor diagrams for a loaded ideal transformer

Finally from relation 8,

$$\frac{V_1}{V_2} = \frac{I_2}{I_1}$$

$$V_1 I_1 = V_2 I_2$$

$$S_1 = S_2$$

The primary circuit phase angle is the same as the secondary circuit phase angle, i.e. each circuit has the phase angle ϕ, hence

$$S_1 \cos\phi = S_2 \cos\phi$$

but $\quad S_1 \cos\phi = P_1$

and $\quad S_2 \cos\phi = P_2$

hence $\quad P_1 = P_2 \qquad (9)$

Thus the input power is equal to the output power. This is a result which we would expect from an ideal transformer in which there are no losses.

Example 4 An ideal 25-kVA transformer has 500 turns in the primary winding and 40 turns in the secondary winding. The primary winding is connected to a 3·0-kV, 50-Hz supply. Calculate

(*a*) the primary and secondary currents on full load;
(*b*) the secondary e.m.f.;
(*c*) the maximum core flux.

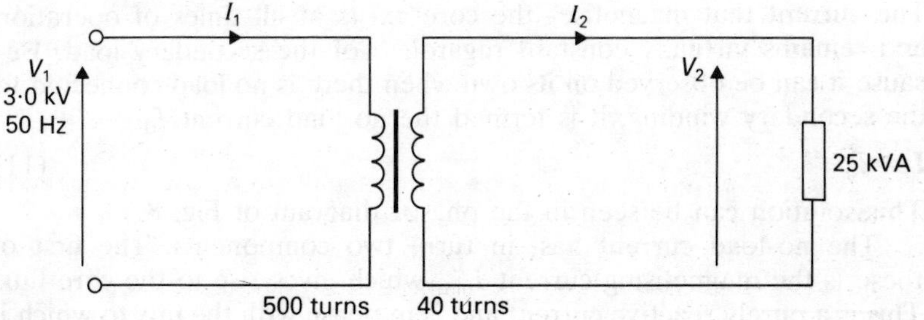

Fig. 7

$$I_1 = \frac{S}{V_1} = \frac{25 \times 10^3}{3 \times 10^3} = \underline{8{\cdot}33\ \text{A}}$$

$$I_2 = \frac{I_1 N_1}{N_2} = \frac{8{\cdot}33 \times 500}{40} = \underline{104{\cdot}2\ \text{A}}$$

$$E_2 = E_1 \frac{N_2}{N_1} = V_1 \frac{N_2}{N_1} = 3000 \times \frac{40}{500} = \underline{240\ \text{V}}$$

$$E_1 = 4{\cdot}44 f N_1 \Phi_\text{m}$$

$$\Phi_\text{m} = \frac{E_1}{4{\cdot}44 f N_1} = \frac{3000}{4{\cdot}44 \times 50 \times 500} = 27 \times 10^{-3}\ \text{Wb}$$

$$= \underline{27\ \text{mWb}}$$

---

3 Practical transformer

The practical transformer differs from the ideal transformer in three ways:

(*a*) it requires a current to magnetise the core;
(*b*) it has electric losses;
(*c*) it has magnetic losses.

In a practical transformer, there is a current in the primary winding to magnetise the core. When a load is connected to the secondary winding of the transformer, an additional current flows in the primary winding to maintain the m.m.f. (ampere-turn) balance in the core. Thus the total primary current consists of two components – the current to magnetise the core and the current to maintain the m.m.f. balance. Since the primary current is designated I_1, some other symbol must be given to the component current required to maintain the m.m.f. balance. Let it be $I_{2'}$, which is described as I_2 referred. From relation (7), it follows that

$$\frac{N_1}{N_2} = \frac{I_2}{I_{2'}} \tag{10}$$

The current that magnetises the core exists at all times of operation and remains virtually constant regardless of the secondary load. Because it can be observed on its own when there is no load connected to the secondary winding, it is termed the no-load current I_0.

$$\boldsymbol{I}_1 = \boldsymbol{I}_0 + \boldsymbol{I}_{2'} \tag{11}$$

This solution can be seen in the phasor diagram of Fig. 8.

The no-load current has, in turn, two components. The first of these is the magnetising current $I_{0\text{m}}$, which gives rise to the core flux. This is a purely reactive current and is in phase with the flux to which it

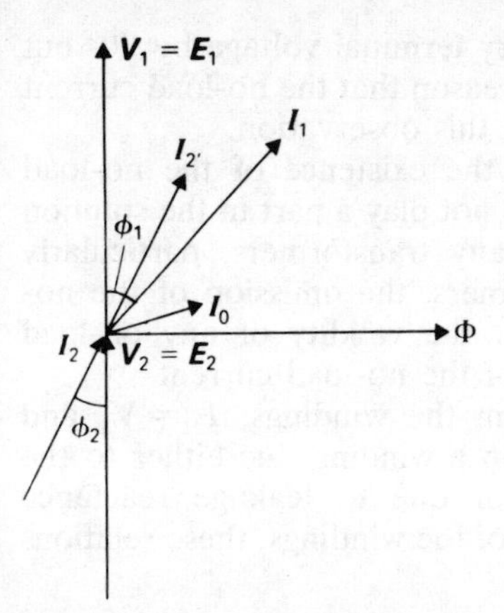

Fig. 8 Phasor diagram for a practical transformer with no winding

gives rise. The second component is an active one due to the hysteresis and eddy-current losses which occur in a practical core. This component is designated I_{0l}.

In the first component, the subscript has the 0 and m on the same level, indicating the magnetising component of the no-load current. Only when the m is dropped relative to the number or symbol does it indicate a maximum value. In the second component, the subscript includes the l for losses.

The total no-load current is given by the phasor sum of the components.

$$\boldsymbol{I}_0 = \boldsymbol{I}_{0m} + \boldsymbol{I}_{0l} \tag{12}$$

This relationship is illustrated in Fig. 9. You will see that the resulting

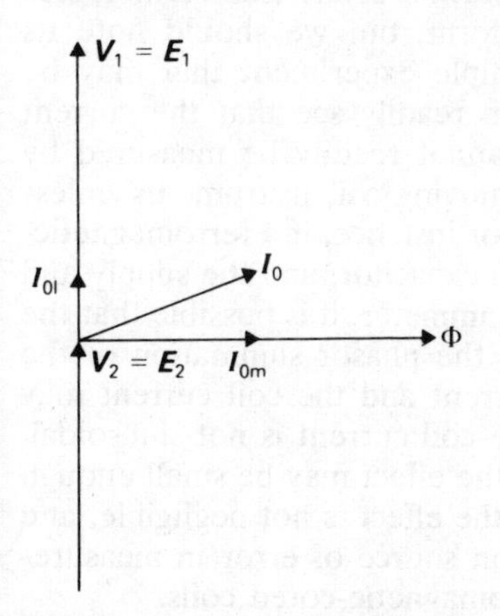

Fig. 9 Phasor diagram of a practical transformer on no-load and showing the components of the no-load current

no-load current does not lag the primary terminal voltage by 90° but by an angle somewhat less. It is for this reason that the no-load current in Fig. 8 was drawn to correspond with this observation.

You should remember that, whilst the existence of the no-load current should be acknowledged, it does not play a part in the solution of the accompanying problems. So many transformers, particularly power transformers and signal transformers, the omission of the no-load current does not materially affect the validity of any on-load solution, due to the relative smallness of the no-load current.

Provided that there are no losses in the windings, $E_1 = V_1$ and $E_2 = V_2$. If there were a voltage drop in a winding due either to the resistance of the winding conductor or due to leakage reactance created by imperfect magnetic coupling of the windings, these relations would not hold.

4 Magnetisation of the core

You have already noted that if the applied voltage across a winding wound on a ferromagnetic core is sinusoidal, then the rate of change of the core flux must also be sinusoidal. However it does not follow that the current which creates this flux varies sinusoidally. If the B/H characteristic is redrawn as a ϕ/i characteristic, the relation between flux and current can be derived as indicated in Fig. 10. The change of scale for the axes is derived from $\phi = BA$, where the cross-sectional area A of the core is constant, and from $i = Hl/N$, where the core length l and the number of turns in the winding N are constant. Using this ϕ/i characteristic, the waveform of the current can be obtained by plotting instantaneous values of flux against current.

At this introductory stage in the operation of the transformer, it is not necessary to analyse such a waveform, but we should note its significance which will arise in any simple experiment that may be carried out with a transformer. We can readily see that the current waveform is not sinusoidal, hence it cannot readily be measured by standard instruments such as rectifier moving-coil instruments unless the readings are correctly interpreted. For instance, if a ferromagnetic-cored coil is connected in parallel with a capacitor, and the supply and the branch currents are measured by an ammeter, it is possible that the supply current will not be obtained by the phasor summation of the branch currents because the supply current and the coil current may not be sinusoidal. It is the case that the coil current is not sinusoidal, for the reason observed in Fig. 10, but the effect may be small enough to be negligible. In most transformers, the effect is not negligible, and non-sinusoidal waveforms are a common source of error in measurements made on circuits containing ferromagnetic-cored coils.

When testing a transformer, the no-load current is better measured at least by using a moving-iron ammeter, which attempts to indicate

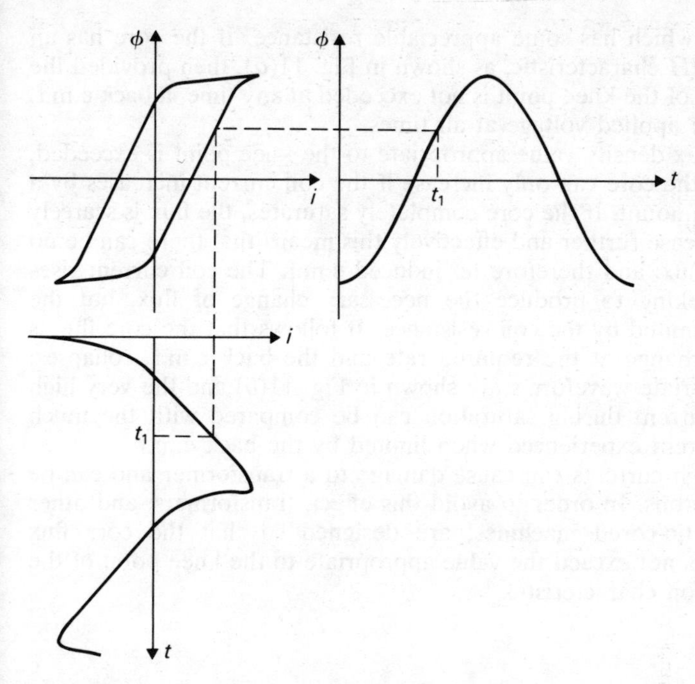

Fig. 10 Waveform of magnetising current

the true r.m.s. value regardless of waveform. However, the waveform of the no-load current does cause some error of indication, yet the moving-iron ammeter is less affected by this factor than is the rectifier moving-coil ammeter.

Finally the effect of saturation of the core on the back e.m.f. requires some comment. Consider a ferromagnetic-cored coil, the

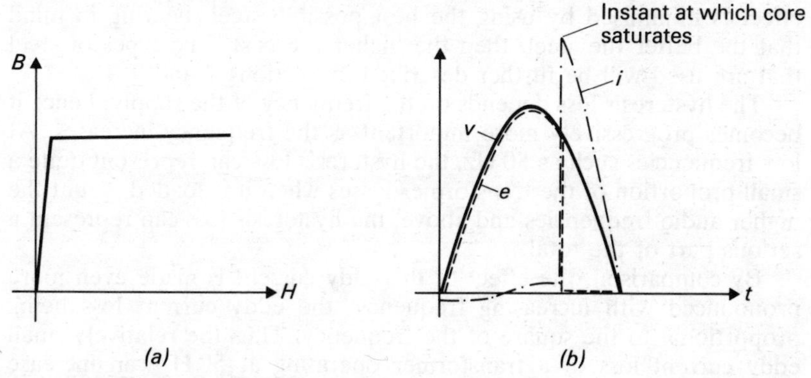

Fig. 11 Effect of excessive flux in a ferromagnetic core: (*a*) *B*/*H* characteristic; (*b*) waveform diagram

winding of which has some appreciable resistance. If the core has an idealised B/H characteristic, as shown in Fig. 11(a), then provided the flux density of the knee point is not exceeded at any time, a back e.m.f. opposes the applied voltage at all times.

If the flux density value appropriate to the knee point is exceeded, the flux in the core can only increase if the coil current increases by a very large amount. If the core completely saturates, the flux is scarcely able to increase further and effectively this means that there can be no change of flux, and therefore no induced e.m.f. The coil current rises rapidly, seeking to produce the necessary change of flux, but the current is limited by the coil resistance. It follows that the core flux is unable to change at the required rate and the back e.m.f. collapses. The appropriate waveforms are shown in Fig. 11(b) and the very high value of current during saturation can be compared with the much smaller current experienced when limited by the back e.m.f.

Such high currents can cause damage to a transformer and can be most dangerous. In order to avoid this effect, transformers, and other ferromagnetic-cored machines, are designed so that the core flux density does not exceed the value appropriate to the knee point of the magnetisation characteristic.

5 Transformer losses

There are three sources of active power loss in a transformer and all of these have already been mentioned. They are

(a) hysteresis loss;
(b) eddy-current loss;
(c) winding i^2R losses.

The hysteresis loss arises from the magnetic cycling of the core and its effect is minimised by using the best possible steel, bearing in mind that the better the steel, then the higher the cost. The types of steel that are used will be further described in Sections 7 and 9.

The hysteresis loss depends on the frequency of the supply; hence it becomes progressively more important as the frequency increases. At low frequencies such as 50 Hz, the hysteresis loss can represent quite a small proportion of the transformer losses when it is loaded, yet at the higher audio frequencies and above, the hysteresis loss can represent a serious part of the total.

By comparison, the effect of the eddy current is made even more pronounced with increasing frequency, the eddy current loss being proportional to the square of the frequency. Thus the relatively small eddy current loss of a transformer operating at 50 Hz can increase drastically if operated at audio frequencies.

The eddy current loss is minimised by the choice of steel or other ferromagnetic material used in the core. It is necessary to use a material of high resistance, whereby the eddy current induced is minimised. The loss is also reduced either by lamination or by the use of pulverised material in the core.

The combined effects of the hysteresis loss and the eddy current loss are termed the core loss. Formerly the core loss was termed the iron loss, but this name has fallen from favour because cores can be made from materials other than iron, e.g. cobalt and nickel.

The variation of core loss with frequency is shown in Fig. 12. It is difficult to show this in a satisfactory manner due to the rapid growth of the eddy current loss with increase in frequency. Because of this, it is common practice to draw the characteristic of the dissipation factor to a base of frequency, where the dissipation factor is the ratio of loss to frequency. This characteristic is also shown in Fig. 12.

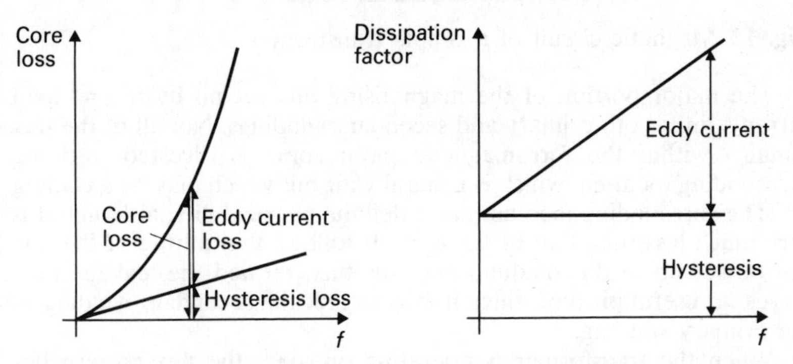

Fig. 12 Core loss/frequency and dissipation factor/frequency characteristics of a transformer core

The windings of the transformer have resistance and the currents in the windings give rise to i^2R losses. Even though the windings are made of copper or aluminium, there remains sufficient resistance to give rise to an appreciable total loss. In a modern transformer operating at full load and a frequency of, say, 50 Hz, the i^2R losses can account for about 90 per cent of the total transformer losses. Thus the core materials have been improved for this type of operation to the point that the principal loss arises in the windings.

By comparison, as the frequency rises, the core losses become more significant until the roles are reversed and the core loss accounts for about 90 per cent of the total losses.

The i^2R loss was formerly described as the copper loss, but this term is now rarely used as it is quite inappropriate to transformers with aluminium windings.

6 Magnetic losses

The term 'magnetic losses' is somewhat misleading, because it does not involve the loss of energy (or power). Instead it refers to the magnetic circuit flux which fails to link both coils of the transformer and which is therefore thought of as having been lost; consider the magnetic circuit of a simple transformer as shown in Fig. 13.

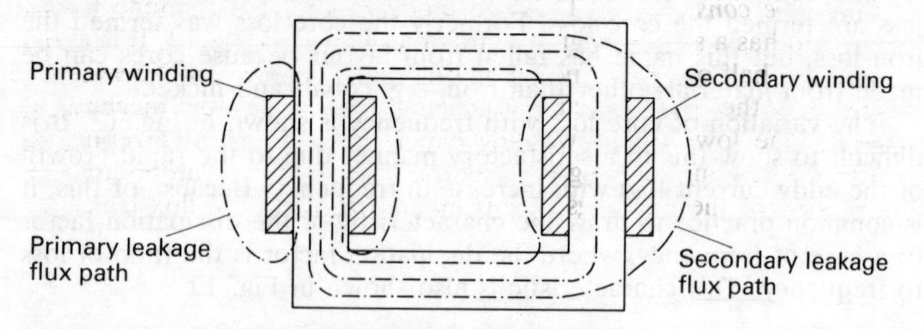

Fig. 13 Magnetic circuit of a simple transformer

The major portion of the magnetising flux set up by the no-load currents links both primary and secondary windings. Not all of the flux remains within the ferromagnetic path: some is diverted into the surrounding medium, which is generally air but which may be a cooling oil. The surrounding medium has a definite permeability, although it is very much less than that of the core. It follows that a little of the flux can be set up in this medium and this flux, termed the leakage flux, serves no useful purpose since it fails to link the secondary winding to the primary winding.

When the transformer is operating on load, the flux pattern becomes that shown in Fig. 13. The leakage flux due to the primary winding is termed the primary leakage flux. However, the secondary current sets up an m.m.f. which opposes the main flux and causes a portion of the flux to be diverted into secondary leakage paths. There is thus formed a flux which links only the secondary winding, and this is termed the secondary leakage flux. In Fig. 13, the secondary flux does not appear separately but combines with the main flux, which thus appears to be diverted outside the secondary winding as indicated.

The effect of these leakage fluxes is purely self inductive, hence both windings appear to have self-inductive reactance. This serves no useful purpose but acts as effective voltage drops in the windings. The main core flux decreases very slightly as the load increases, but the leakage fluxes are practically proportional to the currents in the respective windings. The effect of flux leakage upon the ratio of transformation is thus to reduce the secondary terminal voltage for a given primary applied voltage.

To minimise the effects of flux leakage, various forms of transformer construction are employed according to the operating frequency. These transformers operating at 50 or 60 Hz are used for power applications, whilst those operating at higher frequencies are classified as audio- or high-frequency transformers.

Power transformers generally operate at 50 Hz or 60 Hz and are constructed in one of the following two manners.

Core-type construction. This is a modification of the simple transformer and has a single magnetic circuit as shown in Fig. 14. It is usual to wind one-half of each winding on one limb, the low-voltage winding (which has the greater current) being innermost for mechanical strength. The low-voltage winding experiences the greater repulsion between its current-carrying conductors, because of the larger current. The placing of the windings in this manner reduces the flux leakage.

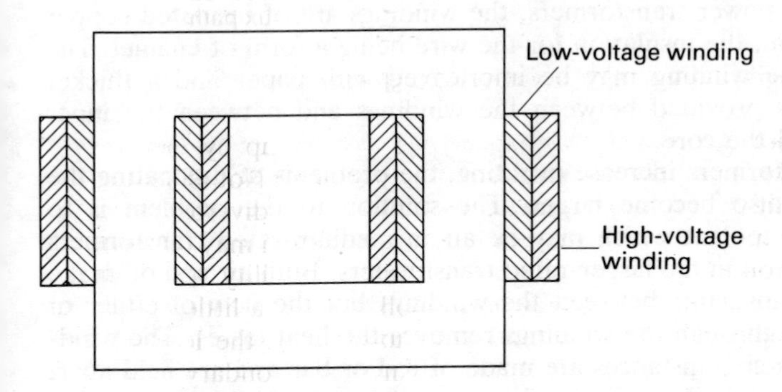

Fig. 14 Core-type transformer construction

Shell-type construction. This method of construction involves the use of a double magnetic circuit as shown in Fig. 15. The windings are again wound concentrically, i.e. within one another, but they are completely around the centre limb. In this form, the leakage flux forms only a very small part of the total flux and coupling coefficients well in excess of 0·99 are quite normal.

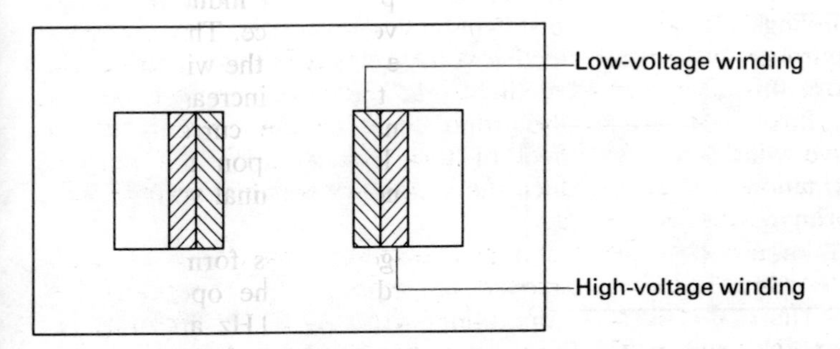

Fig. 15 Shell-type transformer

In both forms of core construction, the material is likely to be a silicon steel chosen for its high permeability and high resistivity. The steel is laminated and, except in the cheapest transformers, it is cold rolled to give grain orientation, thus giving better working flux densities. Stalloy is commonly used in cores.

The introduction of nickel to the steel gives high values of permeability at low magnetic field strengths. This introduces less distortion in the magnetising current but this benefit is rarely essential at supply frequencies except when the transformer is being used for measurement purposes. High nickel content (about 50 per cent) gives the group of alloys termed Radiometal which have the advantages of reducing the core size and of low hysteresis loss.

In small power transformers, the windings are of insulated copper or aluminium, the insulation for the wire being a form of enamel. The layers of the winding may be interleaved with paper and a thicker insulation is provided between the windings and between the inner winding and the core.

As transformers increase in rating, the problems of dissipating the i^2R losses also become larger. The solution to the problem is to introduce a coolant which may be air in medium-rated transformers but which is oil in the larger-rated transformers. Both air and oil make satisfactory insulants between the windings but the flow of either of these media through the windings removes the heat losses. The windings in these circumstances are made of rod or bar and are held apart by wooden or similar solid insulant material wedges.

In order to minimise the amount of copper (or aluminium) used in the construction of the windings, a form of transformer called an autotransformer has been evolved, and this only has a single winding as shown in Fig. 16. The single winding serves as both primary and secondary, and it is tapped at different points for the input and output. As with the normal form of transformer, the voltage ratio depends on the numbers of turns involved between the input points and the output points, thus in the instance given, the autotransformer steps down the supply voltage. Unlike the two-winding transformer, the autotransformer transfers the electrical power from one circuit to the other

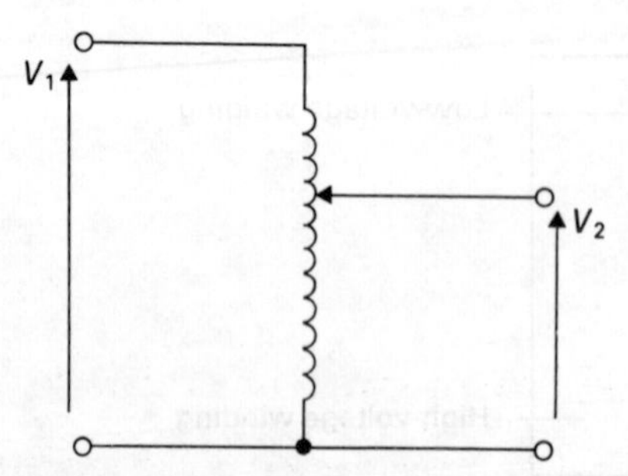

Fig. 16 1-ph auto-transformer

partly by the direct electrical connection between the circuits, the remainder of the power being transferred by the usual magnetic interchange process.

The form of construction may be either core type or shell type, but the former is more common. Often the connection tapping the winding on the secondary side can be varied by a sliding contact, whereby the output voltage can easily be adjusted. This form of autotransformer is commonly termed a variac.

The advantages of the autotransformer are that it requires less conductor material and as a consequence there are less I^2R losses and greater operating efficiency. The disdavantages mainly occur under fault conditions whereby the higher supply voltage may inadvertently come through the low-voltage circuit, which can give rise to quite dangerous situations. Also the common connection should operate at a low voltage (say that of the neutral supply wire) or preferably it should be earthed. In this way the voltages of the secondary conductors are limited with respect to their surroundings.

Although this chapter deals with the 1-ph transformer, some mention of the 3-ph transformer is in order. Most transformers rated above 25 kVA are of the 3-ph core type illustrated in Fig. 17. In this instance both the primary windings and the secondary windings are star-connected. On the secondary side, this has the advantage that the neutral conductor can be brought out from the star point. Also it is convenient to have a star point that may be earthed, thus determining the voltages of the line conductors with respect to the surrounding earth. Although the windings are shown separately, they would in practice be concentrically wound as previously described.

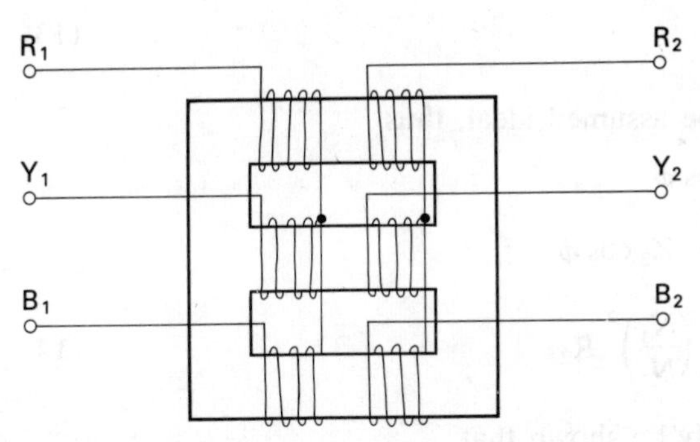

Fig. 17 3-ph core-type transformer

8 Impedance transformation and matching

Consider the effect of the secondary load when seen from the primary winding. This may be analysed by considering the arrangement shown

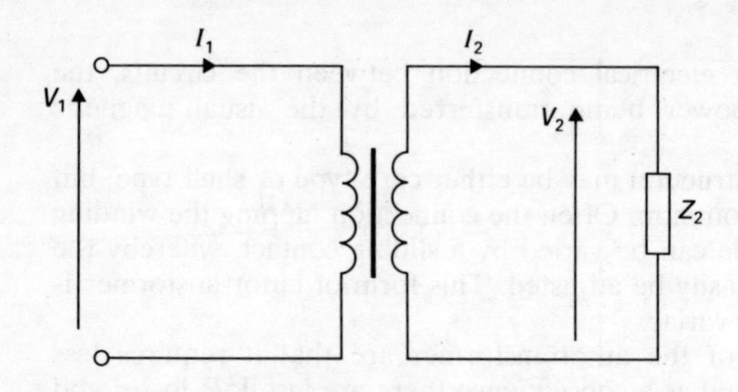

Fig. 18 Transformer supplying a secondary load

in Fig. 18. As before, the subscript 2′ indicates that the quantity has been referred to the primary winding of the transformer.

$$Z_2 = \frac{V_2}{I_2}$$

$$Z_1 = \frac{V_1}{I_1}$$

$$= V_2 \frac{N_1}{N_2} \frac{1}{I_2} \frac{N_1}{N_2}$$

$$= \left(\frac{N_1}{N_2}\right)^2 Z_2$$

$$= Z_{2'}$$

$$Z_1 = Z_{2'} = \left(\frac{N_1}{N_2}\right)^2 \cdot Z_2 \tag{13}$$

The transformer may be assumed ideal, thus

$$Z_1 \cos \phi = Z_2 \cos \phi$$

$$= \left(\frac{N_1}{N_2}\right)^2 Z_2 \cos \phi$$

hence $$R_1 = R_{2'} = \left(\frac{N_1}{N_2}\right)^2 R_2 \tag{14}$$

and also it may similarly be shown that

$$X_1 = X_{2'} = \left(\frac{N_1}{N_2}\right)^2 X_2 \tag{15}$$

Example 5 A 2 : 1 step-down transformer supplies a resistor of resistance 100 Ω. Find the equivalent input resistance of the transformer.

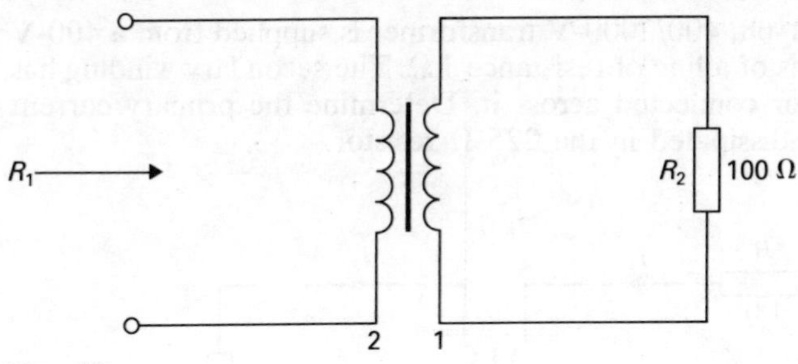

Fig. 19

$$R_1 = \left(\frac{N_1}{N_2}\right)^2 R_2 = \left(\frac{2}{1}\right)^2 100 = \underline{400\ \Omega}$$

Example 6 A 2:1 step-down transformer supplies a capacitor of capacitance 4 μF. Find the equivalent capacitance as seen from the input to the transformer.

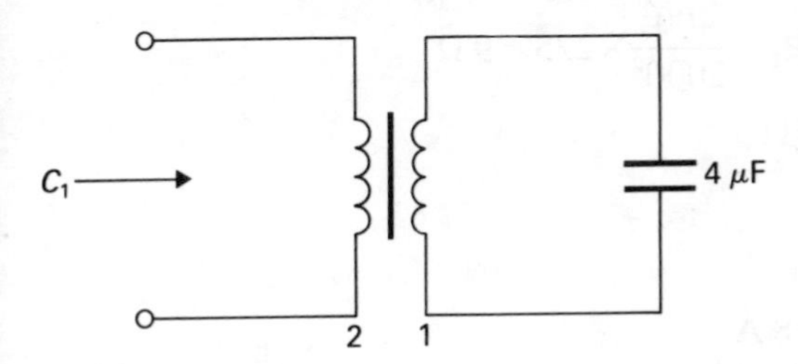

Fig. 20

By comparison with the previous example, it is easy to fall into the trap of saying four times the secondary capacitance, i.e. 16 μF. However the transformation of the capacitance causes its reactance to increase by the factor $(N_1/N_2)^2$, and if the reactance is to increase, it follows that the capacitance must decrease by that factor. The input capacitance is therefore $\underline{1\ \mu\text{F}}$.

A common problem that requires solution is that of a transformer connected to a load, yet supplied from a source to which it is joined by a line of appreciable impedance. This form of problem is illustrated in Example 7 and you should note that the supply voltage of 400 V is not that applied to the primary winding of the transformer. Instead there is a voltage drop in the line due to the 1-Ω resistor, and the voltage across the primary winding is reduced. Instead the method of analysis is that now shown.

Example 7 A 1-ph, 400/2000-V transformer is supplied from a 400-V source by means of a line of resistance 1 Ω. The secondary winding has a 225-Ω resistor connected across it. Determine the primary current and the power dissipated in the 225-Ω resistor.

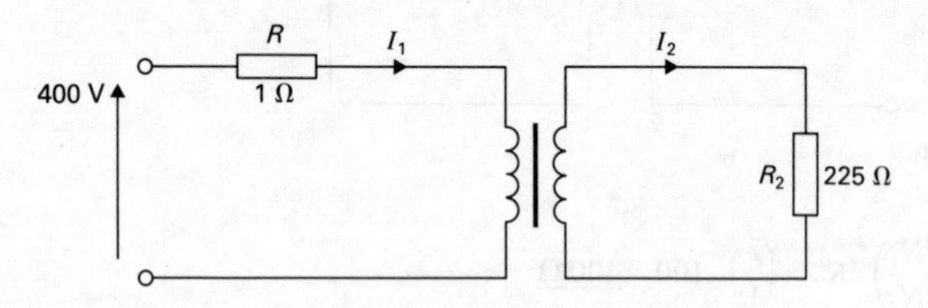

Fig. 21

$$R_2 = 225\ \Omega$$

$$R_{2'} = \left(\frac{N_1}{N_2}\right)^2 R_2 = \left(\frac{E_1}{E_2}\right)^2 R_2 = \frac{400^2}{2000^2} \times 225 = 9\ \Omega$$

$$R_{in} = R + R_{2'} = 1 + 9 = 10\ \Omega$$

$$I_1 = \frac{V}{R_{in}} = \frac{400}{10} = \underline{40\ \text{A}}$$

$$I_2 = I_1 \frac{E_1}{E_2} = 40 \times \frac{400}{2000} = 8\ \text{A}$$

$$P_2 = I_2^2 R_2 = 8^2 \times 225 = 14\,400\ \text{W}$$

$$= \underline{14{\cdot}4\ \text{kW}}$$

This example draws to your attention that the rating of a transformer does not imply that that is the operating condition. Thus in this case the 400/2000-V transformer is operating with a primary voltage of 360 V and a secondary voltage of 1800 V. However, the e.m.f. ratio of the transformer remains the same and it is that ratio which we require to use.

In earlier studies, we considered the implications of the maximum power transfer theorem. One of the possibilities was to adjust the load to match the source whereby maximum power could be dissipated in the load. Since a load seen through a transformer appears to be changed in resistance (or impedance), then it follows that a transformer can be used to match a load to a source.

This form of approach is useful in circuits involving small quantities of power, e.g. electronics circuits. Although the efficiency of power transfer is low (not more than 50 per cent), the matching of the load to the source (probably an amplifier or a transmission line) ensures the highest possible power dissipated in the load.

Example 8 A load of resistance 600 Ω is to be matched to an a.c. source of 30 mV and internal resistance 6 Ω. Determine the turns ratio of the coupling transformer.

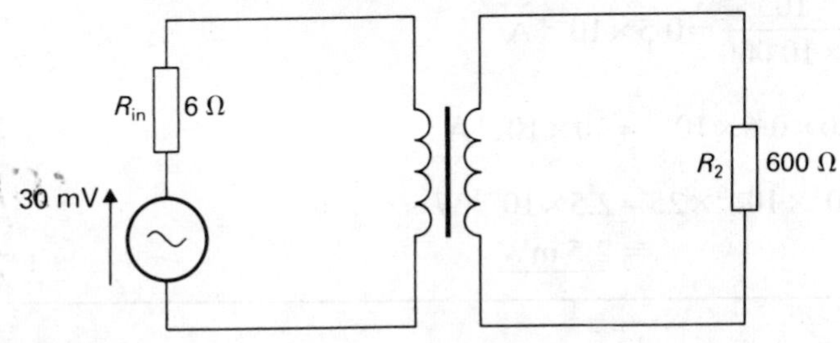

Fig. 22

The input resistance to the transformer must equal the internal resistance of the source if matching is to be attained.

$$R_{in} = 6\ \Omega$$

$$= \left(\frac{N_1}{N_2}\right)^2 R_2 = \left(\frac{N_1}{N_2}\right)^2 \times 600$$

$$\left(\frac{N_1}{N_2}\right)^2 = \frac{6}{600} = \frac{1}{100}$$

$$\frac{N_1}{N_2} = \frac{1}{10}$$

The turns ratio of the matching transformer is $\underline{1:10}$

Example 9 An a.c. source of 10 V and internal resistance 10 kΩ is matched to a load by a 20 : 1 transformer. Determine the resistance of the load and the power dissipated by it.

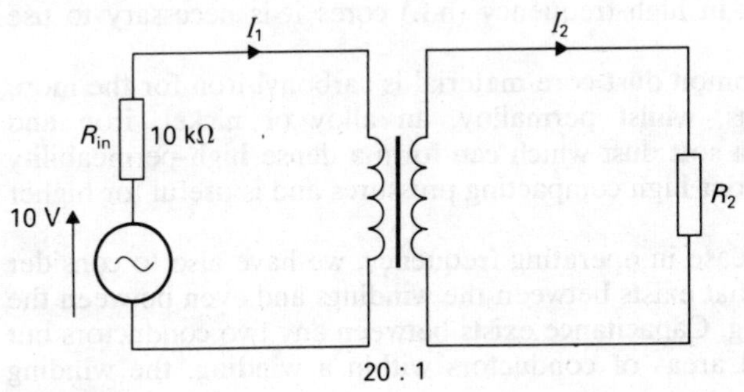

Fig. 23

$$R_{in} = 10\,000\ \Omega$$

$$= \left(\frac{N_1}{N_2}\right)^2 R_2 = \left(\frac{20}{1}\right)^2 R_2$$

$$R_2 = \frac{10\,000}{400} = \underline{25\ \Omega}$$

$$I_1 = \frac{V}{2R_{in}} = \frac{10}{2 \times 10\,000} = 0{\cdot}5 \times 10^{-3}\ \text{A}$$

$$I_2 = \frac{N_1}{N_2} I_1 = 20 \times 0{\cdot}5 \times 10^{-3} = 10 \times 10^{-3}\ \text{A}$$

$$P_2 = I_2^2 R_2 = 10^2 \times 10^{-6} \times 25 = 2.5 \times 10^{-3}\ \text{W}$$

$$= \underline{2{\cdot}5\ \text{mW}}$$

---

9 Audio-frequency and high-frequency transformers

Coupling transformers used for the purpose of matching operate at audio frequencies and even higher frequencies. Such transformers require to have high values of inductance because they operate in circuits containing high values of impedance. It follows that the inductance of the primary winding may be between 1 H and 80 H.

Such high values of inductance are an indication of the need to use core materials which have high permeability. The core material therefore is different from that which would be found in a power transformer. At the same time, the hysteresis loss incurred by the material must be small; the hysteresis loss is proportional to the frequency; thus as the frequency rises, it is necessary to counteract the increased loss by the use of a material with small hysteresis.

Ferromagnetic materials also suffer from the eddy current loss which becomes increasingly formidable as the frequency rises. In audio-frequency (a.f.) cores, it can be acceptable to use laminated core construction, but in high-frequency (h.f.) cores it is necessary to use dust cores.

The most common dust core material is carbonyl iron for the more basic applications, whilst permalloy, an alloy of nickel, iron and molybdenum, is a soft dust which can form a dense high-permeability core structure under high compacting pressures and is useful for higher frequency cores.

With the increase in operating frequency, we have also to consider the capacitance that exists between the windings and even between the turns of a winding. Capacitance exists between any two conductors but due to the small areas of conductors within a winding, the winding capacitance is very small. However, at high frequencies, this small capacitance becomes appreciable.

The electric field gives rise to a dielectric power loss which is the electric equivalent of the magnetic hysteresis loss. At frequencies up to about 15 kHz, this power loss is negligible; thus at audio frequencies the dielectric loss can be neglected. At high frequencies, the dielectric loss becomes appreciable and the variation of the resulting dissipation factor is shown in Fig. 24.

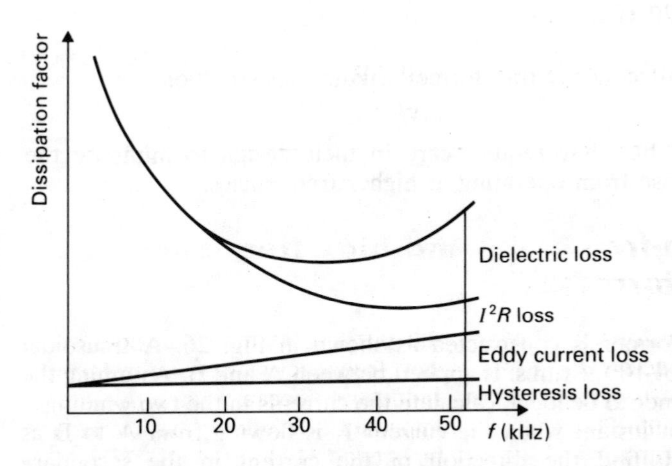

Fig. 24 Variation of dissipation factor with frequency for a ferromagnetic core

The combination of hysteresis loss, eddy current loss and dielectric loss becomes increasingly more important to a transformer core as the frequency increases, thus the criterion for such transformers is the need to use materials incorporating small losses.

The capacitance between turns is effectively in parallel with the winding inductance and care must be taken that this does not produce resonance at any of the operating frequencies. To minimise the capacitance, core windings are wound in two parts as shown in Fig. 25.

A transformer is intended normally to ensure that the primary and secondary windings are electrically separate. If the inter-winding capacitance effect becomes too large, then effectively the windings become electrically coupled. To minimise this effect, an earthed screen is inserted between the windings. This increases the capacitance to earth but decreases the inter-winding capacitance. A typical screen is a simple inter-winding copper sheet which does not meet round the core; if it did, it would effectively become a short circuit around the core.

Thus, a.f. and h.f. transformers are similar to the more basic power

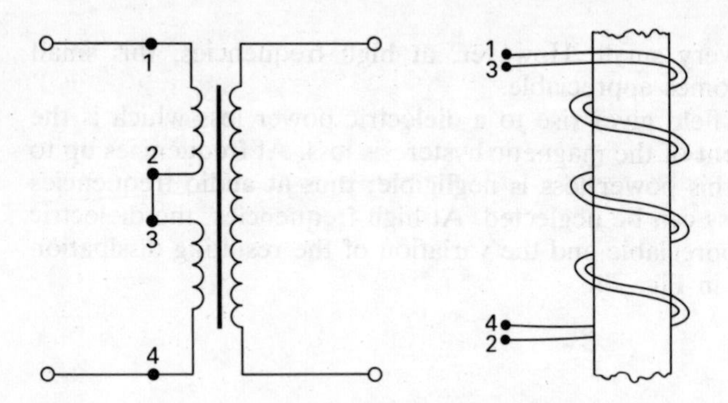

Fig. 25 High-frequency transformer winding construction

transformers, but they require care in their design to minimise the losses that arise from operating at higher frequencies.

Problems

1. A transformer is constructed as shown in Fig. 26. A sinusoidal voltage of 100 V r.m.s. is applied between A and B. Assuming the transformer to be ideal, calculate the currents in the two windings.

   At an instant when the current I_1 is flowing from A to B as indicated, find the direction of the current in the secondary winding.

   Why is a practical transformer not normally constructed with primary and secondary windings on separate limbs as shown in Fig. 26?

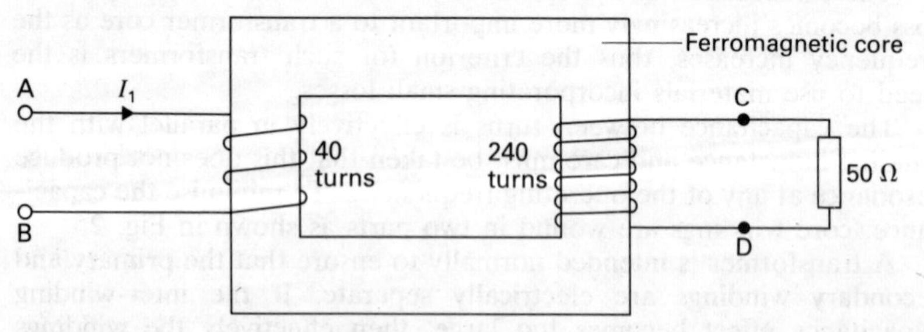

Fig. 26

2. The high-voltage winding of a transformer has 360 turns and is rated at 220 V. The low-voltage winding is rated at 36 V. How many turns are there in the low-voltage winding?

3. A 2-kVA, 3300/110-V transformer operates at 11 V per turn. Find the number of turns in each winding and the maximum rated current in each winding.
4. A load of pure resistance 10 Ω is connected across the 240-V secondary winding of a step-down transformer and the resulting primary current is 3·0 A. Find the primary voltage.
5. An ideal transformer is supplied from a constant-voltage source rated at 240 V and 50 Hz. It has a primary winding of 200 turns. When a load resistor is connected across the secondary winding, the voltage across the load is 480 V and the input power to the transformer is 240 W. Determine the resistance of the load and the number of turns in the secondary winding.
6. An ideal transformer supplies 500 W to a resistive load of resistance 200 Ω. The transformer is supplied from a 240-V, a.c. source. Determine:
   (*a*) the voltage across the load;
   (*b*) the ratio of the transformer;
   (*c*) the current taken from the supply.
7. A 3300/250-V, 50-Hz, 1-ph transformer has a core of effective cross-sectional area 13 000 mm^2 and its low-voltage winding has 80 turns. Determine:
   (*a*) the number of turns in the high-voltage winding;
   (*b*) the maximum flux density in the core.
   If the transformer supplies a load of 25 kW at 0·8 power factor lagging when connected to a 3·3-kV source, calculate the approximate values of the primary and secondary currents.
8. A 20-kVA, 1-ph transformer has a turns ratio of 44:1. The primary winding has 4000 turns and is connected to an 11-kV, 50-Hz sinusoidal supply. For full-load conditions, calculate:
   (*a*) the approximate values of primary and secondary currents;
   (*b*) the maximum instantaneous core flux.
9. A 1-ph transformer has a turns ratio 1:10 and a secondary winding of 1000 turns. The primary winding is connected to a 25-V sinusoidal supply, and the maximum instantaneous core flux is 2·25 mWb. Find:
   (*a*) the supply frequency;
   (*b*) the number of primary turns;
   (*c*) the open-circuit secondary voltage.
10. A transformer on no load requires a magnetising current but is otherwise ideal. When the secondary current is 100 A at a power factor of 0·866 lagging, the primary current is 11·0 A at a power factor 0·788 lagging. The primary voltage is 200 V. Find the secondary voltage.
11. An ideal transformer has its primary winding of 50 turns connected to a 250-V, 50-Hz sinusoidal a.c. supply and the secondary winding is open circuited. Calculate the maximum flux density in the core if the effective cross-sectional area of the core is 0·02 m^2.

The secondary winding supplies a 2-kW, unity-power factor load at 15 V. Under these conditions, find the number of turns in the secondary winding and calculate the primary current. With the stated load, calculate the primary input resistance.

12. A load of constant resistance $R = 16\ \Omega$ and inductance $L = 38$ mH is supplied from a constant-voltage source through the primary winding of an ideal transformer of turns ratio 1:10. What power is dissipated in the load, shown in Fig. 27, with terminals P and Q:
(*a*) short-circuited;
(*b*) open-circuited.

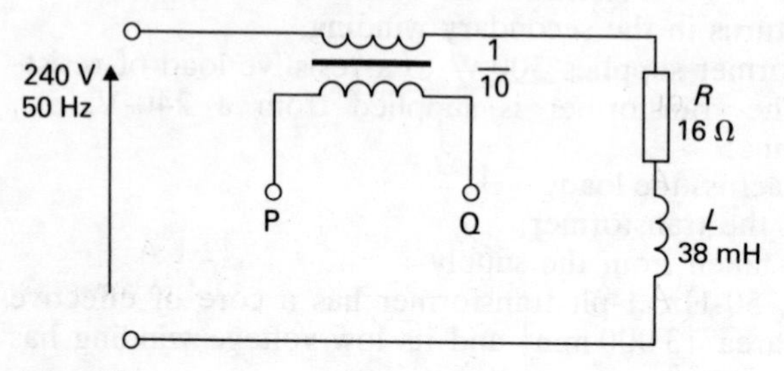

Fig. 27

13. A 12-Ω resistor is supplied from a 110-V, 50-Hz source as shown in Fig. 28 through the primary winding of an ideal transformer of turns ratio 1:10. The secondary winding of the transformer is connected to a capacitor of capacitance 2·5 μF. Determine the current in each winding of the transformer.

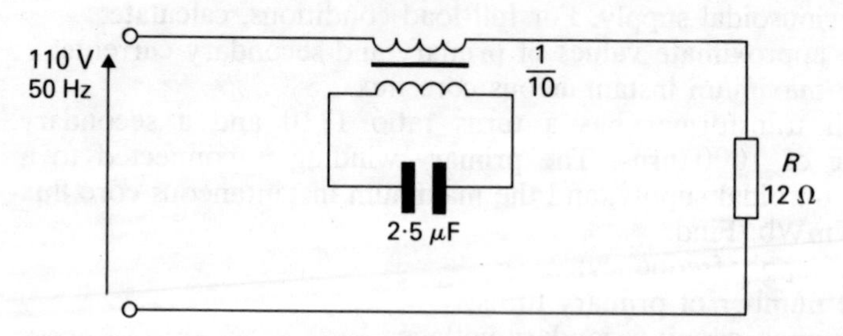

Fig. 28

14. A 15-Ω resistive load is to be matched to an amplifier which has an output of effective internal resistance 5·0 kΩ. Determine the turns ratio of a transformer that will match the source to the load in order to obtain maximum power transfer.

15. An ideal transformer has 100 primary turns and 200 secondary turns, the secondary winding being centre-tapped as shown in Fig. 29. Calculate the power dissipated in the two 10-Ω resistors.

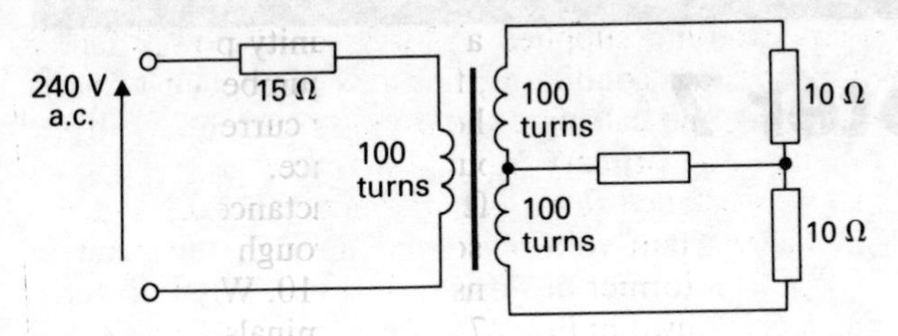

Fig. 29

Answers

1. 12 A; 72 A; D to C through load
2. 59 turns
3. 300 turns; 10 turns; 0·6 A; 18·2 A
4. 1920 V
5. 960 Ω; 400 turns
6. 316 V; 1:1·32; 2·1 A
7. 1056 turns; 1·08 T; 125 A; 9·5 A
8. 1·8 A; 80 A; 12·4 mWb
9. 25 Hz; 100 turns; 250 V
10. 20 V
11. 1·13 T; 3 turns; 8·0 A; 31·3
12. 2300 W; 0 W
13. 6·3 A; 0·63 A
14. 18·25 : 1
15. 720 W

Chapter 7

Electrical machines

The electrical energy in circuits and in networks mostly comes from machines which take in mechanical energy and convert it to electrical energy. Some of that electrical energy leaves the circuits and networks by means of machines which convert it back to mechanical energy. The conversion of energy by such electromechanical machines plays an important part in the operation of electrical systems and it is helpful to have an understanding of the ways in which energy conversion can take place within electro-mechanical machines.

Such machines can operate with either direct current or alternating current, but the latter systems are complicated by the effects of inductance and capacitance. To avoid these difficulties, let us start by considering the simpler d.c. machine operation, and then look at some aspects of a.c. machines.

1 Motors and generators

A machine that takes in electrical energy and gives out mechanical energy is said to be a motor. Engineers prefer to consider the rates of energy transfer, hence it is usual to think that a motor converts electrical power to mechanical power.

A machine that takes in mechanical energy and gives out electrical energy is said to be a generator. Again it is usual to express this statement in terms of power, hence a generator converts mechanical power to electrical power.

In a d.c. system, the electrical power is given by

$$P = VI \tag{1}$$

where V is the terminal voltage to the machine and I is the current passing between the terminals as indicated in Fig. 1. In a mechanical system, the energy is usually transmitted by a rotational arrangement, and the mechanical power is given by

$$P = T\omega \tag{2}$$

where T is the shaft torque (in newton metres) and ω is the angular velocity (in radians per second). It is common practice to express the angular velocity as a rotational speed N (in revolutions per minute); hence

$$P = \frac{2\pi NT}{60} \tag{3}$$

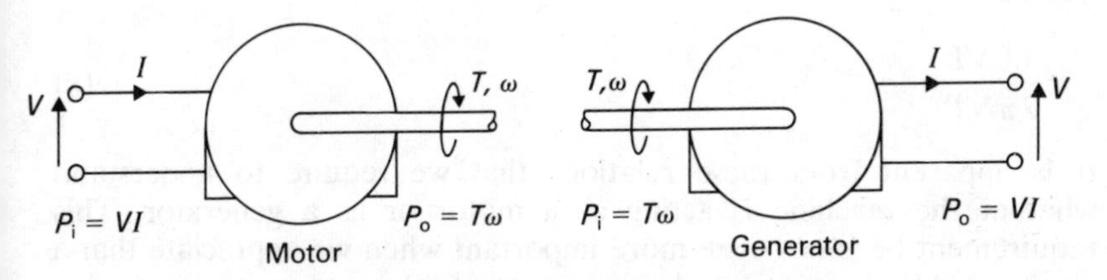

Fig. 1 Motor and generator power relationships

In an ideal machine, the input power P_i to a machine would be equal to the output power P_o. In practice, the machine experiences a loss of power due to such factors as I^2R losses in the windings, hysteresis and eddy-current losses in the core, and friction and windage losses arising from the movement of the rotating parts. The ratio of output power to input power is termed the efficiency of the machine.

Efficiency **Symbol: η** **Unit: none**

Since the efficiency is the ratio of two quantities, both measured in the same units, it follows that efficiency cannot have a unit. It is either expressed as a percentage value, e.g. 75 per cent, or as a per-unit value, e.g. 0·75. The latter form is generally used by engineers whilst the percentage value is retained to express such values to the general public.

Symbolically therefore the efficiency is given by

$$\eta = \frac{P_o}{P_i} \tag{4}$$

For a motor,

$$\eta = \frac{P_o}{P_i}$$

$$= \frac{T\omega}{VI} \tag{5}$$

$$= \frac{2\pi NT}{60VI} \tag{6}$$

For a generator,

$$\eta = \frac{P_o}{P_i}$$

$$= \frac{VI}{T\omega} \tag{7}$$

$$= \frac{60VI}{2\pi NT} \tag{8}$$

It is apparent from these relations that we require to understand whether the machine is acting as a motor or as a generator. This requirement becomes even more important when we appreciate that a given machine can be made to operate both as a motor and as a generator. For instance, a machine can drive a train up a hill, yet when the train comes back down, the machine can be used as a brake by generating electrical energy from the potential energy of the train. This interchangeability of the mode of operation can be summarised in the form of Fig. 2. The difference between the input power P_i and the

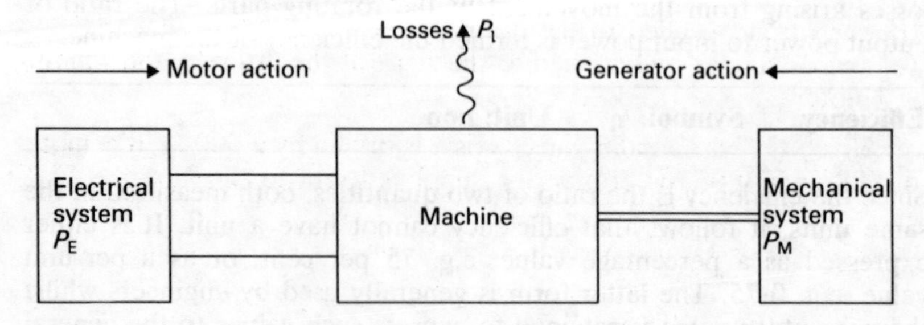

Fig. 2 Energy interchange in a d.c. machine

output power P_o is the power loss P_l within the machine, thus

$$P_l = P_i - P_o \quad (9)$$

$$= P_i - \eta P_i$$

$$= P_i(1-\eta) \quad (10)$$

Example 1 A 500-V, d.c. motor is loaded to operate at 850 rev/min while developing a torque 12·5 N m. The supply current to the motor is 2·5 A. Find the efficiency of the motor.

$$\eta = \frac{2\pi NT}{60VI} = \frac{2\pi \times 850 \times 12{\cdot}5}{60 \times 500 \times 2{\cdot}5} = \underline{0{\cdot}89}$$

Example 2 A 650-V, d.c. generator supplies a current 125 A and operates at 1250 rev/min. Given that the efficiency of the generator is 0·95, find:

(*a*) the torque required to drive the generator;
(*b*) the power loss in the generator.

$$\eta = \frac{60VI}{2\pi NT}$$

$$T = \frac{60VI}{2\pi N\eta} = \frac{60 \times 650 \times 125}{2\pi \times 1250 \times 0{\cdot}95} = \underline{653 \text{ N m}}$$

$$P_o = VI = 650 \times 125 = 81\,250 \text{ W}$$

$$P_i = \frac{P_o}{\eta} = \frac{81\,250}{0{\cdot}95} = 85\,530 \text{ W}$$

$$P_l = P_i - P_o = 85\,530 - 81\,250 = \underline{4280 \text{ W}}$$

Example 3 A lift of mass 250 kg is raised with a velocity 5·0 m/s. The rope by which the lift is suspended is wound onto a drum of diameter 60 cm and the efficiency of the winding mechanism is 0·85. Determine:

(*a*) the rotational speed of the winding drum;
(*b*) the torque applied to the drum by the rope;
(*c*) the driving torque applied to the shaft of the drum by the winding motor.

Given that the winding mechanism is driven by a 500-V, d.c. motor of efficiency 0·91, determine the supply current to the motor.

(*a*) $$u = \omega r = \frac{2\pi Nr}{60}$$

$$N = \frac{60u}{2\pi r} = \frac{60 \times 5}{2\pi \times 0{\cdot}3} = \underline{159 \text{ rev/min}}$$

(*b*) $T = Fr = mgr = 250 \times 9{\cdot}81 \times 0{\cdot}3 = \underline{736\ \text{N m}}$

(*c*) $$P_o = \frac{2\pi NT}{60} = \frac{2\pi \times 159 \times 736}{60} = 12\,260\ \text{W}$$

$$P_i = \frac{P_o}{\eta} = \frac{12\,260}{0{\cdot}85} = 14\,420\ \text{W}$$

Let the shaft torque be T_s, hence the shaft power in the input power to the winding mechanism and

$$T_s = \frac{60P_i}{2\pi N} = \frac{60 \times 14\,420}{2\pi \times 159} = \underline{866\ \text{N m}}$$

The shaft power is the output power of the motor which has an efficiency 0·91; hence the electrical power is given by

$$P_e = \frac{14\,420}{0{\cdot}91} = 15\,850\ \text{W}$$

$$= VI$$

$$I = \frac{15\,850}{500} = \underline{31{\cdot}7\ \text{A}}$$

---

2 Types of electrical machines

There is a considerable variety of types of machine acting either as electrical motors or electrical generators. However, most machines can be assigned to one of the following five types:

(*a*) direct-current machines;
(*b*) three-phase induction machines;
(*c*) three-phase synchronous machines;
(*d*) single-phase induction machines;
(*e*) relays and contactors.

The last type mentioned produces linear motion and does not concern our present studies.

Direct-current machines are now manufactured in very large numbers mainly to operate as motors. The d.c. motor can be very easily controlled for either speed adjustment or torque adjustment, and these properties make the machine very suitable in automation systems where control to obtain optimum operation is essential. The d.c. motor can also produce very high torques at low speeds and is an ideal machine for starting systems into motion, e.g. starting a train.

The 3-ph induction machine is usually found acting as a motor. It is the work-horse of most industrial drives, providing a convenient method of producing a rotating drive for a wide range of production machines in which the act choice of operating speed is not critical.

Three-phase synchronous machines are numerically the least common, but they are the large generators which provide almost all our electrical supplies. Most synchronous generators are large machines and a typical rating is about 660 MW, although even larger machines are being made.

Finally there is the 1-ph induction machine which operates as a motor. There are quite a few different forms of 1-ph induction motor and these lie outside the scope of this book. However, this is a machine with which everyone is familiar, being the motor which drives such things as washing machines, vacuum cleaners, in fact most things which turn electrically within your house.

At first glance, most rotating electrical machines look the same and in fact they all do have a great deal in common. If we consider only the d.c. machines, the 3-ph induction machines and the 3-ph synchronous machines, we find that:

1. Each machine can act either as a motor or as a generator.
2. Each machine has a fixed part (the stator) within which is a rotating part (the rotor).
3. The stator is a cylinder with either a smooth inner surface or with salient pieces, termed poles, attached to the inner surface.
4. The rotor is a cylinder with either a smooth outer surface or with salient pieces, again termed poles, attached to the outer surface.
5. Conductors connected to form windings are distributed either along the smooth surfaces or around the poles.
6. These windings pass electric currents which create magnetic fluxes both in the rotor and the stator and also in the space, termed the air gap, between the rotor and the stator.
7. Each machine has the same form of mounting and enclosure with bearings to retain the rotor, and with cooling arrangements. These vary in detail according to the conditions in which the machine is to operate.

There is however one essential difference between the various types of machine, and that is the form of magnetic flux set up by the currents in the windings. Thus for instance the magnetic field may be fixed in space and remain constant in magnitude as in the d.c. machine, or it may move in space and possibly fluctuate in magnitude as in the 3-ph induction machine, whilst a sort of combination of these two possibilities arises in the 3-ph synchronous machine.

The reasons for these differences lie in the forms of windings constructed within the machine and the types of current used to excite the windings. The effects of these differences will become apparent in the subsequent chapters dealing with the three forms of electrical machine considered above.

However, before progressing to consider specific machines, there remain some common principles to be observed.

3 Some basic principles of electrical machines

We have noted that a generator is a device that accepts mechanical energy and converts it into electrical energy. It follows that the generator provides a source of e.m.f. whereby current is made to flow in a circuit. But this action requires some explanation.

First of all, the mechanical energy cannot be directly changed into electrical energy. Instead, the mechanical energy is changed into magnetic energy, which is then transformed into electrical energy. The electrical machines therefore all involve magnetic fluxes which interact with the windings into which are induced e.m.f.s.

We are already familiar with the principle of Faraday's Law, i.e. if the flux linking a circuit is varied, then an e.m.f. is induced in that circuit. It follows that to induce an e.m.f. in the generator winding, all we require to do is vary the flux linking the winding. This can be done in one of the following ways:

(*a*) the winding can be moved relative to the magnetic field;
(*b*) the magnitude of the flux linking the winding can be varied;
(*c*) these two actions can occur simultaneously.

Considering each of these in turn, the first method gives rise to an e.m.f. termed a motional e.m.f. It is so called because it arises out of the relative motion between winding and field. There are various ways of achieving such a situation: thus, for instance, the field can be stationary in space and a winding made to rotate in that field. This is the arrangement used in the d.c. machine. Alternatively, the winding can be held fixed in space and the magnetic field rotated past the winding. This is the arrangement used in the three-phase synchronous machine.

In any situation involving a motional e.m.f., the e.m.f. is given by

$$e = Blu \tag{11}$$

Since instances of both d.c. and a.c. machines have been given, we can observe that this relation can apply to both forms of current system.

Instead of moving either the winding or the field, it is possible to have both fixed in space and to vary the magnitude of the magnetic flux. This gives rise to an induced e.m.f. due to the change of flux linkage, and this is the form of induced e.m.f. already encountered in a transformer. For this reason, this form of e.m.f. is termed a transformer e.m.f. and is given by

$$e = \frac{d\psi}{dt} = N\frac{d\phi}{dt} \tag{12}$$

Finally, machines can be made in which the e.m.f.s induced are partly motional and partly transformer, and the 3-ph induction machine may be included in this category.

It is generally acceptable to look on the function of an e.m.f. in a generator as being the force induced into the winding whereby current is made to flow in the electrical circuit. After all, the function of a generator is to provide a supply of e.m.f. When we come to the motor, the function of the e.m.f. is not so obvious. However let us look to the comparable situation that occurs in mechanics.

You will know that for every action there is an equal and opposite reaction. Thus if a man leans against a wall, the wall pushes back with an equal but opposite force. The weight of the man pushes down against the floor which in turn pushes up with an equal but opposite force. However if the man pushes a barrow, the force he applies to the barrow is opposed by a frictional force, and if the barrow moves with uniform velocity, then these two forces are equal but opposite.

In electrical circuits, a similar situation arises. The generator produces an e.m.f. which is equal in magnitude to the volt drops around the circuit but which acts in the opposite direction, a fact summarised by Kirchhoff's Second Law. However if we apply a voltage to a motor, it reacts by inducing an e.m.f. which is equal but opposite to the applied voltage. This is the action we have already seen in the primary winding of a transformer, but it also occurs in electrical machines. The e.m.f. in an electrical machine is the force in the circuit associated with the interchange of form of energy, i.e. the interchange from electrical energy to magnetic energy or vice versa.

Again thinking of the man pushing the barrow, the rate at which he transfers energy to the barrow is given by the product of the reaction of the barrow and the velocity of movement. In the same way, the rate of energy interchange electrically is given by the product of the reaction force (i.e. the e.m.f.) and the current, thus at any instant the rate of energy interchange, i.e. power, is given by

$$p = ei \tag{13}$$

The e.m.f. induced in the machine windings varies according to the type of machine considered, and for this reason, we are better to continue thinking about induced e.m.f.s in the various specific cases, but you should remember that the e.m.f. is generally given by $e = Blu$ and that it is the electrical force arising from the interchange of form of energy.

Having thought about the electrical force, it follows that we should also give consideration to the mechanical force which arises in electrical machines. For the basic three forms of machine already described, the mechanical force generally arises from the principle that a conductor carrying a current and lying in a magnetic field experiences a mechanical force. This is determined by the relation

$$F = Bli \tag{14}$$

In the case of a motor, we can expect that a system of current-carrying conductors will be situated with a magnetic field and the resulting forces will cause the system to move, generally to rotate. However again we must remember that for each action there is a reaction. It follows that whilst the rotating system is pushed in one direction, the magnetic field system is pushed in the opposite direction.

In most machines, the rotor is made to rotate by the mechanical action of the current-carrying conductors, and it is easy to forget that there is a consequent reaction on the mountings of the stator which experiences force trying to make it turn in the opposite direction. This applies whether the machine acts as a motor or as a generator. The only difference between these two situations is the direction of the motion relative to the direction of the applied force.

This matter of direction of force does give difficulties, so again let us turn to the example of the man with the barrow. If he pushes the barrow along the ground, then he is acting as a motor, i.e. giving his energy to the mechanical load. However, if the man comes to a steep slope, he may get to a point that although he is pushing the barrow, it is too heavy for him and he starts to move back down the slope. In this case the force he is applying still acts in the forward direction but the motion is now backwards instead of forwards. In this case, the load is generating energy into the man. Whether acting as a motor or as a generator, force is still involved at the point of interchange of energy form and the rate of energy interchange, i.e. power, is given by

$$p = Fu \qquad (15)$$

If the machine is perfect, i.e. there are no losses of energy in the magnetic system, then

$$p = ei = Fu \qquad (16)$$

When discussed in this abstract manner, the principles of e.m.f. and mechanical force can seem rather difficult, but when observed in specific cases, they are much easier to understand. Let us therefore continue to consider individually the d.c. machine, the 3-ph induction machine and the 3-ph synchronous machine.

Problems

1. A 250-V d.c. motor is loaded to operate at 900 rev/min and develops a torque 15 N m. The supply current to the motor is 6·5 A. Find the efficiency of the motor.
2. A 500-V d.c. generator supplies a current 80 A and operates at 1150 rev/min. Given that the efficiency of the generator is 0·83, find

   (*a*) the torque required to drive the generator;
   (*b*) the power loss in the generator.

3. A lift of mass 450 kg is raised at a velocity 4·0 m/s. The rope by which the lift is suspended is wound onto a drum of diameter 65 cm and the efficiency of the winding mechanism is 0·81. Determine
   (*a*) the rotational speed of the winding drum;
   (*b*) the torque applied to the drum by the rope;
   (*c*) the driving torque applied to the shaft of the drum by the winding motor.

   Given that the winding mechanism is driven by a 500-V d.c. motor of efficiency 0·89, determine the supply current to the motor.
4. An electric crane has an overall efficiency 0·60 while it is lifting a load of mass 100 kg with a steady velocity 0·5 m/s. Calculate the current taken by the motor from a 400-V d.c. supply.
5. A car starter motor, developing a torque of 18 N m at 10 rev/s, works at an efficiency 0·85. Find
   (*a*) its input power;
   (*b*) the current that it takes from a 12-V battery.
6. The maximum demand on a hydroelectric generating station is 10 MW and the annual load factor is 20 per cent. Calculate the total energy supplied in one year.

   A total quantity 750×10^6 kg of water at an average head of 40 m passed through the turbines during a 36-h period. Given that the overall operating efficiency was 0·88, calculate for this period
   (*a*) the energy (in kilowatthours) generated;
   (*b*) the average power generated.
7. A pump driven by a 500-V d.c. motor delivers 600 tonnes of oil per hour against a total head of 18 m. At this load the pump operates with an efficiency 0·75 and the motor runs with an efficiency 0·90. Calculate the output power of the motor and its input current.
8. The motor in Problem 7 is replaced by a 415-V, 3-ph induction motor which operates with an efficiency 0·91 and a power factor 0·84. Determine the output power of the motor and its line supply current.

Answers

1. 0·87
2. 400 N m, 8·2 kW
3. 117·5 rev/min, 1430 N m, 1770 N m, 49 A
4. 2·0 A
5. 1330 W, 110 A
6. $63 \cdot 1 \times 10^{12}$ J or $17 \cdot 5 \times 10^6$ kWh, $7 \cdot 2 \times 10^6$ kWh, 2·0 MW
7. 39·2 kW, 87·2 A
8. 39·2 kW, 71·5 A

Chapter 8

D.C. machines

Although d.c. supplies are not made available by the electricity supply authorities, it is nevertheless a relatively simple process to rectify the alternating current to direct current whereby d.c. machines may be operated. It is also convenient to control the direct current as part of the process of rectification, and this ability to control when coupled with the suitability of the d.c. machine for many forms of industrial drive has made the d.c. machine one of the most common electrical motors.

It is also a machine which readily demonstrates the application of basic electrical principles and magnetic principles to the production of movement. this demonstration is more simple than those associated with a.c. machines because the element of time variation associated with alternating current is not present. Given therefore the choice of machine to be first investigated, the d.c. machine is preferable since it is the most simple.

In order to understand the operation of the d.c. machine, we must start by looking in greater detail at the construction, and in particular at the method of connecting the windings.

1 Principle of a commutator

To generate an e.m.f., it is possible to move a conductor which is part of a circuit across a magnetic field. This action cannot be continuous as eventually the conductor will come to the end of the field, so we must seek an alternative method of continuously inducing an e.m.f. into a circuit. The most simple solution is that of rotating a coil in a uniform magnetic field. A suitable arrangement is shown in Fig. 1. The corners

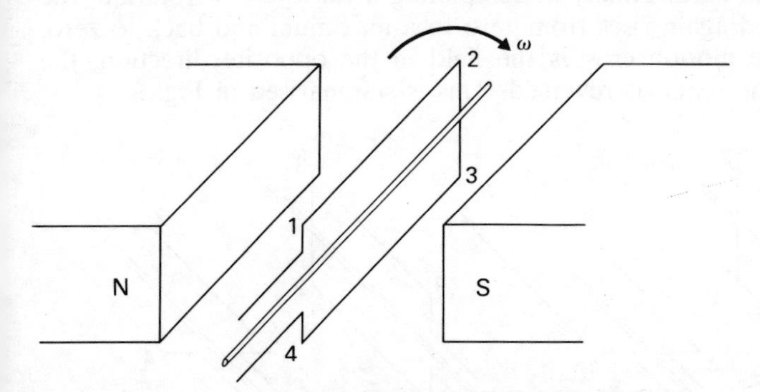

Fig. 1 Simple rotating coil in a magnetic field

of the coil are numbered 1234 so that we can observe the effects of rotating the coil in subsequent diagrams. The e.m.f. induced in the coil depends on the length of the sides of the coil, on the rotational speed and on the density of the magnetic field. It also depends on the direction of the motion of the coil sides relative to the magnetic field, and in the instance shown in Fig. 1, the sides are moving in the direction of the magnetic field; thus, at that instant, no e.m.f. is induced. When the coil is in this position, it is said to be in the neutral plane.

When the coil moves on, its sides start to move across the magnetic field until they move with maximum velocity across the field when the coil has rotated 90° from the neutral plane. This second position is illustrated in Fig. 2.

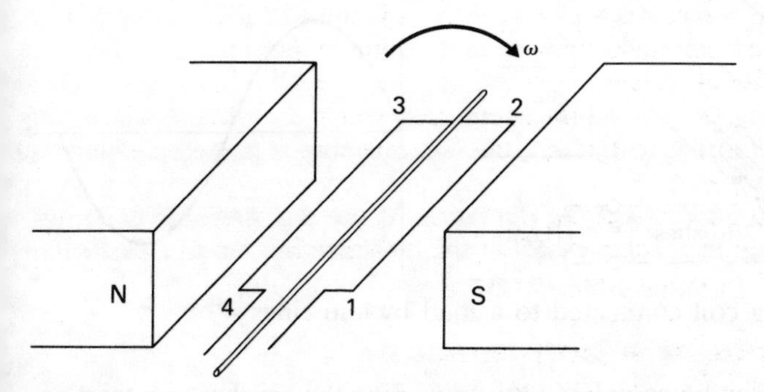

Fig. 2 Coil rotated 90° from the neutral plane

After the coil passes the position of 90° displacement from the neutral plane, the speed with which the sides continue to cut the magnetic field starts to fall until eventually the coil again lies in the neutral plane having rotated through 180°. At this point, the induced e.m.f. is again zero. Finally in completing a full cycle of rotation, the e.m.f. induced again rises from zero to a maximum and back to zero, but since the motion crosses the field in the opposite direction, the polarity of the e.m.f. is reversed. This is summarised in Fig. 3.

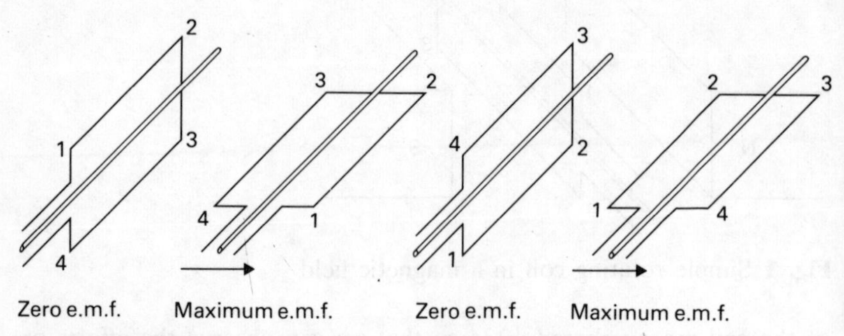

Fig. 3 Polarities of e.m.f. induced in a rotated coil

The problem with the induced e.m.f. is to obtain access to it in order that we may observe it. One method of connection is to connect the two coil terminals to two continuous and insulated rings mounted on the shaft as shown in Fig. 4. Such rings are called slip rings. Electrical contact to the rings is made through two blocks of carbon that are held against the rings under the pressure of suitable springs. The blocks are called brushes, and a brush rubs against each slip ring.

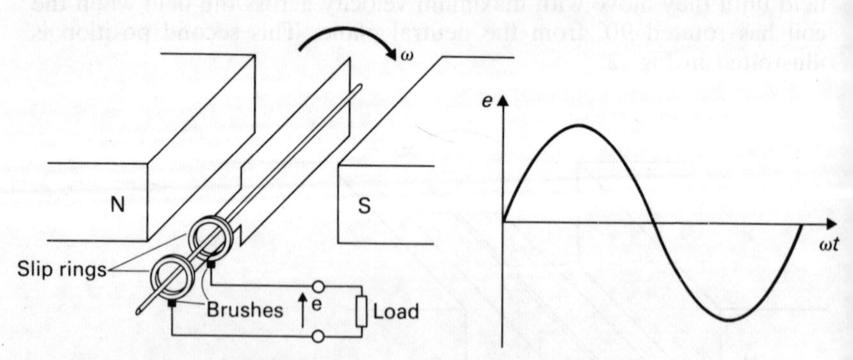

Fig. 4 Rotating coil connected to a load by slip rings

The circuit can be completed by connecting the brushes to a load or to suitable test equipment. In this way, we could observe that when the coil is rotated with uniform speed, the e.m.f. induced in it is a sinusoidal alternating voltage.

The alternating e.m.f. can be converted into a direct one by using a different form of slip ring. Instead of having two separate rings, we use one ring which is split into two parts insulated from one another. Such an arrangement is called a commutator and it requires that the connections are made through two brushes placed on opposite sides as shown in Fig. 5.

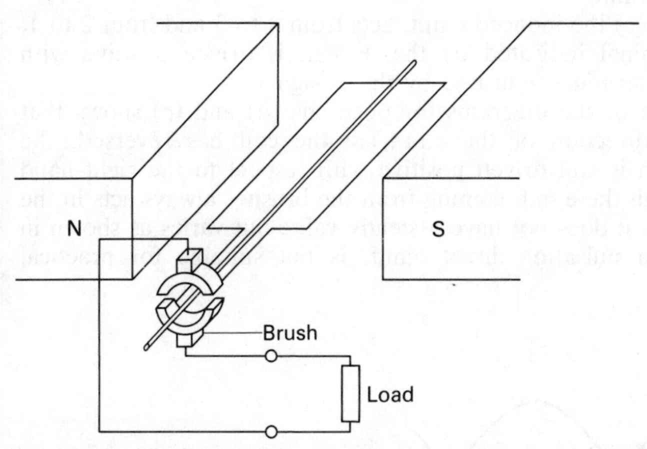

Fig. 5 Rotating coil connected to a load by a commutator

The action of the commutator is to reverse the connections from the coil at the instant when the coil passes through the neutral plane. In this way, the e.m.f. transmitted to the load always acts in the same direction round the circuit and it is therefore a direct e.m.f.

Let us consider this action in greater detail. Figure 6 shows the coil in three positions, which correspond to 90° before passing through the neutral plane, passing through the neutral plane, and 90° after passing through the neutral plane.

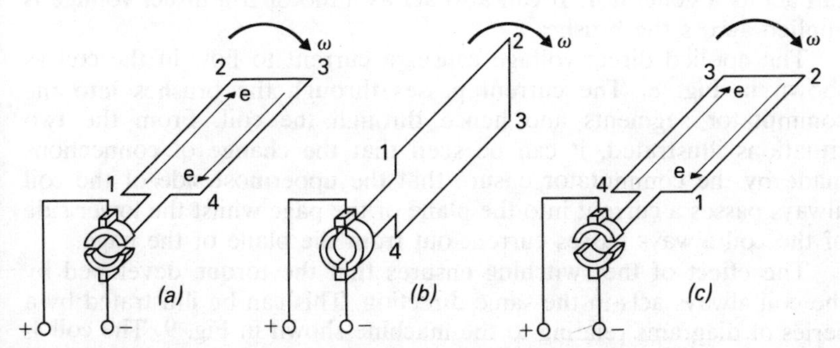

Fig. 6 Action of a commutator in a simple d.c. machine

In position (a), the induced e.m.f. acts from 1 to 2 and from 3 to 4. Thus the terminal indicated by the + sign is driven positive with respect to the terminal indicated by the − sign.

In position (b), the coil has moved into the neutral plane and the induced e.m.f. is zero. At this point the brushes make contact with both parts of the commutator ring. These parts are called segments; thus the brushes make contact with both segments. Effectively the brushes therefore short circuit the coil but this does not matter as there is no induced e.m.f.

In position (c) the induced e.m.f. acts from 4 to 3 and from 2 to 1. Thus the terminal indicated by the + sign is driven positive with respect to the terminal indicated by the − sign.

Comparison of the diagrams for positions (a) and (c) shows that although the direction of the e.m.f. in the coil has reversed, the left-hand brush is still driven positive with respect to the right-hand brush. Although the e.m.f. coming from the brushes always acts in the same direction, it does not have a steady value but varies as shown in Fig. 7. Such a pulsating direct e.m.f. is not suitable for practical

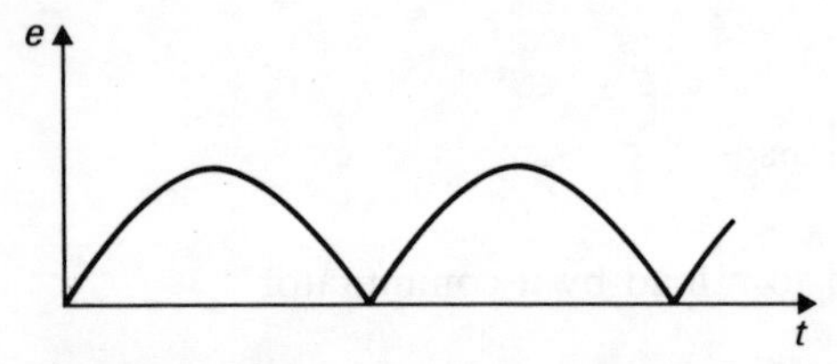

Fig. 7 Variation of e.m.f. between the brushes

applications, but we have obtained the basis whereby we can generate an e.m.f. that essentially gives rise to a direct voltage.

Before proceeding to consider the form of commutator machine found in general commercial practice, it is worth further exploring the possibilities of the simple commutator machine which we have seen can act as a generator. It can also act as a motor if a direct voltage is applied across the brushes.

The applied direct voltage causes a current to flow in the coil as shown in Fig. 8. The current passes through the brushes into the commutator segments and hence through the coil. From the two situations illustrated, it can be seen that the change of connections made by the commutator ensure that the uppermost side of the coil always passes a current into the plane of the page whilst the lower side of the coil always passes current out from the plane of the page.

The effect of the switching ensures that the torque developed by the coil always acts in the same direction. This can be illustrated by a series of diagrams relating to the machine shown in Fig. 9. The coil is shown at various positions and we can see that although the torque

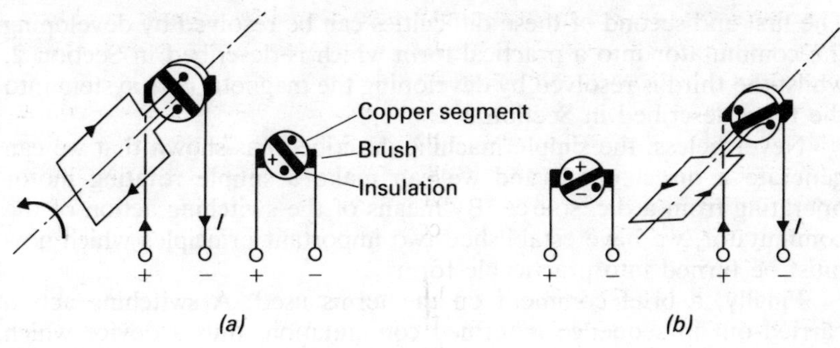

Fig. 8 Simple d.c. commutator motor

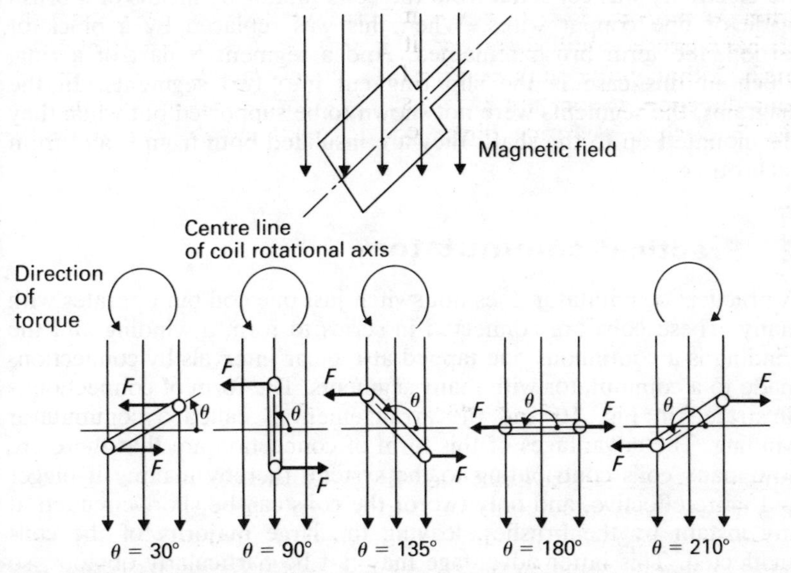

Fig. 9 Forces and torque on the coil of the simple d.c. motor

always acts in the same direction, it varies in magnitude according to the angle between the plane of the coil and the neutral plane.

Such a simple motor can be made to operate but it is not very satisfactory for practical purposes for three reasons:

1. When the brushes overlap the joins of the segments, the supply is short-circuited momentarily.
2. The torque developed pulsates according to the position of the coil. Further there are two positions in which the coil experiences no torque and from which the motor could not start.
3. The torque developed is weak, partly because of the small number of current-carrying conductors and partly because of the large air gap between the poles of the magnetic core.

The first and second of these difficulties can be resolved by developing the commutator into a practical form which is described in Section 2, whilst the third is resolved by developing the magnetic core system into the form described in Section 3.

Nevertheless, the simple machine described has shown that we can generate a direct e.m.f. and we can make a simple rotating motor operating from a d.c. source. By means of the switching action of the commutator, we have established two important principles which now must be turned into practicable forms.

Finally, a brief comment on the terms used. A switching action carried out in sequence is termed commutation, thus a device which switches in sequence is a commutator. The term brush originated when the electricity was collected from the commutator by means of a brush made of fine copper wires. When this was replaced by a block of carbon, the term brush remained. And a segment is part of a ring, which in this case is the slip ring cut into two segments. In the diagrams, the segments were not shown to be supported but when they are mounted on to the shaft, they are insulated both from it and from each other.

2 Practical commutator

A practical commutator does not switch just one coil but operates with many. These coils are connected in series to form a winding and the winding is a continuous one tapped at regular intervals by connections made to a commutator with many segments. The form of connection is illustrated in Fig. 10 and the arrangement is called a commutator winding. The advantages of this form of connection are that there are now many coils contributing to the system, thereby making it bigger and more effective, and only two of the coils can be short-circuited at any instant by the brushes, leaving the large majority of the coils unaffected. This latter advantage may not be particularly obvious, so

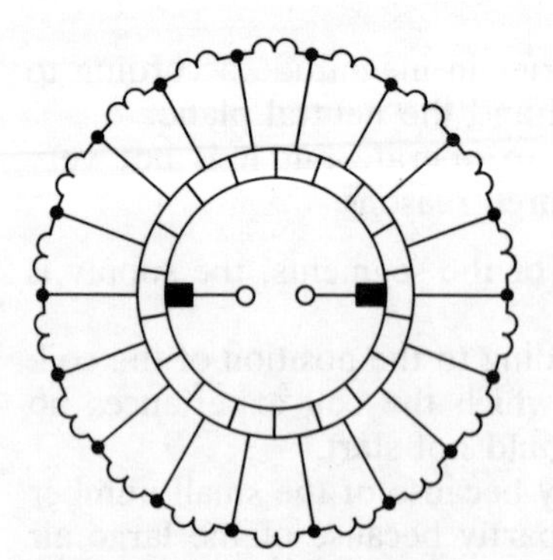

Fig. 10 Schematic diagram of a simple commutator winding

we had better consider in some detail the commutator action assuming the commutator winding to be involved in a machine used as a motor. In order to indicate the rotation of the winding, three of the segments are numbered as indicated in Fig. 11.

Because there is a continuous winding, the current passing from the positive brush to the negative brush finds that there are two parallel

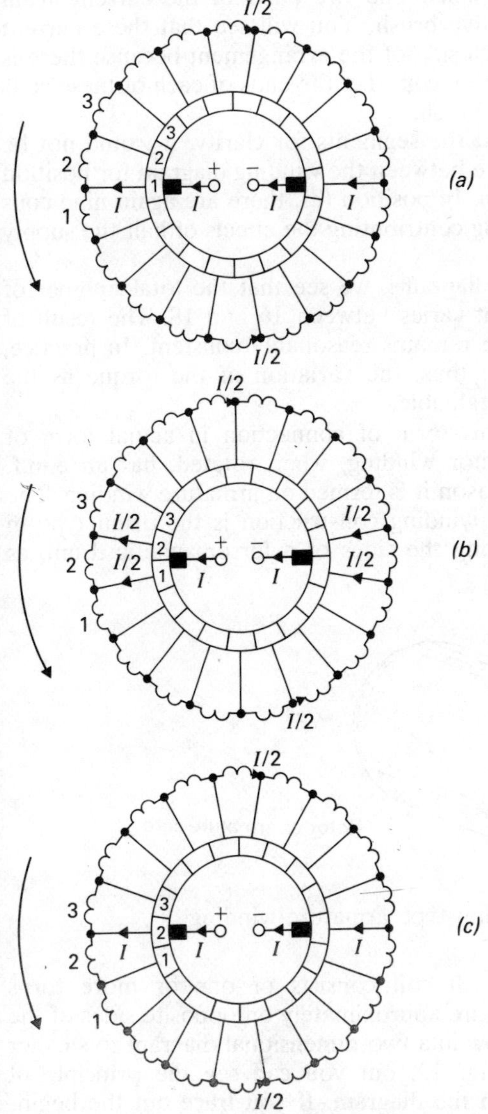

Fig. 11 Action of multi-segment commutator

paths available to it. In position (*a*), the current entering at the positive brush divides equally, half of it passing through the top nine coils and the other half passing through the bottom nine coils. The two parts of the current come together again at the negative brush.

When the winding rotates to position (*b*), the positive brush supplies current to two segments. As a result, half the current passes through the top seven coils and the other half of the current passes through the bottom seven coils. The two parts of the current again come together at the negative brush. You will see that these current paths ignore one coil at each side of the arrangement because there is no applied voltage across these coils, i.e. the ends of each of these coils are connected to the same brush.

If we had not numbered the segments for clarity, it would not be possible to tell the difference between the winding diagram for position (*c*) and that for position (*a*). In position (*c*), there are again nine coils on either side of the winding contributing the effects of half the supply current,

From this sequence of diagrams, we see that the total number of effective coils at any instant varies between 16 and 18. The result of this is that the total torque remains reasonably constant. In practice, many more coils are used; thus, the variation of the torque as the winding rotates becomes negligible.

It remains to realise this form of connection in actual form of construction. The commutator winding, when rotated, has an e.m.f. induced in it and for this reason it is termed an armature winding. The common form of armature winding construction is the drum type in which the coils are laid along the sides of a ferromagnetic drum, as shown in Fig. 12.

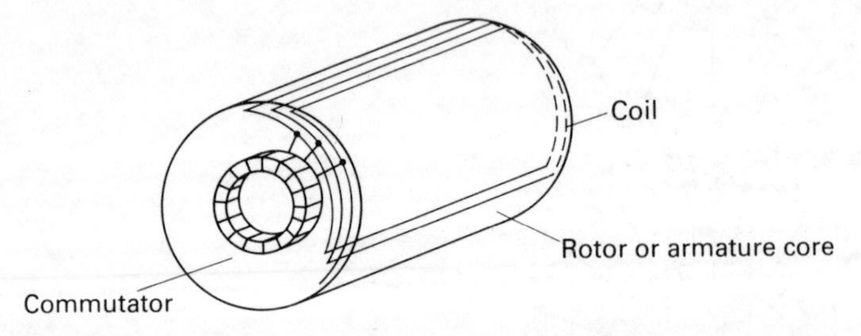

Fig. 12 Construction of drum-type armature winding

In this arrangement, each coil consists of one or more turns constructed so that its sides are approximately on opposite sides of the drum. This is difficult to show in a two-dimensional diagram so smaller coils have been used in Fig. 12, but you can see the principle of construction quite clearly in the diagram. If you trace out the beginnings of the current paths, you will see that the coil sides on one side of the drum all carry current in one direction whilst those on the other side of the drum all carry current in the opposite direction.

The coils are held in slots cut along the sides of the drum and usually there are two coil sides in each slot. The resulting form of connection is indicated in Fig. 13. You can see that half the current passes from the 'a' terminal to the 'b' terminal through half the coils and the remainder of the current through the remaining coils. However as the remaining half of the coils are effectively wound in the opposite direction, the result is that the two effects are cumulative, thus we obtain a strong motor action.

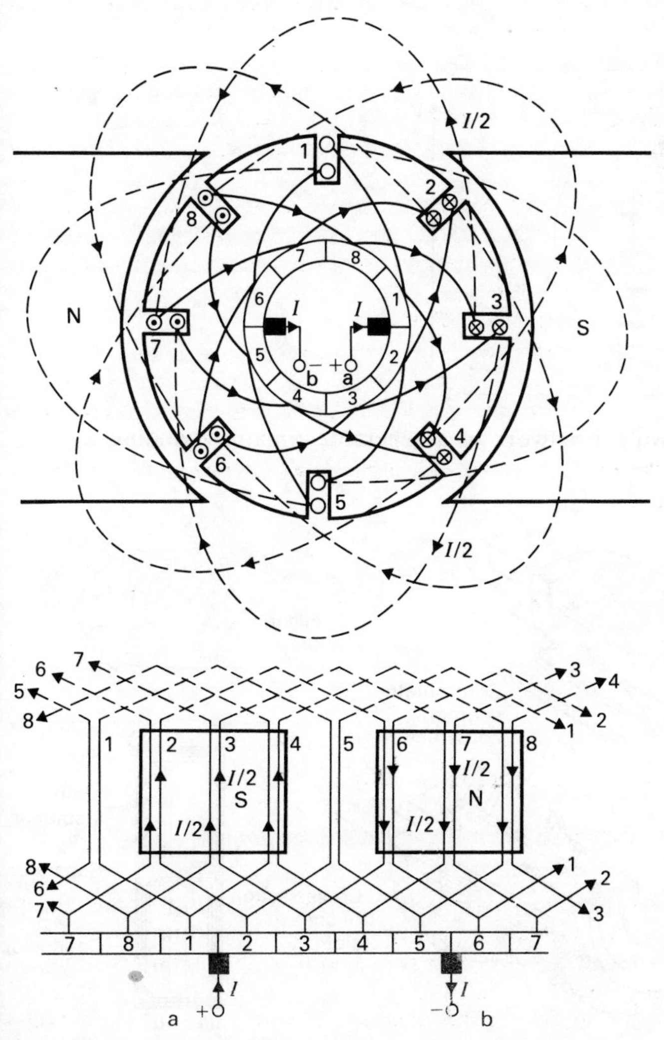

Fig. 13 Arrangement of a double-layer winding in a motor

Just as the force and torque effects of the coils are cumulative when the winding is arranged to act in a motor, so the e.m.f.s are cumulative when the winding is made to rotate in a magnetic field. Again one-half of the coils contribute their number of individual e.m.f.s which are series connected and therefore may be added together. The other half of the coils being both effectively connected the other way round and moving back across the magnetic field also give rise to an identical e.m.f. acting in parallel with the former e.m.f. and with the same polarity. The equivalent network of a generator armature winding is shown in Fig. 14.

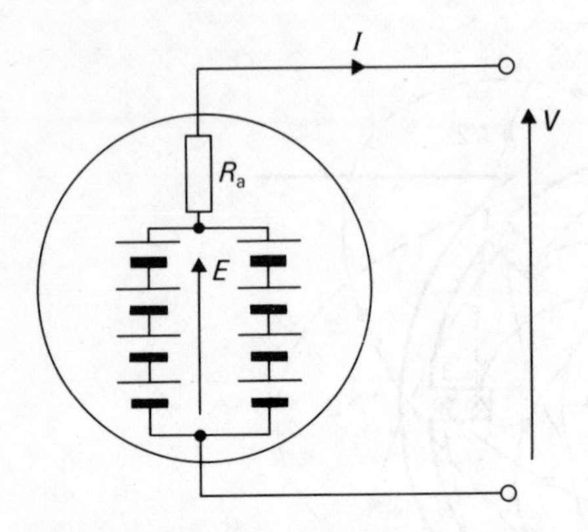

Fig. 14 Equivalent network of a generator armature winding

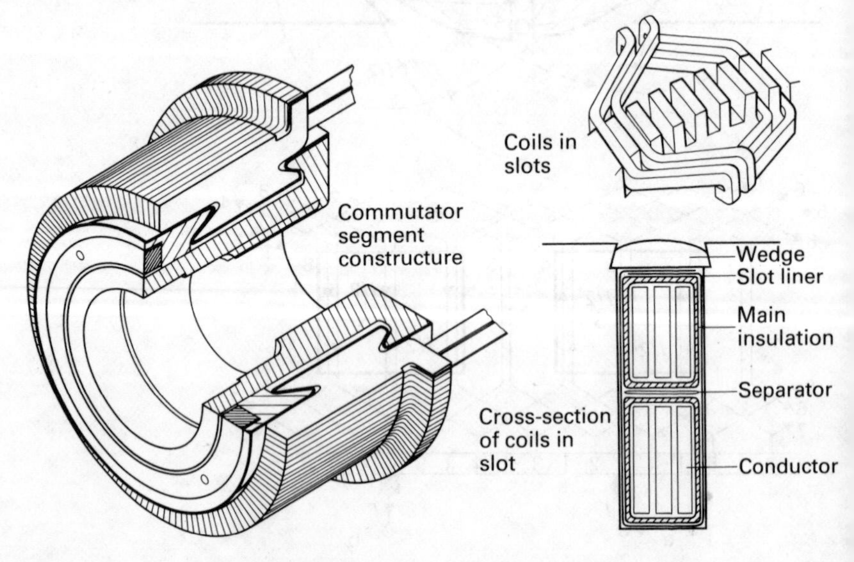

Fig. 15 Commutator winding construction details

The complete rotating part of a d.c. machine is termed the rotor of the machine and consists of the ferromagnetic drum mounted on a shaft. Along the sides of the drum are the coils which are insulated from the drum and from each other. Because of the forces which they experience, the coils must be firmly wedged into position. The ends of the coils are connected together and to the segments of the commutator. The segments are made of copper and each segment is insulated from the next. The segments are held in position by a dove-tail form of construction. Details of the winding and the commutator arrangements are shown in Fig. 15.

3 D.C. machine construction

When we first considered the principle of the d.c. machine, the arrangement consisted of a coil rotating in a uniform magnetic field emanating from two rectangular poles. However we have now evolved a rotor consisting of a ferromagnetic drum. If this is inserted into the previous arrangement for a uniform magnetic field, the field becomes distorted, and it is necessary to redesign the shape of the poles to maintain the uniformity of distribution of the magnetic field. A basic suitable arrangement is shown in Fig. 16.

The poles are now shaped to ensure that the air gaps between them and the core of the rotor are of uniform length. In this way, the magnetic field between the poles and the core of the rotor is uniformly distributed as indicated. The poles are mounted onto the outer body of the machine which is termed the yoke of the machine. In order to excite this magnetic circuit, windings are placed round the poles as indicated.

You will see that only the coils under the poles experience the effects of the magnetic field. You may think that this has an adverse

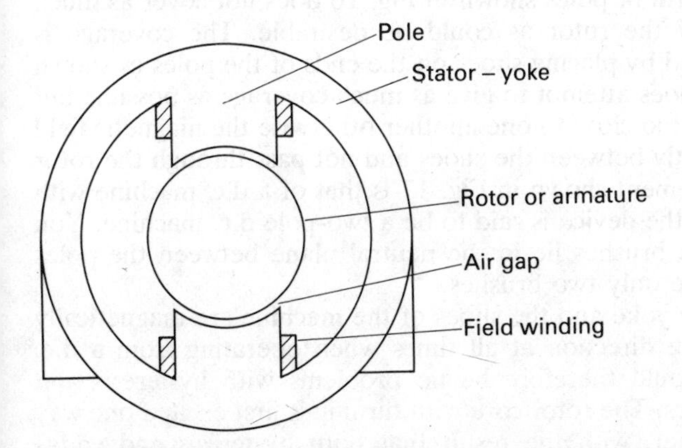

Fig. 16 Essential magnetic circuit of a d.c. machine

effect on the machine but in fact it is beneficial to its operation. The coils outwith the main magnetic field lie near to the neutral plane and you will recall that these coils contribute little to either the induction of an e.m.f. or to the production of torque. It is, however, in passing the neutral plane, that the coil experiences commutation. Because there is little e.m.f. induced in the coil, it does not matter that it is short-circuited for an instant by the brush. Also by the time the coil is entering into the field of the next pole, the action of its commutation has passed. The only difficulty arising is that the current in the coil has to change its direction, and, by Lenz's Law, we may anticipate that such a change is not readily acceptable to a magnetic system. Nevertheless local magnetic fields introduced by interpoles, which are small poles interposed between the main poles, help the current to reverse and thus aid commutation. The interpoles are indicated in Fig. 17.

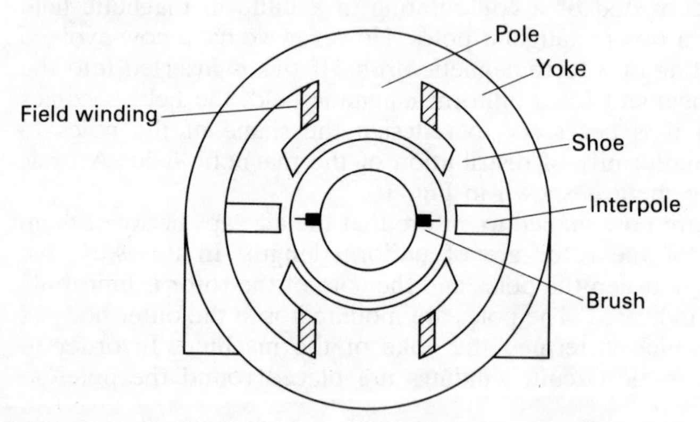

Fig. 17 Arrangement of a two-pole d.c. machine

The simple form of poles shown in Fig. 16 does not cover as much of the surface of the rotor as could be desirable. The coverage is therefore extended by placing shoes on the ends of the poles as shown in Fig. 17. The shoes attempt to give as much coverage as possible but they must not be too close to one another otherwise the magnetic field would cross directly between the shoes and not pass through the rotor core. The arrangement shown in Fig. 17 is that of a d.c. machine with two poles, hence the device is said to be a two-pole d.c. machine. You will note that the brushes lie in the neutral plane between the poles and that there are only two brushes.

The poles, the yoke and the shoes of the machine are magnetically excited in the one direction at all times when operating from a d.c. supply. There should therefore be no problems with hysteresis and eddy-current losses. The rotor core with turning is first excited one way and then the other, with the result that both hysteresis and eddy-current losses can arise. For these reasons, the rotor is usually laminated and made from a good quality steel, whilst the poles, the yoke and the shoes can be made of cast steel. However, in recent years, many d.c. machines are operated from rectified alternating current which gives rise to a pulsating direct current that fluctuates. This fluctuation gives rise to hysteresis and eddy-current losses particularly in the poles and the shoes and as a result many d.c. machines also have laminated poles and shoes.

On to the ends of the yoke, there are mounted end housings which are called bell housings. These carry the bearings in which the shaft turns. The bell housings have air vents cut into them and a fan is mounted on the shaft so that air is blown through the machine when it is operating, thus keeping it cool. The fan is mounted at the other end of the rotor from the commutator, as indicated in Fig. 18. The fixed parts of the machine are termed the stator, as opposed to the moving parts which make up the rotor.

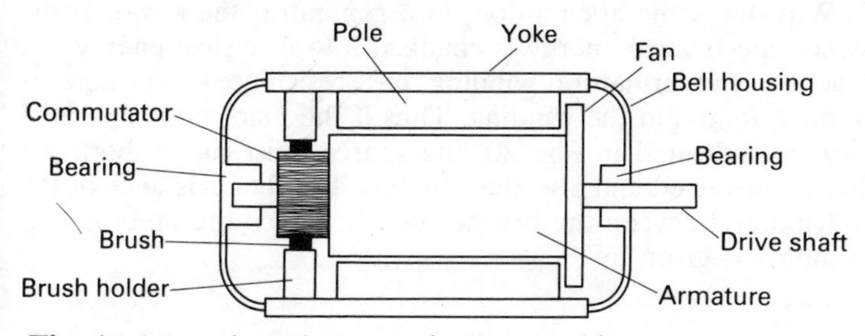

Fig. 18 Mounting of a rotor in d.c. machine

For a variety of reasons mainly stemming from the difficulties associated with commutation, many d.c. machines have more than two poles. Figure 19 shows a four-pole d.c. machine and you can see in it that the number of brushes remains equal to the number of poles. As you move round the stator, the poles are met in the order N–S–N–S.

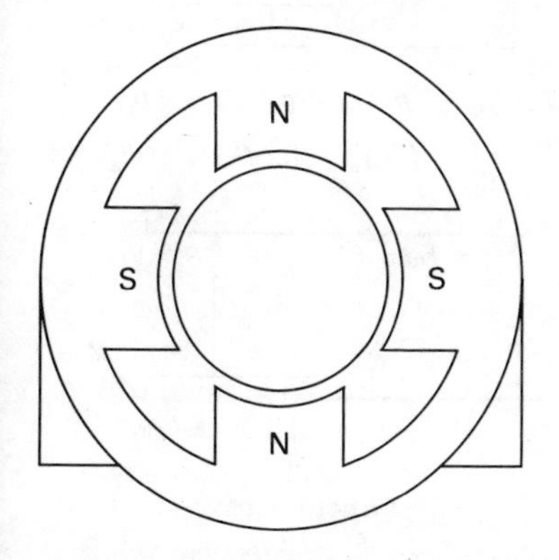

Fig. 19 Arrangement of a four-pole d.c. machine

So far, we have treated the induced e.m.f. as though it were only a part of the principle of a generator. However, when any rotor is made to rotate in a magnetic field, then an e.m.f. is induced in the armature winding, no matter whether the machine acts as a motor or as a generator. It was merely convenient to think of the e.m.f. as the basis of a generator but now we must remind ourselves that an e.m.f. is induced whenever there is an interchange of energy. In relation to the current I_a in the armature winding we can therefore say that the rate of energy interchange is given by

$$P = EI_a \tag{1}$$

In a motor, this power is the rate at which electrical energy is changed into mechanical energy (assuming that there are no losses in the transfer). With the same assumption, in a generator, the power is the rate at which mechanical energy is changed into electrical energy.

In practice, the armature winding has resistance and there is therefore an I^2R loss in the winding. Thus if the machine is operated as a motor, as indicated in Fig. 20, the source must supply both the power that is converted and also the I^2R loss. Let the resistance of the armature winding between the brushes be R_a; hence the input power P_i to the motor is given by

$$P_i = EI_a + I_a^2 R_a \tag{2}$$

The input power is given by the product of the terminal voltage and the supply current, which in this instance is equal to the armature

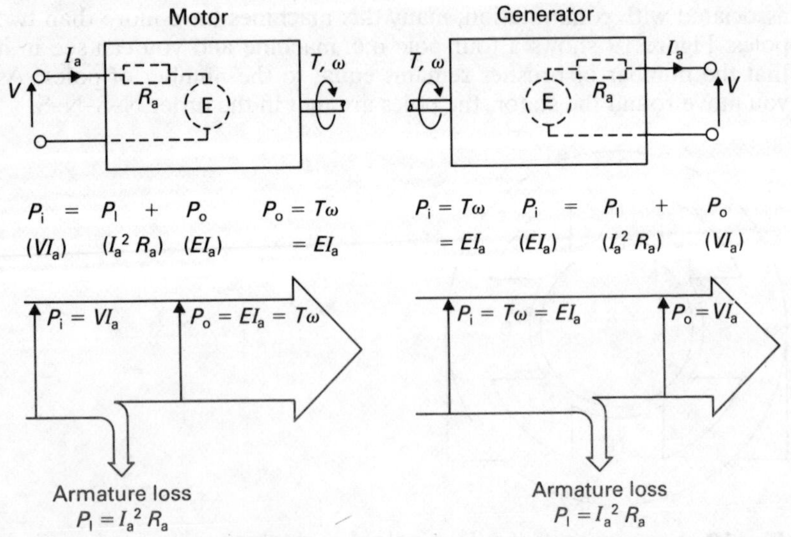

Fig. 20 Power flow diagrams for a d.c. motor and a d.c. generator

current I_a; thus

$$P_i = VI_a$$

and $VI_a = EI_a + I_a^2 R_a$

thus $V = E + I_a R_a$ (3)

If the machine is operated as a generator, the power input to the electrical system is the rate of mechanical energy conversion, which is

$$P_i = EI_a$$

The I^2R loss has to be deducted from this and what remains is the power output of the generator which can be measured at the output terminals as VI_a; thus

$$P_i = I_a^2 R_a + VI_a = EI_a \quad (4)$$

and $V = E - I_a R_a$ (5)

From relations (3) and (5), we may conclude that the e.m.f. induced in a motor is less than the terminal voltage, yet the e.m.f. induced in a generator is greater than the terminal voltage.

Example 1 A 250-V, d.c. motor is loaded to operate at 1250 rev/min and the armature current is 5·0 A. Given that the resistance of the armature is 2·0 Ω, determine the output torque of the motor.

$$V = E + I_a R_a$$

$$E = V - I_a R_a = 250 - 5 \times 2 = 240 \text{ V}$$

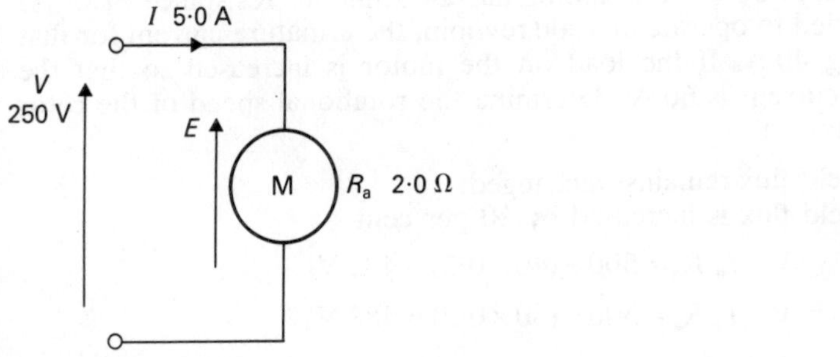

Fig. 21

It follows that the converted power is

$$P = EI_a = 240 \times 5 = 1200 \text{ W}$$

$$= \frac{2\pi NT}{60}$$

$$T = \frac{60 \times 1200}{2\pi \times 1250} = \underline{9{\cdot}2 \text{ N m}}$$

The e.m.f. induced in an armature winding can be derived from the relation that $E = Blu$. In the case of a winding, the length l has already been fixed, so it remains that

$$E \propto Bu$$

Again the flux, of density B, emanates from a pole of fixed area A; thus the e.m.f. in a d.c. machine is proportional to the total flux cut by the side of coil when passing from one neutral plane to the next; hence

$$E \propto \Phi u$$

Finally the velocity with which the sides of the coils cut across the magnetic field depends on the rotational speed of the rotor. Thus

$$E \propto \Phi N \tag{6}$$

Let k be a suitable constant of proportionality; hence in general

$$E = k\Phi N \tag{7}$$

Let us consider two sets of conditions, thus

$$E_1 = k\Phi_1 N_1$$

and $$E_2 = k\Phi_2 N_2$$

thus $$\frac{E_1}{E_2} = \frac{\Phi_1 N_1}{\Phi_2 N_2} \tag{8}$$

This is a useful relationship whereby we may predict the operation of a d.c. machine.

Example 2 A 500-V d.c. motor has an armature resistance of 0·2 Ω and is loaded to operate at 1200 rev/min, the armature current for that load being 40 A. If the load on the motor is increased so that the armature current is 60 A, determine the rotational speed of the rotor given that:

(a) the field flux remains unchanged;
(b) the field flux is increased by 10 per cent.

$$E_1 = V - I_{a_1} R_a = 500 - (40 \times 0{\cdot}2) = 492 \text{ V}$$

$$E_2 = V - I_{a_2} R_a = 500 - (60 \times 0{\cdot}2) = 488 \text{ V}$$

(a) $$\frac{E_1}{E_2} = \frac{\Phi_1 N_1}{\Phi_2 N_2} \quad \text{where} \quad \Phi_1 = \Phi_2$$

hence $$N_2 = \frac{E_2 N_1}{E_1} = \frac{488 \times 1200}{492} = \underline{1190 \text{ rev/min}}$$

(b) $$\frac{E_1}{E_2} = \frac{\Phi_1 N_1}{\Phi_2 N_2} \quad \text{where} \quad \Phi_2 = 1{\cdot}10 \times \Phi_1$$

$$N_2 = \frac{E_2 \Phi_1 N_1}{1{\cdot}1\Phi_1 \times E_1} = \frac{488 \times \Phi_1 \times 1200}{1{\cdot}1 \times \Phi_1 \times 492} = \underline{1082 \text{ rev/min}}$$

The earlier discussions concerning torque were made in the context of the action of a motor. However the current passing through the coils of a generator also produces a torque. In a generator, the shaft torque overcomes the torque of the armature and the machine is driven to produce electrical energy. By comparison, a motor produces a torque which overcomes the load and hence gives rise to the mechanical energy.

The torque can be derived from the relation $F = BlI$. In the case of a winding the length l has already been fixed, so it remains that

$$F \propto BI_a$$

As before, the flux density is directly related to the flux per pole, hence

$$F \propto \Phi I_a$$

Finally, the coil sides all act at fixed distances about the shaft of the rotor; thus the force produced is proportional to the torque and

$$T \propto \Phi I_a \tag{9}$$

It follows that

$$\frac{T_1}{T_2} = \frac{\Phi_1 I_{a_1}}{\Phi_2 I_{a_2}} \tag{10}$$

Again this is a most useful relationship whereby we may predict the operation of a d.c. machine.

Example 3 A 500-V d.c. generator is driven by a shaft torque of 100 N m and has an output of 15 kW. When the output is increased to 25 kW, the shaft torque increases to 150 N m. Assuming the generator to be free of losses, determine the necessary change in flux.

$$\frac{T_1}{T_2} = \frac{\Phi_1 I_{a_1}}{\Phi_2 I_{a_2}}$$

and $\quad P_{o_1} = VI_{a_1}$

hence $\quad I_{a_1} = \dfrac{15\,000}{500} = 30\ \text{A}$

also $\quad I_{a_2} = \dfrac{P_o}{V} = \dfrac{25\,000}{500} = 50\ \text{A}$

thus $\quad \dfrac{\Phi_2}{\Phi_1} = \dfrac{T_2 I_{a_1}}{T_1 I_{a_2}} = \dfrac{150 \times 30}{100 \times 50} = 0{\cdot}9$

It follows that the necessary change in flux is a 10 per cent reduction.

Example 4 A 250-V, d.c. motor develops an output power 4·5 kW. Given that the armature resistance is 1·0 Ω, find the armature current.

$$VI_a = EI_a + I_a^2 R_a$$
$$= P_o + I_a^2 R_a$$
$$0 = I_a^2 - 250 I_a + 4500$$

hence $I_a = \underline{19{\cdot}5\ \text{A}}$

Of the various relations that we have examined, three are of particular importance. These are

$$V = E \pm I_a R_a$$

$$E \propto \Phi N$$

$$T \propto \Phi I_a \qquad (11)$$

We shall use these relations extensively in our examination of the performance of d.c. machines which follows and it is necessary that you should be completely familiar with them before proceeding further.

5 Armature reaction

In our analysis of the d.c. machine, the only flux considered has been that of the field emanating from the windings wound on the stator poles. However, the armature current in the armature winding also gives rise to a rotor field which adds to the main field and modifies it. The details of this change in the total flux are not important at this introductory stage, but the reaction of the armature winding has one significant effect which requires to be noted.

Figure 22(*a*) is a stretched-out diagram of the air gaps between the poles and the rotor surface, and we can see that the field flux is concentrated between the poles and the rotor. The distribution of the magnetic field is illustrated by Fig. 22(*b*). The m.m.f. F only exists across the air gaps and, allowing for fringing at the edges of the poles, the resulting flux density varies in a comparable manner.

The armature winding also produces an m.m.f. which is a maximum along the axis determined by the positions of the brushes. This results in the m.m.f. distribution shown in Fig. 23(*a*). Because of the large air gaps between the poles, the m.m.f. produces little flux and the flux density due to the armature reaction only becomes appreciable under the poles as shown in Fig. 23(*b*).

Combining the two fields results in a total flux density distribution shown in Fig. 24. This diagram shows two characteristics, one being the theoretical combination of the fields and the other being the actual result. The difference between the characteristics arises in those parts

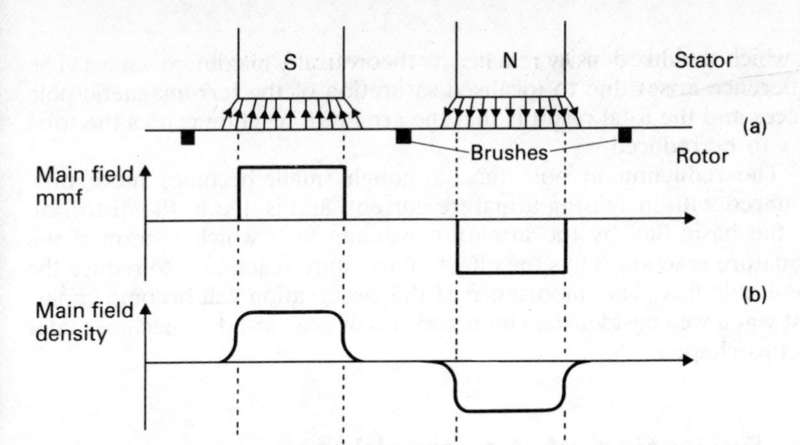

Fig. 22 Distribution of main field in a d.c. machine

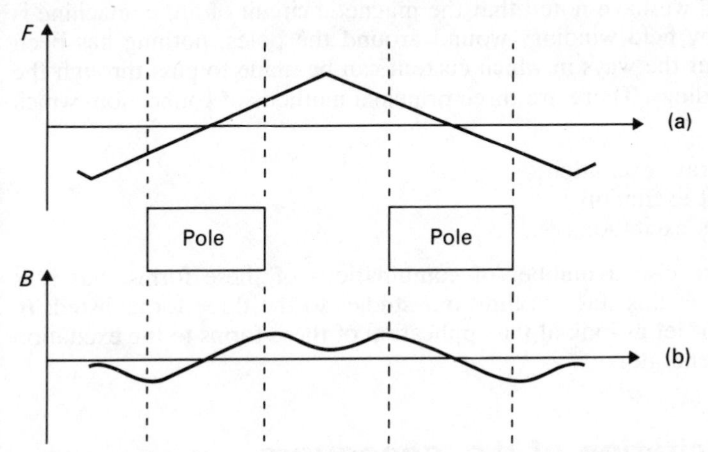

Fig. 23 Distribution of armature field in a d.c. machine

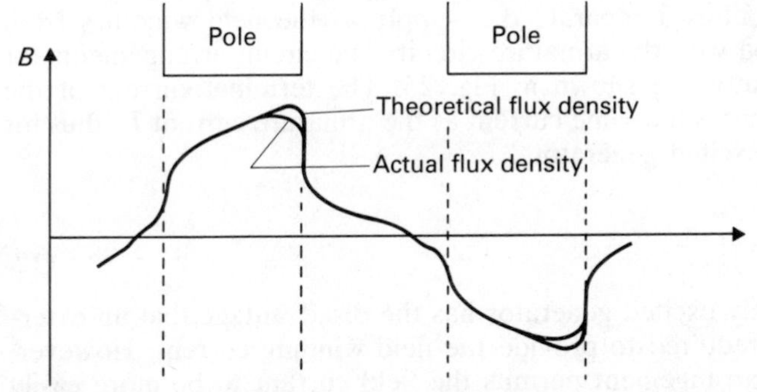

Fig. 24 Flux density distribution in a d.c. machine

at which the flux density reaches its theoretically maximum values. The difference arises due to localised saturation of the ferromagnetic pole pieces and the total result is that the armature reaction causes the total flux to be reduced.

The reduction in pole flux, although small, becomes more pronounced with increasing armature current, and is due to the distortion of the basic flux by the armature winding flux, which is termed the armature reaction. Thus the effect of armature reaction is to reduce the total pole flux. The importance of this observation will become apparent when we consider the characteristics of practical d.c. machines later in this chapter.

6 Excitation of d.c. machines

Although we have noted that the magnetic circuit of a d.c. machine is excited by field windings wound around the poles, nothing has been said about the ways in which current can be made to pass through the field windings. There are three principal methods of connection, which are:

(*a*) separate excitation;
(*b*) shunt excitation;
(*c*) series excitation.

There are also a number of combinations of these forms, but it is sufficient at this stage to limit our studies to the three forms listed. In particular, let us look at the application of these forms to the excitation of d.c. generators.

7 Excitation of d.c. generators

Separate excitation. As the name suggests, a separately-excited generator requires a separate d.c. supply to the field windings from that associated with the armature circuit. The circuit arrangements for such a generator are shown in Fig. 25. The terminal current of the armature circuit is the same current as the armature current I_a, thus for a separately excited generator

$$V = E - I_a R_a$$
$$= E - IR_a \qquad (12)$$

A separately excited generator has the disadvantage that an external supply is required to provide the field winding current. However, this separate arrangement permits the field current to be more easily controlled and, if necessary, the control can be achieved over a wide range. The width of range is from zero to the maximum field current,

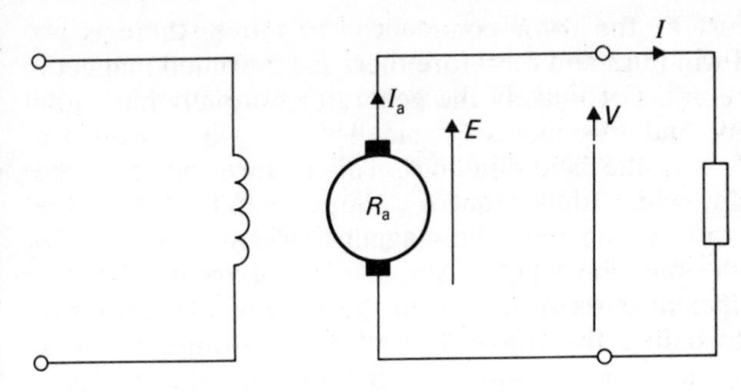

Fig. 25 Separately excited d.c. generator

the polarity of which can be reversed so that the generated e.m.f. may take up either polarity. In control systems, this variation is particularly useful, as it permits good control of the generator terminal e.m.f. The variation of the field current is generally achieved by transistor circuit or by thyristor circuit control.

Shunt excitation. The field winding of the shunt-excited generator is connected in parallel with the load. When a component is connected in parallel with a load, it is said to shunt it; hence the load in this case is shunted by the field winding, which is thus supplied by the armature.

The function of the field winding is to produce the necessary m.m.f. to create the magnetic field. The m.m.f. is given by the product of the winding turns and the current in the winding. For efficiency, we wish as little as possible of the armature current to be diverted into the field circuit and thus away from the load circuit. It is therefore usual to make the field winding with a large number of turns and hence of relatively high resistance, with the result that the field current is small. The form of connection is shown in Fig. 26.

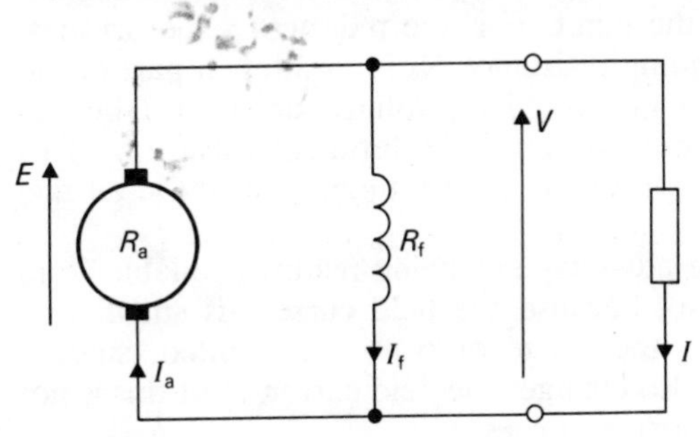

Fig. 26 D.C. shunt generator

A problem with the shunt-excited generator that is not immediately apparent is the induction of the armature e.m.f. when the armature starts to rotate. Just as the rotor commences to move, there is no current in the field windings and therefore there is no excited magnetic field to induce an e.m.f. Fortunately the generator normally has some residual magnetism, and this induces a small e.m.f. which causes a small current to flow in the field windings. This in turn increases the total magnetic field, which thus induces a larger e.m.f. and so the process continues to build up until the magnetic system tends toward saturation, at which point the e.m.f. is stabilised. This means that the e.m.f. causes a sufficient current to flow in the field windings to give enough field flux to induce that e.m.f. The effect of saturation will be further discussed in the section dealing with generator characteristics.

If you are carrying out an experiment in the laboratory with a d.c. generator which is shunt excited and it fails to develop an induced e.m.f., you should check that you have connected the field windings correctly to the armature. If the windings are reverse connected, then the small induced e.m.f. which is initially produced gives rise to a field current which gives rise to a flux that cancels the residual magnetic field, thus reducing the e.m.f. rather than increasing it.

A machine that has just been made requires to be magnetised from an external source in order that the magnetic circuit is excited for the first time. Thereafter the residual magnetism ensures the build up of the induced e.m.f. whenever the generator is correctly operated.

From Fig. 26, we observe three important relations:

$$I_f = \frac{V}{R_f} \tag{13}$$

$$I_a = I + I_f \tag{14}$$

$$E = V + I_a R_a \tag{15}$$

The last relation requires some explanation, because from the diagram of Fig. 26, it appears that the e.m.f. E is the p.d. across the brushes. However, the armature winding resistance R_a is an integral part of the circuit between the brushes and the $I_a R_a$ voltage drop must be deducted from the induced e.m.f. to obtain the terminal voltage V. This difficulty in showing the action of the induced e.m.f. recurs in all d.c. machine circuit diagrams.

Shunt excitation has the advantage of being readily available from the machine terminals. Also, because the field current is small, it is easily controlled. However, the range of control is limited, since a change of terminal voltage also changes the field current, but this is not a serious disadvantage in most instances.

Series excitation. As the name suggests, the field windings are connected in series with the load, and the field current is therefore the load current. It follows that the field current is relatively large and therefore fewer turns are required in the field winding than are required in a shunt machine. Because the current is relatively large, it is also necessary to make the winding from a conductor of comparatively large cross-sectional area in order to minimise the winding resistance and consequent voltage drop.

Like the shunt machine, the series-excited machine requires an amount of residual magnetism whereby it may develop an induced e.m.f. during starting. And again like the shunt machine, if the field winding were reverse connected, the initial field current would create an m.m.f. which would oppose the residual magnetic field and therefore the machine would fail to develop an induced e.m.f.

The circuit diagram for a series-excited generator is shown in Fig. 27. From the diagram, we can observe that

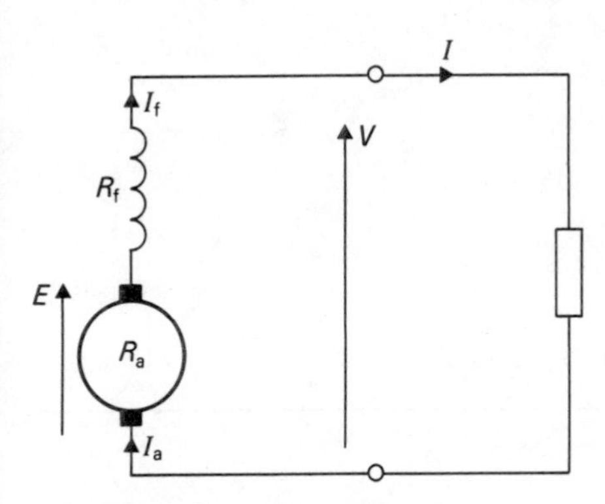

Fig. 27 D.C. series generator

$$I = I_a = I_f \tag{16}$$

and $V = E - IR_a - IR_f$

$$= E - I(R_a + R_f) \tag{17}$$

The series-excited generator has no advantages of importance when compared with the shunt-excited machine except in those very few applications where the operating current is more important than the operating voltage. However, an understanding of series excitation is important in our studies of series-excited motors, and also many practical generators incorporate both shunt- and series-excitation arrangements.

Example 5 A d.c. shunt generator supplies 200 kW at 250 V when fully loaded. The resistance of the field winding is 100 Ω and the resistance of the armature winding is 0·02 Ω. Under full-load conditions, determine

(*a*) the load current;
(*b*) the armature current;
(*c*) the armature e.m.f.

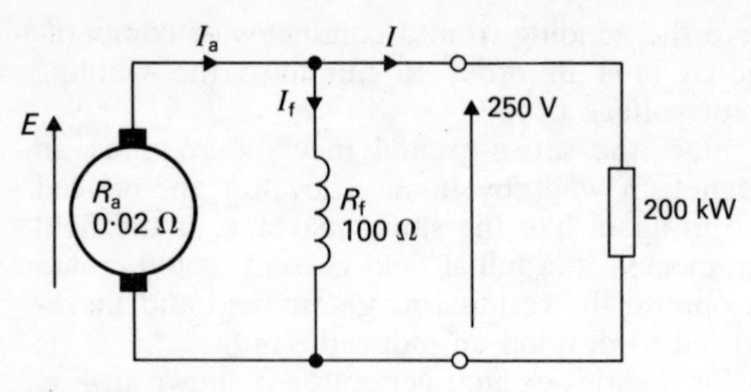

Fig. 28

$$I = \frac{P}{V} = \frac{200 \times 10^3}{250} = \underline{800\ \text{A}}$$

$$I_f = \frac{V}{R_f} = \frac{250}{100} = 2{\cdot}5\ \text{A}$$

$$I_a = I + I_f = 800 + 2{\cdot}5 = \underline{802{\cdot}5\ \text{A}}$$

$$E = V + I_a R_a = 250 + (802{\cdot}5 \times 0{\cdot}02) = \underline{266\ \text{V}}$$

---

8 Characteristics of d.c. generators

For each form of generator, there are two groups of characteristics to be considered. These are:

(*a*) open-circuit characteristics;
(*b*) load characteristics.

Separately excited generator. Under open-circuit conditions, the armature winding carries no current, yet it has an e.m.f. induced in it due to its rotation in the magnetic field, which is energised from a separate source. The e.m.f. is given by

$$E \propto \Phi N$$

and we may consider this relation for variation of speed when the field flux is maintained constant, and for variation of the field flux when the speed is maintained constant. In the first instance, the field flux is constant provided the field current I_f is constant, thus

$$E \propto N$$

Initially, let the field current be I_{f_1}, thus the e.m.f./speed characteristic takes the form shown in Fig. 29(*a*). Higher values of field current I_{f_2} and I_{f_3} increase the field flux and therefore give rise to proportionally

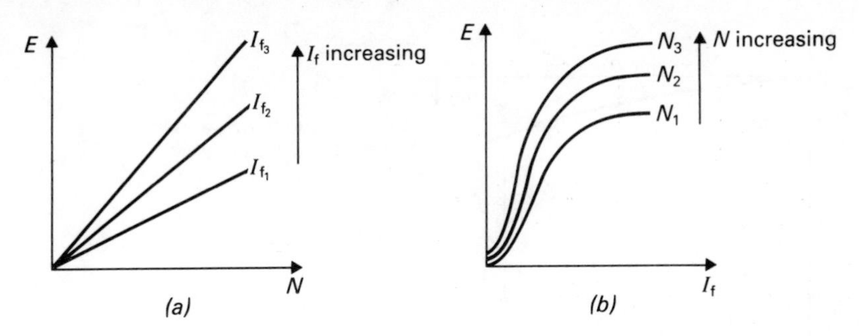

Fig. 29 Open-circuit characteristics of a separately excited d.c. generator

greater e.m.f.s for any given speed, but in each case the characteristics are essentially linear. Alternatively, let the speed be constant and the field current varied. Since the speed N is constant,

$$E \propto \Phi$$

The flux Φ is varied by the field current I_f and the relation takes the form defined by the B/H characteristic of the magnetic circuit. It follows that the E/I_f characteristic also takes a similar form as illustrated in Fig. 29(*b*).

By increasing the rotor speed, the e.m.f. characteristic is also increased; thus a family of characteristics can be created with a separate characteristic for each particular speed. None of the characteristics passes through the origin of the graph because even when I_f is zero, there is still a small e.m.f. induced by the residual magnetism.

The function of a generator is to supply current to a load, and generally this is undertaken at a specific supply voltage. Suppose that the generator is excited to give the required terminal voltage and that a load is introduced, thus increasing the armature current from zero. Provided that the field current and the rotor speed are maintained constant, the terminal voltage/load current characteristic takes the form shown in Fig. 30. This characteristic is termed the external characteristic of the generator.

This characteristic indicates that the terminal voltage drops with increase in load current. There are two reasons for this drop:

1. The armature reaction decreases the field flux and hence reduces the induced e.m.f.
2. There is the I_aR_a voltage drop in the armature.

In practice, the drop in output voltage at full load compared with the no-load terminal voltage is small although appreciable. If necessary, the terminal voltage may be maintained constant by increasing the field current and/or the rotor speed.

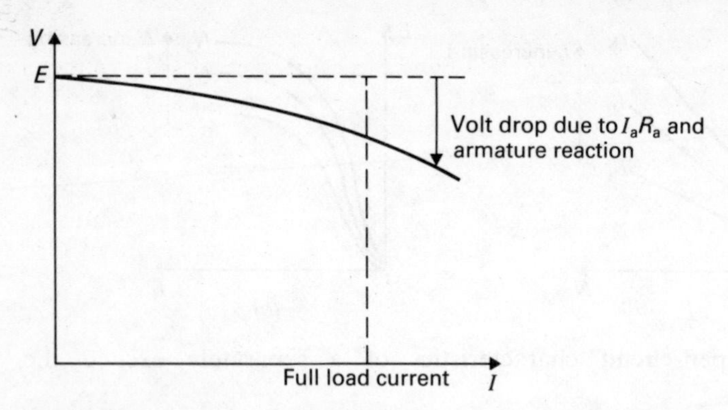

Fig. 30 External or load characteristic of a separately excited d.c. generator

Shunt generator. When a shunt generator is operated under no-load conditions, it is the output terminals that are open-circuited as shown in Fig. 31(a). The armature still requires to supply current to the field winding. However to obtain the open-circuit characteristic, it is a simpler test to operate the generator at constant speed with the field separately excited, thus obtaining the characteristic shown in Fig. 31(b). Normally because the field current I_f is comparatively small, as is the armature resistance R_a, it follows that this characteristic is almost the open-circuit characteristic of the shunt generator and it is not necessary to allow for the very small I_fR_a voltage drop.

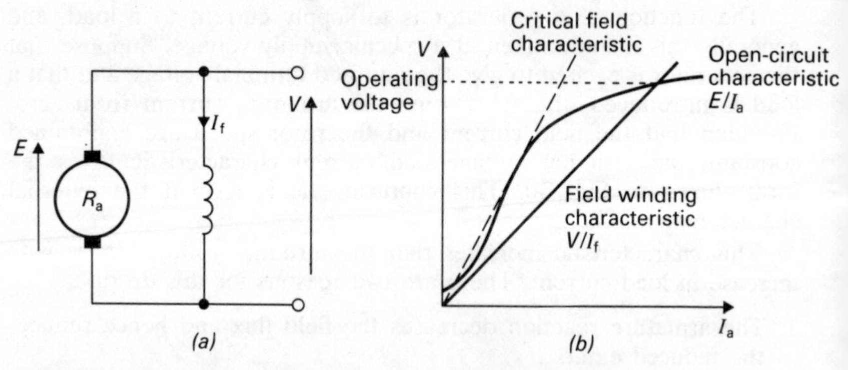

Fig. 31 D.C. shunt generator under no-load conditions

The V/I_f characteristic of the field windings is a straight line, its gradient depending on the resistance of the windings. We have already noted that the generator starts by inducing an e.m.f. due to the residual magnetism. This e.m.f. causes a current to flow in the field windings which enlarges the magnetic field, which in turn increases the induced e.m.f. This process continues until the e.m.f. causes just sufficient current to flow to produce a sufficient field to exactly induce the e.m.f. This condition occurs when the open-circuit characteristic meets the V/I_f characteristic of the field windings. Once the steady-state condition has been achieved, there are several possible changes of condition that we may apply to the generator. For instance, if the resistance of the field winding circuit is increased, then the slope of the field characteristic increases and the characteristic intersects at a lower value of E. However, if the line of the field-resistance characteristic becomes tangential to the open-circuit characteristic, then a critical point is reached at which the generator is unable to sustain the induced e.m.f., which suddenly falls to a very low value. This limitation means that the variation of field resistance is not a suitable method of obtaining control of the output voltage over a large range. Nevertheless, it is often quite sufficient to control the output voltage over a small range near the normal operating voltage.

If a wider range of output voltage variation is required, it is necessary to vary the speed of the rotor.

Finally, we should note that if the field-circuit resistance is too high, the shunt generator will fail to excite itself, thus a shunt generator can fail to excite because it has no residual magnetism, because the field windings are reverse connected or because the resistance of the field circuit is too high. Other reasons for failure to excite originate from malfunction of components, e.g. dirty brushes on the commutator or a winding being open- or short-circuited.

When the shunt generator is operated under load, the load characteristic shown in Fig. 32 is similar to that of the separately excited generator. In this instance however, the characteristic is drawn on the basis that the speed N is constant, the field current I_f now depending

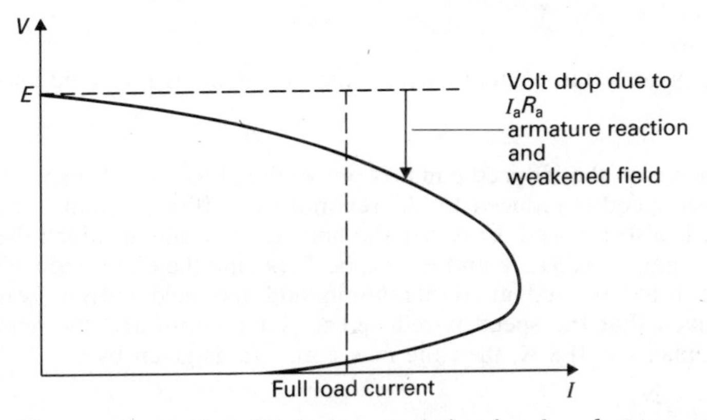

Fig. 32 External or load characteristic of a d.c. shunt generator

on the terminal voltage of the generator. Because this voltage tends to decline with increase of load current I, it follows that I_f also declines, thus reducing the field flux Φ and hence the induced e.m.f. E. This in turn further reduces the terminal voltage; thus the droop of the characteristic is slightly more pronounced in the shunt generator than in the separately excited generator. The decline in the terminal voltage is therefore caused by:

(*a*) the effect of armature reaction;
(*b*) the $I_a R_a$ voltage drop in the armature winding;
(*c*) the reduction in the field current I_f mainly due to the other two causes.

Example 6 An open-circuit test was carried out on a d.c. shunt generator driven at 1000 rev/min, the field being separately excited.

Terminal voltage V (volts)	312	357	390	414	435
Field current I_f (amperes)	0·8	1·2	1·6	2·0	2·4

The resistance of the field circuit is 200 Ω. Find the open-circuit voltage of the generator when:

(*a*) it is driven at 1000 rev/min;
(*b*) it is driven at 800 rev/min.

The results of the test are illustrated in Fig. 33. The V/I_f characteristic is constructed by considering a voltage, say 400 V, and hence determining the field current as $400/200 = 2{\cdot}0$ A. By plotting this condition on the graph and joining the point to the origin, as shown, the V/I_f characteristic is constructed.

The intersection of the open-circuit characteristic of the armature when operating at 1000 rev/min and the V/I_f characteristic occurs at a point corresponding to a terminal voltage of 418 V

$$E \propto \Phi N$$

For any given value of field current, the flux remains the same and

$$E \propto N$$

thus in each case the induced e.m.f. is proportional to the rotor speed. If the rotor speed is reduced to 800 rev/min from 1000 rev/min, then the e.m.f. is also reduced. Consider the first set of results in which the terminal voltage was 312 V and let this be E_1, being the e.m.f. induced when the rotor rotated at 1000 rev/min and the field current was 0·8 A. Given that the speed is reduced to 800 rev/min and the field current remains at 0·8 A, then the new e.m.f. E_2 is given by

$$\frac{E_1}{E_2} = \frac{N_1}{N_2}$$

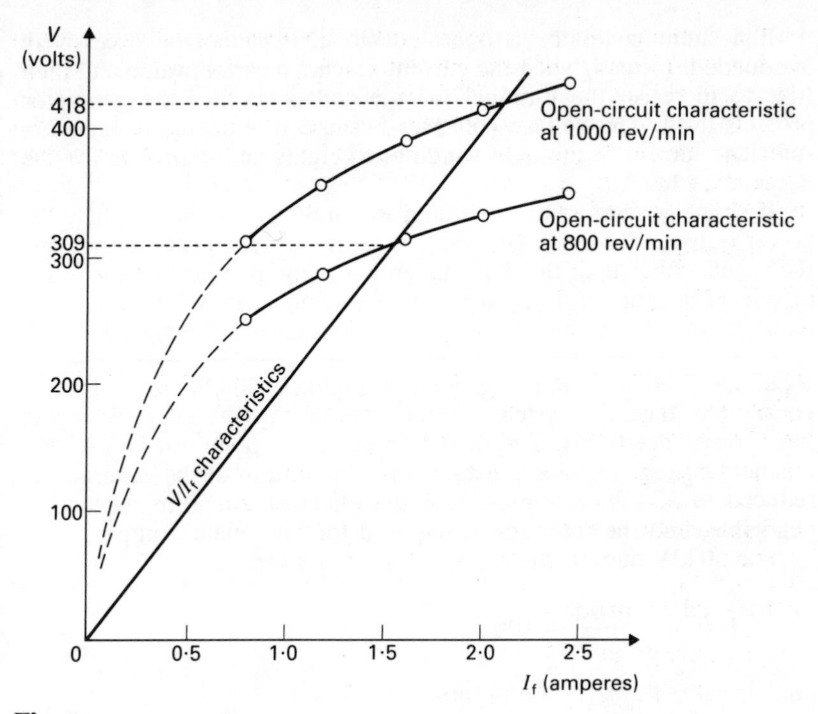

Fig. 33

hence $E_2 = E_1 \dfrac{N_2}{N_1} = 312 \times \dfrac{800}{1000} = 250\ \text{V}$

By repeating this procedure, we obtain the open-circuit characteristic of the generator operating at 800 rev/min as being

Terminal voltage V (volts)	250	286	312	331	348
Field current I_f (amperes)	0·8	1·2	1·6	2·0	2·4

This characteristic is drawn on Fig. 33 and its intersection with the V/I_f characteristic indicates that the open-circuit voltage of the generator at 800 rev/min is <u>309 V</u>

Over a normal working range, i.e. from no load to full load, the drop in output voltage of a shunt-excited generator is generally less than 5 per cent, which is much smaller than is suggested by the characteristic shown in Fig. 32. The characteristic of that diagram was exaggerated to emphasise that there is a voltage drop with increase of load.

If a shunt generator is operated alone, it cannot be excessively overloaded. Instead, when the current reaches a certain value and then tries to increase, the terminal voltage collapses; thus the generator protects itself. The voltage collapses because it is unable to maintain sufficient current in the field windings to create sufficient field for the necessary e.m.f.

If shunt generators are operated in parallel with one another, the power output of a single generator is more easily adjusted by varying the speed rather than the field current, bearing in mind that the effect of one generator on a group of parallel-connected generators is less likely to vary the terminal voltage, which is essentially constant.

Example 7 A d.c. shunt generator supplies 50 kW to a 500-V, constant-voltage, d.c. system. The armature rotates at 650 rev/min and has a resistance 0·2 Ω. The field current of the generator is 3 A, and this field current remains constant when the output of the generator is reduced to 35 kW. Assuming that the effect of armature reaction is negligible, find the rotor speed required for this smaller output.

For 50 kW output, the supply current is given by

$$I_1 = \frac{P_1}{V} = \frac{50\,000}{500} = 100\text{ A}$$

and $E_1 = V + I_{a_1}R_a = 500 + (100 + 3)0{\cdot}2 = 520{\cdot}6\text{ V}$

For 35 kW output, the supply current is given by

$$I_2 = \frac{P_2}{V} = \frac{35\,000}{500} = 70\text{ A}$$

and $E_2 = V + I_{a_2}R_a = 500 + (70 + 3)0{\cdot}2 = 514{\cdot}6\text{ V}$

The field current is constant and, since the armature reaction is negligible,

$$\frac{E_1}{E_2} = \frac{N_1}{N_2}$$

$$N_2 = 650 \times \frac{514{\cdot}6}{520{\cdot}6} = \underline{642{\cdot}5\text{ rev/min}}$$

Series generator. When the series generator is operated with no load, there can be no armature current and therefore no current in the field windings to induce an e.m.f. However, a small e.m.f. will be generated due to residual magnetism.

When a load is applied to the generator, current can flow in the field windings and an e.m.f. is induced. The external or load characteristic of the series generator is shown in Fig. 34 and it is similar to that of the open-circuit characteristic, which can be obtained by

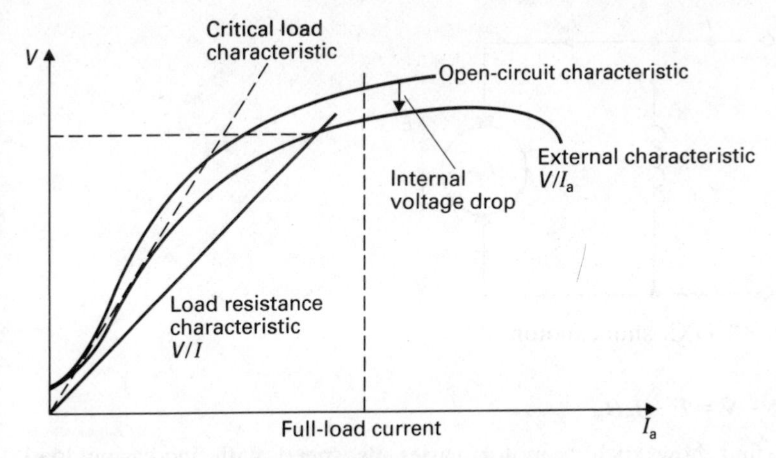

Fig. 34 Characteristics of a d.c. series generator

separately exciting the machine. The difference between the characteristics is caused by a voltage drop due to:

(*a*) armature reaction;
(*b*) the I_aR_a voltage drop;
(*c*) the I_aR_f voltage drop.

The output voltage therefore depends on the resistance of the load, and may be determined by drawing the voltage/current characteristic of the load, as shown, to meet the external characteristic. If the load resistance is too large, the slope of its characteristic becomes too steep and exceeds that of the critical load resistance, in which case the generator fails to excite. On the other hand, if the load resistance is too small, the machine protects itself by letting the output voltage fall, as illustrated.

9 Characteristics of d.c. motors

The methods of excitation of d.c. motors are similar to those of d.c. generators. The shunt motor arrangement is shown in Fig. 35, and effectively the field is supplied from the source and the armature is supplied from the same source. Being in parallel, they are nevertheless independent of one another and merely share the same supply voltage. It would not matter if the field were supplied separately from another source; thus the characteristics which follow for the shunt motor also apply to the separately excited motor.

For the shunt motor shown in Fig. 35,

$$I = I_a + I_f \tag{18}$$

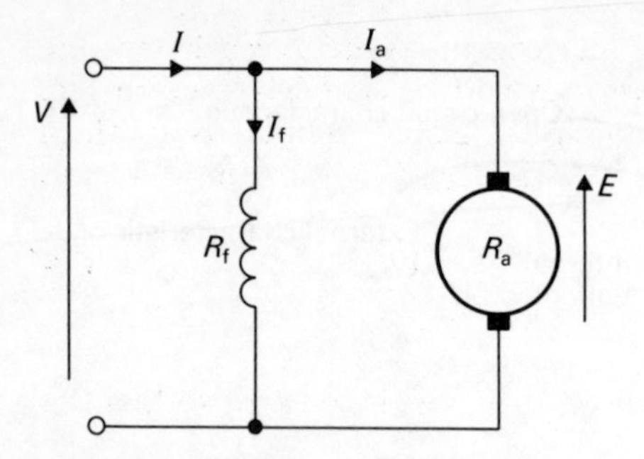

Fig. 35 D.C. shunt motor

also $\quad V = E + I_a R_a$

To find how such a motor varies its speed with increasing load, consider the e.m.f. relationship

$$E \propto \Phi N$$

thus $\quad V - I_a R_a \propto \Phi N$

and
$$N \propto \frac{V - I_a R_a}{\Phi} \tag{19}$$

If the flux is maintained constant, which can be achieved by keeping the field current constant, the speed is proportional to $(V - I_a R_a)$. This means that for any given field current, the speed remains essentially constant although, in practice, it does fall slightly with increase in load due to armature reaction and the $I_a R_a$ voltage drop. The resulting speed/armature current characteristic, showing the small fall in speed with increase of load, is shown in Fig. 36(a).

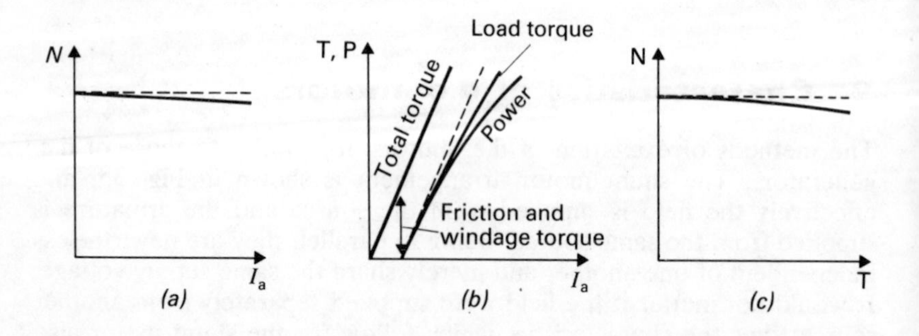

Fig. 36 Characteristics of a d.c. shunt motor: (a) speed/armature current; (b) torque and power/armature current; (c) speed/torque.

If the shunt motor were a constant-speed machine, the speed/armature current characteristic would be that indicated in Fig. 36(*a*) by the dotted line. The droop in the actual characteristic, which is the continuous line, is so small that the shunt motor can be seen to be almost a constant-speed machine.

From relation (9),

$$T \propto \Phi I_a$$

Again suppose that the field current is maintained constant so that the flux remains essentially constant. It follows that

$$T \propto I_a$$

In this relation, the torque is the total torque developed and the relation is illustrated by the torque/armature current characteristic shown in Fig. 36(*b*). In practice, a certain torque is required to overcome friction and windage, even when the motor is not loaded; as a result, some armature current is required under no-load conditions to drive the rotor. This no-load torque has to be deducted from the total torque to obtain the load torque, i.e. that torque applied to the load by the motor driving shaft.

In practice, the load torque/armature current does not increase linearly and this is due to armature reaction. Also the droop in the torque/armature current characteristic reflects that of the speed/armature current characteristic, with the result that the power/armature current characteristic falls away even more noticeably.

By relating the speed/armature current characteristic to the torque/armature current characteristic, we obtain the speed/torque characteristic shown in Fig. 36(*c*). This illustrates that the speed remains reasonably constant with variation of load torque.

The family of characteristics, shown in Fig. 36, was drawn up on the assumption that the field current remained constant. However, the field current can readily be changed by varying the resistance of the field circuit (equally the field current of a separately excited machine can also be readily varied). For most shunt motors, E tends to be equal to the supply voltage; thus, from the relation

$$E \propto \Phi N$$

we may say that approximately

$$N \propto \frac{1}{\Phi}$$

thus as the field current and the field flux increase, the speed is reduced, and vice versa, as shown in Fig. 37.

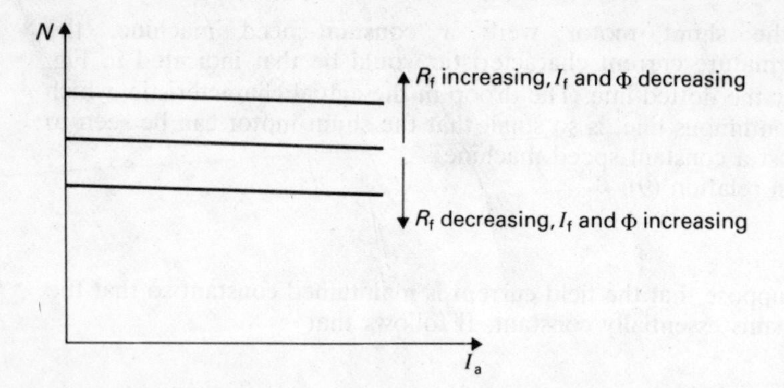

Fig. 37 Effect of field resistance variation on the speed/armature current characteristic of a d.c. shunt motor

---

Example 8 A 240-V, d.c. shunt motor has an armature resistance 0·27 Ω and field winding resistance 160 Ω. On no load, the motor takes a current 2·9 A and its speed is 1250 rev/min. On full load, the motor takes a current 40·0 A and, although the field current remains the same, the flux per pole is reduced by 5 per cent due to armature reaction. Determine the full-load speed of the motor.

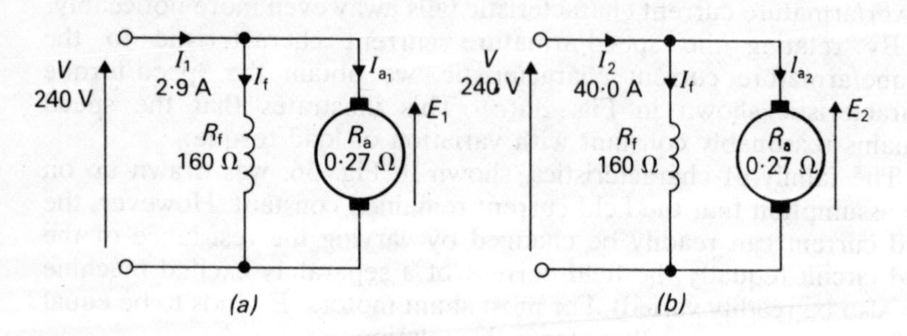

Fig. 38

Let the e.m.f. under no-load conditions be E_1 and the armature current be I_{a_1} as indicated in Fig. 38(a).

$$I_{a_1} = I_1 - I_f = I_1 - \frac{V}{R_f} = 2{\cdot}9 - \frac{240}{160} = 1{\cdot}4 \text{ A}$$

$$E_1 = V - I_{a_1}R_a = 240 - (1{\cdot}4 \times 0{\cdot}27) = 239{\cdot}6 \text{ V}$$

Let the full-load conditions be those defined in Fig. 37(b).

$$I_{a_2} = I_2 - I_f = 40{\cdot}0 - 1{\cdot}5 = 38{\cdot}5 \text{ A}$$

$$E_2 = V - I_{a_2}R_a = 240 - (38{\cdot}5 \times 0{\cdot}27) = 229{\cdot}6 \text{ V}$$

However, $$\frac{E_1}{E_2} = \frac{\Phi_1 N_1}{\Phi_2 N_2}$$

and $$N_2 = \frac{100}{95} \times \frac{229{\cdot}6}{239{\cdot}6} \times 1250 = \underline{1261 \text{ rev/min}}$$

Example 9 A 500-V, d.c. shunt motor has an armature resistance 1·0 Ω and a field winding resistance 500 Ω. When loaded to develop a total torque 100 N m, the motor takes a current of 21 A from the supply. Determine the speed of the rotor.

With the field current remaining unchanged, the motor is further loaded until the total torque is 120 N m. Assuming that the effect of armature reaction is negligible, determine the speed of the rotor.

When the total torque is 100 N m, let the circuit quantities be defined as shown in Fig. 41(a).

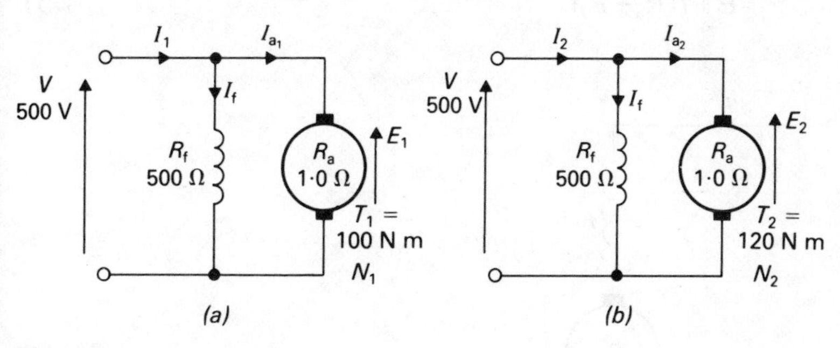

Fig. 39

$$I_f = \frac{V}{R_f} = \frac{500}{500} = 1 \text{ A}$$

$$I_{a_1} = I_1 - I_f = 21 - 1 = 20 \text{ A}$$

$$E_1 = V - I_{a_1}R_a = 500 - (20 \times 1) = 480 \text{ V}$$

$$E_1 I_{a_1} = T_1\omega_1 = \frac{2\pi N_1 T_1}{60}$$

$$N_1 = \frac{60 \times E_1 I_{a_1}}{2\pi T_1} = \frac{60 \times 480 \times 20}{2\pi \times 100} = \underline{917 \text{ rev/min}}$$

When the total torque is 120 N m, let the circuit quantities be defined as shown in Fig. 39(b).

$$\frac{T_1}{T_2}=\frac{I_{a_1}}{I_{a_2}}$$

$$I_{a_2}=\frac{120}{100}\times 20=24\text{ A}$$

$$E_2=V-I_{a_2}R_a=500-(24\times 1)=476\text{ V}$$

$$E_2I_{a_2}=T_2\omega_2=\frac{2\pi N_2T_2}{60}$$

$$N_2=\frac{60\times E_2I_{a_2}}{2\pi T_2}=\frac{60\times 476\times 24}{2\pi\times 120}=\underline{909\text{ rev/min}}$$

The circuit arrangement for the d.c. series motor is shown in Fig. 40. Since there is only the one circuit, it follows that

$$I=I_a=I_f \tag{20}$$

Also $V=E+IR_a+IR_f$

$$=E+I(R_a+R_f) \tag{21}$$

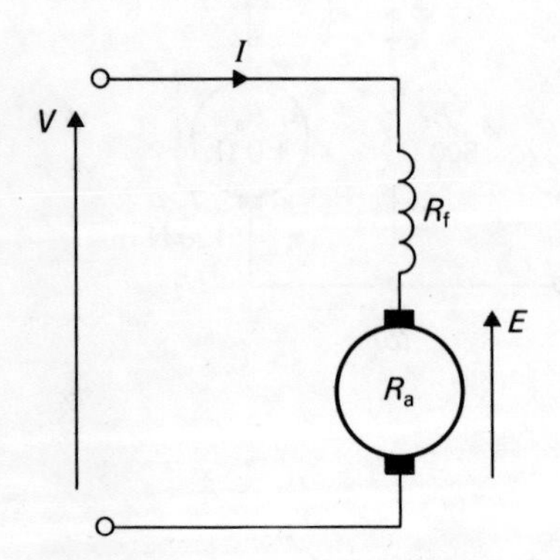

Fig. 40 D.C. series motor

To find how the motor varies its speed with increasing load, consider the e.m.f. relationship

$$E\propto \Phi N$$

thus $V-I(R_a+R_f)\propto \Phi N$

and
$$N\propto\frac{V-I(R_a+R_f)}{\Phi} \tag{22}$$

The supply voltage V can be taken as constant and generally the term $I(R_a + R_f)$ is small; thus it is approximately correct to say that

$$N \propto \frac{1}{\Phi}$$

The flux of the series motor is proportional to the current provided that magnetic saturation does not affect the relationship. It follows that

$$N \propto \frac{1}{I} \tag{23}$$

and the speed/armature current characteristic takes the form shown in Fig. 41(a). It is important to observe the extremes of this characteristic, especially when the load current is small. Under this condition when I is small, then N is large, and if the current becomes too small, then the speed will become so high that the rotor will fly apart under the centrifugal forces applied. It is therefore essential to ensure that the series motor is always loaded so that its speed cannot rise to such dangerous levels, and there is therefore a safe minimum load, as indicated in Fig. 41(a). As the load current increases, the magnetic circuit starts to saturate and the flux cannot continue to increase; hence the speed of the rotor tends to become constant also. The speed does not become constant, however, as E continues to fall due to increasing IR drop in the circuit.

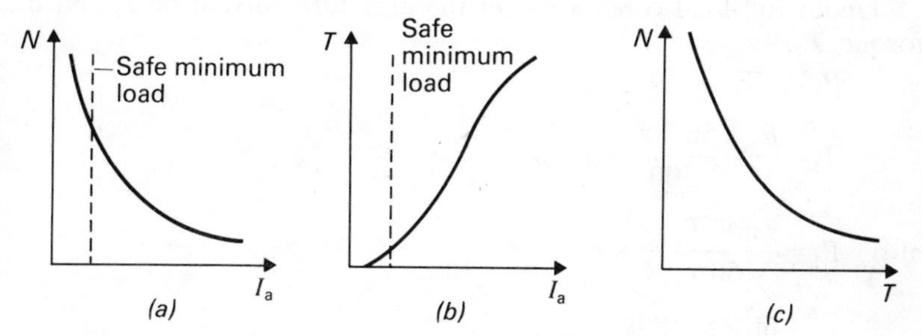

Fig. 41 Characteristics of a d.c. series motor: (a) speed/armature current; (b) torque/armature current; (c) speed/torque

Turning to the torque expression, we remember that

$$T \propto \Phi I_a$$

In the series machine, the flux is proportional to the armature current provided that saturation has not taken place. Thus if

$$\Phi \propto I_a$$

then $T \propto I_a^2$ (24)

As the flux increases, saturation of the magnetic circuit tends to make the flux constant, hence

$$T \propto I_a \tag{25}$$

The resulting torque/armature current characteristic is shown in Fig. 41(*b*). The initial section does not pass through the origin of the graph due to the effects of windage and friction causing a loss between the output torque and the total torque. The characteristic continues to rise parabolically under the influence of relation (24) and then tends to finally become linear with a slight droop due to armature reaction as was also experienced by the shunt motor.

Comparison between the speed/armature current and the torque/armature current characteristics results in the speed/torque characteristic shown in Fig. 41(*c*). This again illustrates the manner in which the speed rises to dangerous levels if the load becomes too small. This gives rise to the requirement that series motors are only used in applications where there is little possibility of the motor being operated without a load, e.g. driving a train or a lift.

Example 10 A 500-V, 50-kW series motor operates at 900 rev/min on full load. If the load is reduced until the output torque is 300 N m, calculate the armature current in the motor. It may be assumed that all losses may be neglected and that the magnetic circuit remains unsaturated.

Under full-load conditions, let the armature current be I_1 and the torque T_1.

$$P_1 = VI_1$$

$$I_1 = \frac{P_1}{V} = \frac{50\,000}{500} = 100\text{ A}$$

also $P_1 = \dfrac{2\pi N_1 T_1}{60}$

and $T_1 = \dfrac{60 \times 50\,000}{2\pi \times 900} = 530\text{ N m}$

Let the torque of 300 N m be T_2 and the corresponding armature current I_2. Since

$$T \propto \Phi I_a \propto I_a^2$$

it follows that $\dfrac{T_1}{T_2} = \dfrac{I_1^2}{I_2^2}$

and

$$I_2 = \sqrt{\left(I_1^2 \times \frac{T_2}{T_1}\right)} = \sqrt{\left(100^2 \times \frac{300}{530}\right)} = \underline{75{\cdot}2\text{ A}}$$

There are few uses remaining for the d.c. generator because most electrical energy is generated by a.c. machines. Nevertheless, d.c. generators are required to excite the a.c. generators, and they are used in certain forms of a.c./d.c. convertors to a diminishing extent. Perhaps the most important use numerically is that of the diesel-electric locomotive in which the diesel engine drives the d.c. generator which provides the direct current for the d.c. series motors.

In almost all other cases, rectifier arrangements are used to convert alternating current into direct current. As the semiconductor rectifiers also permit control of the rectified voltage, their ease of use has reduced the range of applications of the d.c. generator to almost a minimum.

At one time it was thought that a similar fate awaited the d.c. motor and that variable-frequency a.c. drives would replace d.c. drives in many applications. This expectation arose from the development of the thyristor, but instead the thyristor gave added applications to the d.c. motor so that the d.c. motor now leads the field in such applications as variable-speed drives and devices where severe torque variations are experienced. Generally the rectifier and control system comes as one unit and the following resulting applications are to be found:

1. In applications requiring high starting torque and variable speed which does not require accurate control, the series-excited motor is the most suitable. The most common applications are electric locomotives and trains, cranes, car engine starters, fans, etc.
2. In applications where the starting torque is low yet constant operating speed is desirable and perhaps accurately predictable, the shunt-excited motor is the most suitable. The most common applications include conveyor systems, especially those involved with strip mills and paper mills, specialised pumps where speed control is important, fans, blowers, lifts and production machines again where speed control is essential.

Neither form of motor is suitable for applications where the load can fluctuate, e.g. shears and presses. Such applications require motors that have a high starting torque and reasonably constant operating speed. This can be achieved by compound motors which have both series excitation and shunt excitation.

Problems

1. A shunt generator generates a direct voltage of 240 V when operating at 800 rev/min. If the field flux remains unchanged, what e.m.f. does it generate when operating at 1100 rev/min?

2. If the e.m.f. induced in a d.c. generator is 260 V and the armature circuit voltage drop due to $I_a R_a$ is 15 V, what is the terminal voltage of the generator?
3. A 150-kW, 500-V d.c. generator has a field circuit resistance 250 Ω. Calculate, for full-load conditions:
   (*a*) the current in the load;
   (*b*) the field current;
   (*c*) the armature current.
4. A separately excited d.c. generator supplies a load of resistance 6·0 Ω. The armature circuit resistance is 0·5 Ω and armature reaction is negligible. If the e.m.f. induced in the armature winding is 250 V, what is the armature current? Also determine the terminal voltage of the generator.
5. Under no-load conditions, the terminal voltage of a separately excited d.c. generator is 260 V. When the generator is loaded to supply 50 A to a load, the speed and the field current remain unchanged and the terminal voltage falls to 230 V. Given that the resistance of the armature winding is 0·4 Ω, how much of the terminal voltage drop is due to armature reaction?
6. A d.c. shunt generator has a terminal voltage of 250 V and supplies a current 35 A. The armature circuit resistance is 0·38 Ω and the field winding resistance is 125 Ω. The rotational losses of the machine are 325 W. Find:
   (*a*) the armature and field currents;
   (*b*) the I^2R losses in the machine;
   (*c*) the efficiency of the generator.
7. The armature resistance of a 200-V d.c. motor is 0·4 Ω and the no-load armature current is 2·0 A. When loaded and taking an armature current of 50·0 A, the rotor speed is 1200 rev/min. Find the no-load speed stating any assumptions made.
8. A 230-V d.c. motor is separately excited, the field current being 2·2 A. The machine is operated without load but with constant field current and with varying voltage applied to the armature. The corresponding values of armature voltage and current are as follows:

Armature voltage (V)	230	180	130	80
Armature current (A)	4·9	4·5	4·0	3·6

   The armature resistance is 0·15 Ω and the rotor speed is 1150 rev/min when the voltage 230 V is applied to the armature. Estimate the rotor speeds for the other values of applied voltage.
9. The armature resistance of a 250-V d.c. shunt motor is 0·7 Ω and the motor takes an armature current of 2·0 A when operating under no-load conditions. When the motor is loaded such that the

armature current is 60·0 A, the rotor speed is 1000 rev/min. Determine the approximate no-load speed.

10. A d.c. shunt machine has an armature resistance 0·05 Ω. When its brushes are connected to a 100-V supply and the rotor is driven by an engine at 100 rev/min, the armature current falls to zero. The machine is uncoupled from the engine and operated as a motor developing 7·5 kW output. Determine the speed at which the machine operates as a motor.
11. A 15-kW, d.c. shunt motor has a full-load efficiency of 0·88 when supplied from a 200-V source. The armature resistance of the motor is 0·06 Ω and its field resistance is 80 Ω. Determine the approximate change in speed between no-load and full-load conditions, expressing your answer as a percentage of the full-load speed.
12. The resistance of the field winding of a d.c. shunt-excited generator is 250 Ω. The generator is loaded to give an output of 100 kW at 500 V and the corresponding e.m.f. generated in the machine is 522 V. Determine:
    (*a*) the armature resistance;
    (*b*) the generated e.m.f. if the generator supplies 100 kW at 525 V;
    (*c*) the generated e.m.f. if the generator supplies 50 kW at 525 V.
13. A 30-kW, 250-V d.c. shunt generator has a field winding resistance of 125 Ω and its armature resistance is 0·15 Ω. A voltage drop of 2·0 V appears across the brushes. Determine the generated e.m.f. of the generator when operated at full-load.
14. The open-circuit characteristic of a d.c. generator made to rotate at 300 rev/min is as follows:

Field current (A)	0	2	3	4	5	6	7
Voltage (V)	7·5	93	135	165	186	202	215

    Plot the open-circuit characteristic if the rotor were to rotate at 375 rev/min and determine the voltage to which the machine would then excite if the resistance of the field winding is 40 Ω.
15. The magnetisation characteristic of a separately excited d.c. generator operating at 800 rev/min and excited from a 230-V d.c. supply is as follows:

Open-circuit voltage (V)	4	66	128	179	213	240	257	271	279	285	291
Field current (A)	0	0·2	0·4	0·6	0·8	1·0	1·2	1·4	1·6	1·8	2·0

    Determine the resistance of the field winding circuit required to give a no-load voltage of 275 V, the machine still being separately excited.

    What no-load voltage would be obtained if the same value of

field circuit resistance were used when the generator is connected for shunt excitation?

16. A 230-V, d.c. shunt motor has a field winding resistance of 230 Ω and the armature circuit resistance is 0·3 Ω. When operating as a motor on no-load, the supply current is 2·5 A. If the motor is loaded such that the supply current is 35·0 A, calculate the output power and the efficiency of the motor.

    If the motor is used as a 230-V shunt generator, find the input power and the efficiency of the machine when loaded to give a supply current of 35·0 A.

17. The d.c. motor of a crane can be made to act as a brake by generating electric power which is dissipated in a resistor bank. A d.c. series motor has a circuit resistance of 1·0 Ω and the load on it causes the rotor to have a speed of 600 rev/min and applies a torque of 250 N m. Calculate the approximate value of the resistance to be connected across the motor terminals, given that the magnetisation characteristic for the machine when operating at 600 rev/min is as follows:

E.M.F. (V)	383	390	397	403
Current (A)	42	43	44	45

18. A d.c. shunt motor is connected to a constant-voltage supply. When driving load *P*, it runs at 960 rev/min, takes an armature current of 20 A and has a generated e.m.f. of 480 V. When driving load *Q*, it runs at 940 rev/min and takes an armature current of 30 A. When driving load *S*, it takes an armature current of 45 A. Calculate:
    (*a*) the armature resistance;
    (*b*) the supply voltage;
    (*c*) the speed at which it drives load *S*;
    (*d*) the ratio of the gross power output for the three loads.

19. Explain why a resistor must normally be connected in series with the armature winding of a d.c. motor during the starting period.

    A 400-V, 30-kW shunt motor has a full-load efficiency of 0·83. The armature circuit resistance is 0·13 Ω and the field circuit resistance is 200 Ω. When the rated voltage is applied, the armature current at start is to be limited to 1·5 times the full-load value. Calculate:
    (*a*) the starting resistance required in the armature circuit;
    (*b*) the maximum power associated with the starting resistance.

    If the motor, at rest, were connected across the 400-V supply with the starting resistor short-circuited, what initial current would be taken by the armature circuit?

20. Sketch a typical speed/torque characteristic for each of the following d.c. motors:

(*a*) series-excited;
(*b*) shunt-excited.

Explain why the series motor is particularly suited to traction and crane applications.

A 500-V d.c. series motor takes a current of 15 A and develops a quarter full-load torque when operating at 1600 rev/min. The resistance of the armature, brushes and field coils is 1·0 Ω. Assuming the flux per pole to be proportional to the current, calculate the motor speed at half full-load torque.

21. A d.c. series motor operates from a 400-V supply. The armature and field windings have a total resistance of 0·3 Ω. With a certain load, the current is 30 A and the speed is 1000 rev/min. A series speed-control resistor of 2·7 Ω is now connected into the circuit and the current falls to 25 A whilst the flux per pole is 80 per cent of that at 30 A. For this condition, find:
(*a*) the speed;
(*b*) the power developed;
(*c*) the torque as a fraction of the original torque.
22. The field winding of a d.c. series motor consists of two identical portions that may be connected either in series or in parallel. When they are connected in series, the motor drives a given load at 800 rev/min and takes a current of 30 A from a 250-V d.c. supply. Calculate the current and the speed of the motor when the two parts of the field winding are connected in parallel, the load torque being proportional to the square of the speed. Neglect saturation and all losses.
23. A d.c. series motor, having armature and field resistances of 0·06 Ω and 0·04 Ω respectively, was tested by driving it at 2000 rev/min and measuring the open-circuit voltage across the armature terminals, the field being supplied from a separate source. One of the readings taken was: field current 350 A, armature voltage 1560 V. From this information, obtain a point on the speed/current characteristic, and one on the torque/current characteristic for normal operation at 750 V and at 350 A. Take the torque due to rotational loss as 50 N m. Assume that brush drop, and that field weakening due to armature reaction, can be neglected.
24. A d.c. shunt machine has an armature resistance of 0·5 Ω and a field circuit resistance of 750 Ω. When run under test as a motor, with no load, and with 500 V applied to the terminals, the line current was 3·0 A. Allowing for a voltage drop of 2·0 V at the brushes, estimate the efficiency of the machine when it operates as a generator with an output of 20 kW at 500 V, the field circuit resistance remaining unchanged.
25. A 250-V d.c. shunt motor has an armature resistance of 0·1 Ω and a field resistance of 250 Ω. With a certain load, the armature current is 18·0 A and the rotor speed is 800 rev/min. The load

torque is now doubled and the field is weakened by 10 per cent. Determine:
(*a*) the new armature current;
(*b*) the new speed.

Answers

1. 330 V
2. 245 V
3. 300 A, 2 A, 302 A
4. 38.5 A, 231 V
5. 10 V
6. 37 A, 2 A, 1020 W, 0·87
7. 1330 rev/min with constant field flux.
8. 900 rev/min, 650 rev/min, 400 rev/min
9. 1200 rev/min
10. 96·1 rev/min
11. 2·5 per cent
12. 0·11 Ω, 546 V, 536 V
13. 270 V
14. 263 V
15. 153 Ω, 287 V
16. 7130 W, 0·89, 9010 W, 0·89
17. 8·12 Ω
18. 1·0 Ω, 500 V, 910 rev/min, 1·0 : 1·47 : 2·13
19. 2·9 Ω, 50·5 kW, 3000 A
20. 1615 rev/min
21. 1040 rev/min, 8130 W, 0·66
22. 42·5 A, 1130 rev/min
23. 470 rev/min, 5030 W
24. 0·89
25. 40 A, 880 rev/min

Chapter 9

Three-phase induction motors

The 3-ph induction motor has a simple yet exceedingly robust construction which is comparatively cheap to manufacture. It also has good operating characteristics which make it a suitable drive for many production machines such as lathes or fans. These features are reflected by the fact that the induction motor is the most commonly used type of a.c. motor.

The motor is supplied from a 3-ph source and therefore it requires three supply conductors which are connected to three windings attached to the stator. The rotor also has a conductor system but this is not connected to the supply. Instead the current in the rotor conductors is induced by a transformer action from the stator windings. Because of this transformer action, the stator windings are sometimes termed the primary windings and the rotor windings may be termed the secondary windings. Thus the 3-ph induction motor is a sort of transformer with a rotating secondary winding.

There are two principal methods of connecting the rotor conductors together and this gives rise to two classes of induction motor. These are the cage-rotor induction motor and the wound-rotor induction motor. In either case, the principle of operation is essentially the same and this depends on the action of the stator windings which give rise to a rotating magnetic field.

1 The rotating magnetic field

The most simple form of 3-ph stator winding consists of three windings embedded in the inner surface of the cylindrical stator. This form of construction is indicated in Fig. 1.

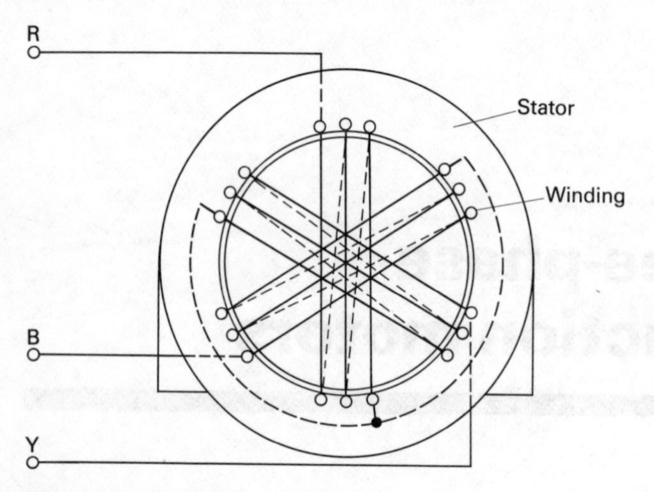

Fig. 1 Simple 3-ph stator construction

You will see that the windings are spaced at 120° intervals around the stator surface. Looking at the diagram, you may be tempted to think that there is only 60° between windings but when describing the angle, we consider the difference to be between the start of each winding, the start being the point at which the supply is connected. Using this description, we can now see that there is 120° between the external connection of the R-phase and that of the Y-phase, whilst there is a further 120° to the connection of the B-phase. Yet a further 120° remains between the B-phase and the R-phase, thus completing a revolution of the stator surface.

The windings are inserted into the stator surface in a manner similar to that already described in the case of the armature winding in the d.c. machine. For convenience, each phase winding is shown distinctly separate from the winding supplied by the next phase. This makes it easy to observe the different windings, but in practice the windings would be spread out to cover the entire surface of the stator.

Each winding has a distinct beginning and an end. Such windings are called phase windings, and are not to be confused with the commutator windings which formed closed loops tapped by means of commutators. The start terminals of the phase windings are connected to the supply lines. In the case shown in Fig. 1, the finish terminals are connected together; thus the windings are star connected, but there is no reason why the windings should not have been delta connected instead.

Each winding sets up a flux which acts along the axis of the winding. This action is shown in Fig. 2 and it shows the field arrangement for only one winding. By considering different instants during a cycle of alternating current flowing in the coil, we see that the flux always acts in the same axis but with varying magnitude and with alternating direction.

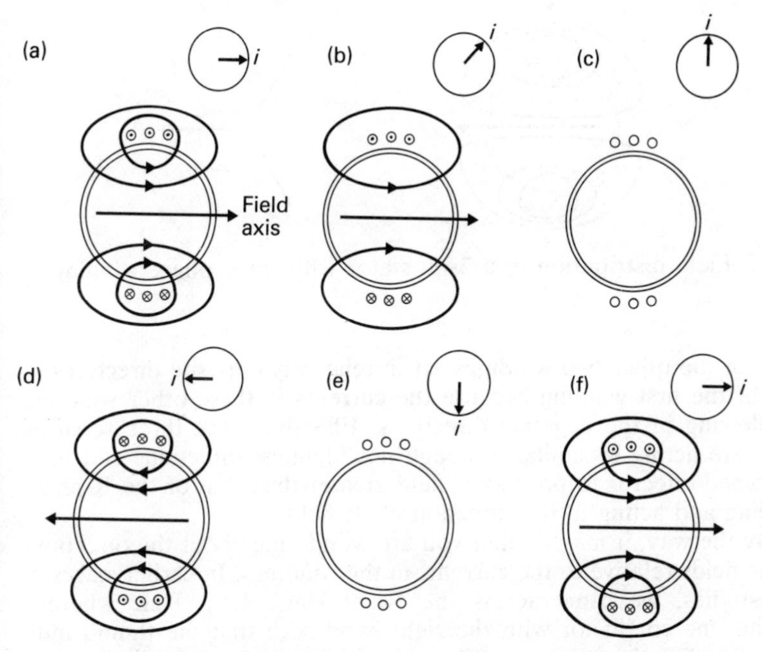

Fig. 2 The pulsating flux due to alternating current in a single coil

In the 3-ph system, there are three windings each producing a flux and each acting in a different direction within the stator. If we combine the three fluxes, we find that a similar magnetic field is set up within the stator, except that the magnitude of the total field is bigger than that associated with any one winding. Actually it is 50 per cent larger, and, for the arrangement shown in Fig. 1, we can see in Fig. 3 how the three fluxes add together to give the larger total flux. This diagram has been drawn for the instant at which the R-phase current is at its maximum value. It follows that the currents in the other two phases at the same instant are half the maximum value and flowing in the opposite direction. By applying our observations of Fig. 2, the R-phase winding produces its maximum flux and the axis of the field is shown by an arrow which also indicates the direction of the field. The

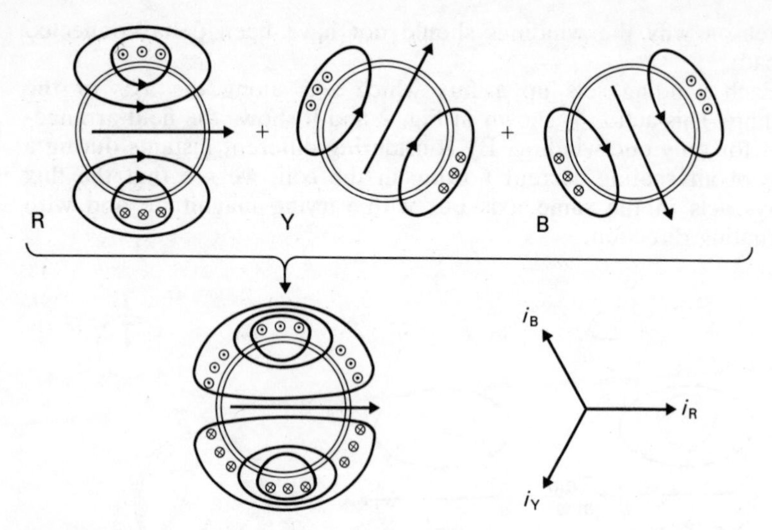

Fig. 3 Field distribution in a 3-ph stator with three phase windings

fields of the other two windings act in relatively opposite directions to that of the first winding because the currents in these other windings are flowing in the negative directions. Effectively the three separate fields are acting in similar although not identical directions and their combined effect is to produce a field greater than that of the R-phase winding and acting in the direction of its field.

By the way, it may be that you are wondering about the directions of the fields relative to the currents in the windings. In earlier stages of our studies, we came across the Right Hand Grip Rule whereby holding the conductor with the right hand such that the thumb indicates the direction of current flow also indicates the field direction by the fingers of that hand. If you wish to check the diagrams, you can use this rule to confirm the directions of the separate phase fields and hence the direction of the combined field.

If we evaluate the combination of winding fields, the result at first sight does not appear to be of much importance. After all, with one winding we could obtain a certain field, yet with three windings, we have only increased the resultant field by 50 per cent, which is not a particularly substantial increase.

The importance of the arrangement only becomes apparent when we consider subsequent instants. For instance, in Fig. 4, we can again look at the flux arrangement as considered in Fig. 3, but at an instant 30° later in the supply cycle. At this instant, the current in the R-phase has fallen to 0·87 of its maximum value, whilst that in the Y-phase is zero. The current in the B-phase has risen in value to 0·87 of its maximum value and it is again flowing in the negative direction. Figure

4 shows the separate phase winding fields and the resultant combined field.

In this case, we again produce a somewhat bigger resultant field than is produced by one winding, but the important observation is that the axis of the resultant field has shifted from that seen in Fig. 3. Actually we choose an instant 30° later in the supply cycle and we now see that the field axis has shifted by 30°. We have therefore made an important observation because we have seen a field set up by three windings fixed in space produce a magnetic field the axis of which has shifted relative to the windings. There is the possibility that this could be a freak condition so let us look at further subsequent instants in the period of the supply cycle.

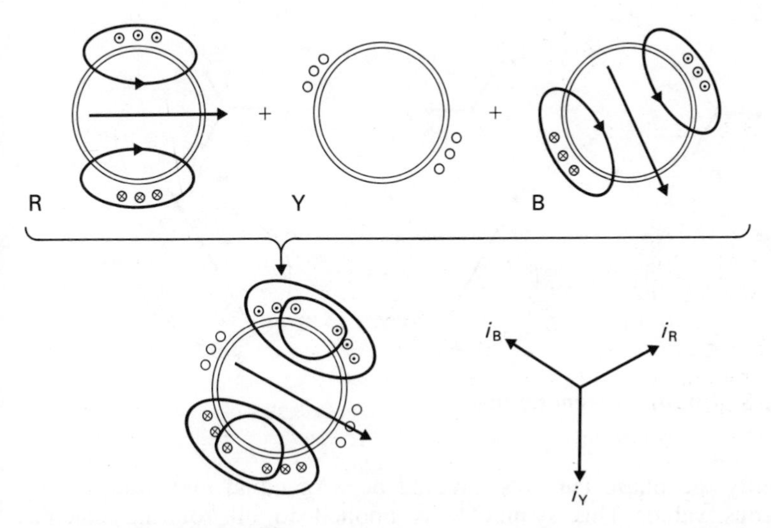

Fig. 4 Field distribution in a 3-ph stator with three phase windings at an instant later than that of Fig. 3.

By now, we are familiar with the production of the resultant field and it should be sufficient to draw the resultant field only for a selection of instants throughout the supply cycle. This is done in Fig. 5 and by taking instants throughout the complete cycle, we find that the resultant field rotates one complete rotation. Thus we may conclude that a system of windings fixed in space and excited from a three-phase a.c. supply can produce a rotating field.

In the production of the rotating magnetic field, one assumption has been made but not stated. It has been assumed that the construction of the 3-ph machine is symmetrical, i.e. the windings are identical although displaced at equal angles from one another. It has also been assumed that the supply is symmetrical and that subse-

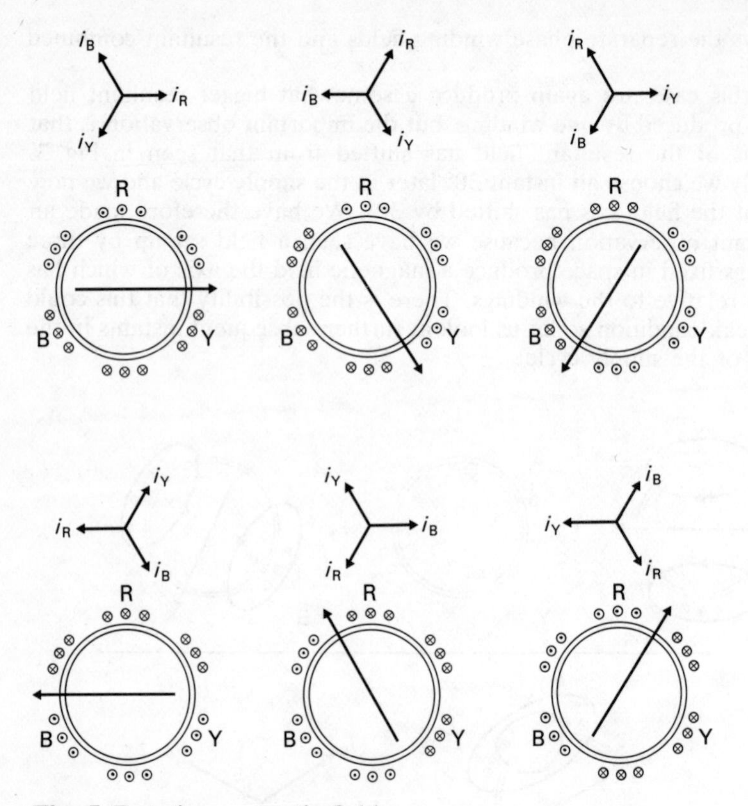

Fig. 5 Rotating magnetic field

quently the phase currents have all been of equal maximum instantaneous value. This symmetry is implicit to all rotating machine studies.

2 Speed of the rotating magnetic field

It has already been noted that spaces have been left between the conductors of one phase winding and the next. This failure to use all the surface space of the stator is wasteful and in practice conductors cover the entire stator surface area as indicated in Fig. 6(*a*). However, drawing the individual conductors is time consuming and it is usual to indicate the winding groups of conductors by shaded areas as shown in Fig. 6(*b*).

Each phase winding has been shown to set up its own field, which acts from one side of the stator across to the other. Seen from the rotor, it would seem that the field emanates from a N-pole at one side

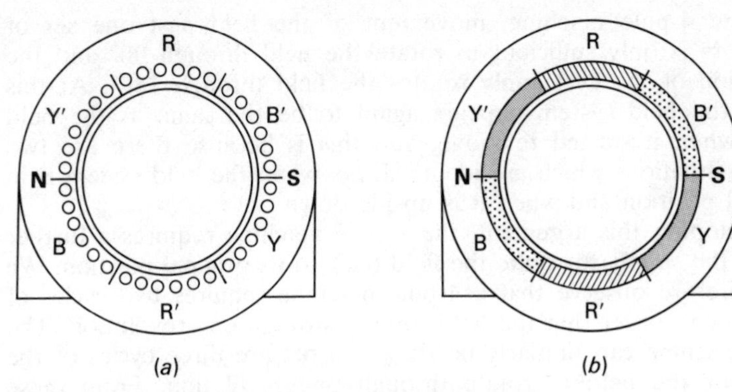

Fig. 6 3-ph, 2-pole winding arrangements

and terminates in a S-pole at the other. The same observation may be made of the resultant field, and for this reason, the winding arrangement is described as a 2-pole winding.

By using six windings instead of three, it is possible to have a 4-pole machine, while nine windings gives a 6-pole machine, and so on. Although unusual, it is quite practicable to have a machine with about a 100 poles. The winding arrangements for 4-pole and 6-pole machines are shown in Fig. 7.

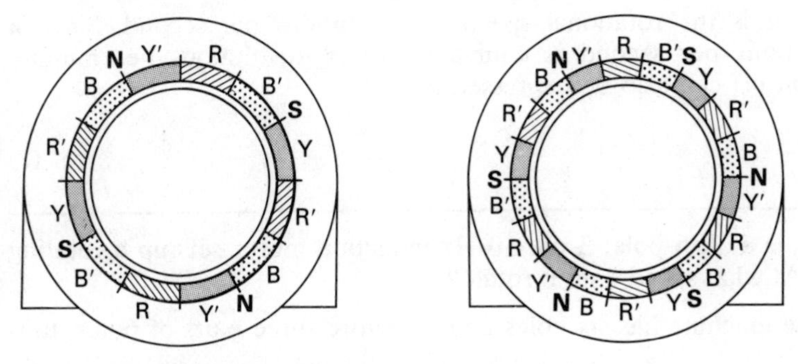

Fig. 7 4- and 6-pole, 3-ph stator windings

Let us now consider the effect of changing the number of poles on the rotation of the magnetic field. In Fig. 5, we saw that one cycle of the supply caused the field to rotate through one revolution. The field axis started half way between the R-phase conductors and after half a cycle had rotated past one set of R-phase conductors, i.e. one side of the R-phase winding, to reach a position again half way between the R-phase conductors yet acting in the opposite direction. A further half cycle completed the movement to reach the original relationship in space between the conductor positions and the field axis.

In the 4-pole machine, movement of the field past one set of conductors is only sufficient to rotate the field through 90° and the completion of the cycle only rotates the field through 180°. At this instant, the field system appears again to be the same as the field system when it started to move, and this is because there are two possible situations which are identical, i.e. when the field system is in its initial position and when it is upside down.

Developing this argument, the 4-pole machine requires a further cycle of the supply to rotate the field back to its original position. We may therefore observe that a 4-pole machine requires two cycles of the supply in order that the field rotates through one revolution. The 6-pole machine can similarly be shown to require three cycles of the supply for the field to rotate through one revolution. From these various cases considered, we can observe that the number of cycles of the supply required for one revolution of the magnetic field is always half the number of poles. However, poles in such machines always come in multiples of two, i.e. in pairs of poles. It follows that the number of pole pairs is equal to the number of cycles of the supply required for one complete rotation of the magnetic field.

If the number of pole pairs is p and the supply frequency is f hertz, then the number of revolution of the magnetic field per second is

$$n = \frac{f}{p} \tag{1}$$

where n is the rotational speed in revolutions per second. Since n revolutions per second is equivalent to N revolutions per minute, relation (1) can also be expressed as

$$N = \frac{60f}{p} \tag{2}$$

Example 1 A 6-pole, 3-ph, 50-Hz induction motor sets up a rotating field. At what speed does it rotate?

The machine has six poles and therefore three pairs of poles, thus $p = 3$.

$$n = \frac{f}{p} = \frac{50}{3} = \underline{16{\cdot}7 \text{ rev/s}}$$

$$\text{or} \quad N = \frac{60f}{p} = \frac{60 \times 50}{3} = \underline{1000 \text{ rev/min}}$$

The speed at which the rotating magnetic field revolves is termed the synchronous speed. In most machines the supply frequency and the number of poles are fixed, and it follows that the synchronous speed is constant for any given machine. However it is possible to produce variable-frequency supplies, and even to change the number of poles, in either of which cases the synchronous speed can be changed but these cases are relatively unusual.

Finally, if the supply connections to any two windings are reversed, then the phase sequence of the stator currents is reversed. The result of this change is to cause the field to rotate in the opposite direction as indicated in Fig. 8. As the common application of the 3-ph induction machine is to serve as a motor, and since the rotor turns in the same direction as the rotating flux, it follows that the direction of rotation of a motor can be reversed simply by interchanging any two of the three supply conductors.

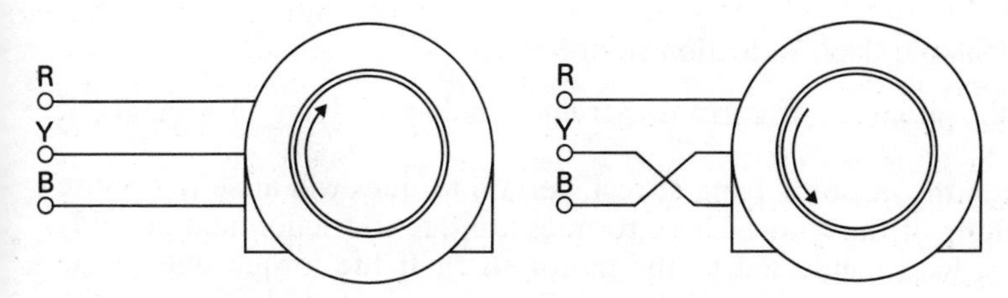

Fig. 8 Direction of rotation of the magnetic field

3 Principle of operation

It has already been mentioned that the currents in the rotor conductors are induced in a fashion similar to that of a transformer. This is quite unlike the d.c. machine, in which the rotor currents pass into and out from the armature winding through the commutator and the brushes. Nevertheless, the action of both machines is similar in that the rotor conductors carry currents while lying in a magnetic field and consequently experience a force tending to make them move at right angles to the magnetic field.

In order to simplify our introduction to the action of the induction motor, consider a machine with the usual 3-ph, 2-pole stator arrangement and with a simple winding mounted on the rotor as shown in Fig. 9. The rotor winding forms a closed loop and does not have terminals.

When the stator supply is switched on, the windings set up a rotating field which turns with synchronous speed. As this field moves past the conductors of the rotor winding, as illustrated in Fig. 10, it induces an e.m.f. in the rotor winding. This is similar to the induced e.m.f. in a transformer winding except that the change of flux linkage occurs due to the motion of the field rather than to the change of its magnitude. Since the rotor winding is a closed loop, the induced e.m.f. causes a current to flow in the rotor conductors, and we thus have

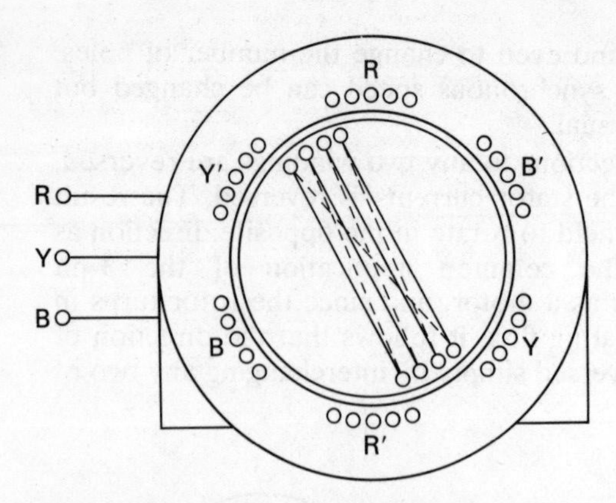

Fig. 9 Simple 3-ph induction motor

again the situation of current-carrying conductors lying in a magnetic field.

When the motor is lying at rest, certain torques will arise to oppose the turning of the rotor. These torques are due to friction and possibly due to a load connected to the motor shaft. If the torque due to the forces on the rotor conductors carrying current and lying in the magnetic field is sufficient to overcome these restraining torques, the rotor will accelerate from rest and commence to turn in the direction of the rotating field.

The driving torque depends on the relation $T = BlIr$, and since the length l of the rotor conductors and the radius r of rotation about the rotor axis are fixed, it follows that the torque is proportional to the flux density and the rotor current. As in the transformer, the flux density is also fixed due to the requirement that the magnetic field be sufficient to induce an e.m.f. in the stator windings equal and opposite to the applied voltage. It follows that the developed torque is essentially proportional to the rotor current.

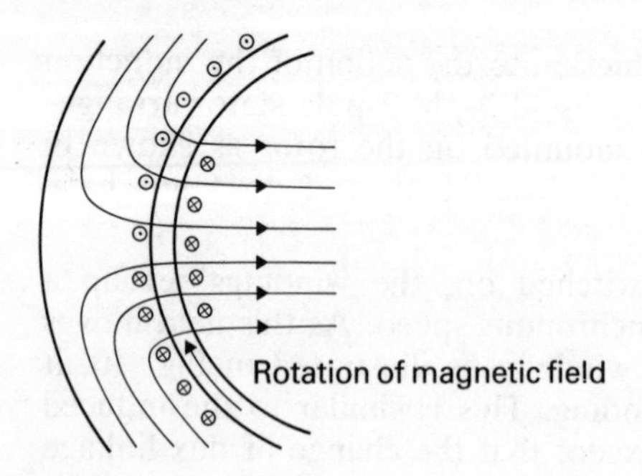

Fig. 10 Interaction between the rotating magnetic field and the rotor winding

The rotor current was induced as a result of the e.m.f. induced in the rotor winding, and the e.m.f. depends on the relation $e = Blu$. Again B and l are constants; thus the rotor e.m.f. depends on the speed with which the field passes the rotor winding. At first, the rotor winding is stationary, and in the 2-pole machine, the field sweeps past the winding with the frequency of the supply, say, 50 times per second. It follows that the rotor e.m.f. starts with the same frequency as the supply frequency.

However, as the rotor starts to rotate in the same direction as the rotating field, the frequency with which the field passes the rotor winding falls away. For instance, if the field were rotating at a synchronous speed of 50 rev/s, then when the rotor winding is at standstill, the frequency of the induced e.m.f. is 50 Hz because the field sweeps past the winding 50 times per second. If the rotor then accelerates to, say, 10 rev/s, then in any second the field rotates 50 times whilst the rotor winding rotates 10 times. As a result, the field only passes the winding 40 times in that second; hence the frequency of the induced e.m.f. is reduced to 40 Hz. Also, because the relative speed of the rotor conductors with respect to the rotating field has been reduced, the magnitude of the induced e.m.f. is reduced and this affects the current in the rotor winding, hence reducing the torque developed by the rotor.

If the rotor were to continue to accelerate until it also rotated at the same speed as the magnetic field, there would be no relative motion between the rotor winding and the magnetic field, with the result that there would be no induced e.m.f., and hence no rotor current and no torque developed by the rotor. If there were no torque the rotor would again slow down, in which case there would again be an induced e.m.f. and again a torque (albeit a small one) would be produced. It follows therefore that the rotor cannot accelerate up to synchronous speed but must always rotate rather more slowly.

The difference between the rotor speed and the synchronous speed is termed the slip speed. This difference is usually expressed as a fraction of the synchronous speed N_1. The rotor speed is represented by N_r and the fractional slip by s; hence

$$s = \frac{N_1 - N_r}{N_1} \tag{3}$$

Example 2 A 4-pole, 50-Hz, 3-ph induction motor operates with a rotor speed 1440 rev/min. Determine the slip for this condition.

$$N_1 = \frac{60f}{p} = \frac{60 \times 50}{2} = 1500 \text{ rev/min}$$

$$s = \frac{N_1 - N_r}{N_1} = \frac{1500 - 1440}{1500} = \underline{0{\cdot}04}$$

---

4 Frequency of the rotor current

For the machine considered in Example 2, there are four poles; thus when the stator field rotates once past a rotor conductor, the conductor experiences the field changing say from that coming from an N-pole to that of an S-pole and back to that from an N-pole, i.e. there is a complete cycle of change, and this is repeated twice because there are four poles. Thus the field of a 4-pole machine in rotating once past a conductor on the rotor will induce two cycles of e.m.f. in that conductor. Extending this argument, the field of a $2p$-pole machine in rotating once past a rotor conductor will induce p cycles of e.m.f. in that conductor.

Also for the machine considered in Example 2, the slip speed measured in revolutions per minute is

$$(N_1 - N_r) = 1500 - 1440 = 60 \text{ rev/min}$$

This is equivalent to 1 rev/s; thus the magnetic field passes a rotor conductor once every second. However, we have already seen that for this four-pole machine, one revolution of the field past a rotor conductor causes two cycles of e.m.f. to be induced; thus the frequency of the induced e.m.f. for a slip speed of 1 rev/s is 2 cycles per second which is 2 Hz.

Had the fractional slip been greater, then the frequency of the induced e.m.f. and hence of the induced current would have been greater. For instance, if the slip had been 0·12, then the slip speed would be

$$N_1 - N_r = sN_1 = 0{\cdot}12 \times 1500 = 180 \text{ rev/min}$$

This slip speed is equivalent to 3 rev/s and the magnetic field would pass a rotor conductor 3 times per second inducing 6 cycles of e.m.f. thereby. The frequency of the induced e.m.f. and of the induced current would therefore be 6 Hz.

In more general terms, the slip speed in revolutions per second is sn_1; let the rotor conductor frequency be f_2. It follows that

$$f_2 = sn_1 p$$

but $n_1 = \dfrac{f}{p}$

where f is the supply frequency, so

$$f_2 = s\frac{f}{p}p$$

$$= sf \qquad (4)$$

The frequency of the rotor e.m.f. and the rotor current in any conductor is therefore directly proportional to the slip. This observation becomes important when we give further consideration to the current set up in the rotor winding due to the induced e.m.f. When discussed earlier, we noted that the induced e.m.f. was reduced as the slip reduced and that this would have an effect on the current. However, being an alternating current, the ratio of the induced e.m.f. to the current is given by the impedance, and this reminds us that the ratio therefore depends not only on the resistance of the winding but also on the inductive reactance. However, the inductive reactance depends on the frequency; thus as the slip reduces, so the inductive reactance also reduces.

When the induction motor is starting, the inductive reactance is high because the slip is high, hence the rotor winding is predominantly inductive and the current lags the induced e.m.f. by almost 90°. When the motor has run up to its normal operating speed, which is near to the synchronous speed and hence the slip is small, the rotor circuit is predominantly resistive and the rotor current is not only almost in phase with the induced e.m.f. but also relatively large in view of the small induced e.m.f. When the slip value tends towards zero, the inductive reactance tends to be negligible and the current is almost proportional to the induced e.m.f., being limited almost entirely by the resistance. The large currents in the rotor at low values of slip give rise to large values of torque, larger than those possible when the slip is high, e.g. $s = 1$ at standstill, and when the current has lower values being limited by the inductive reactance of the rotor winding. The effects of these observations become apparent when we consider the operating characteristics of the motor.

However, before leaving the effects of the rotor frequency, it is as well to mention that the current lagging the induced e.m.f. has an unexpected effect in that those conductors lying at the edge of the field find that the lagging current exists in conductors lying in the part of the field acting in the opposite direction. The result is that some winding conductors produce torque in a direction opposed to the direction of motion, but these conductors are in the minority. The driving torque therefore arises from the majority of the conductors pushing in one direction whilst the minority push in the other. This effect is only important because it makes winding conductors vibrate as they are pushed first in one direction then in the other; hence they must be particularly well made to resist this treatment.

5 Torque/speed characteristics

Earlier, we noted that the entire surface of the stator would be covered in conductors, in order to optimise the operation of the motor. In the same way, the entire surface of the rotor is also covered in conductors. You will also recall that the conductors formed a closed winding and it is therefore possible to make the rotor winding of heavy conductors placed around the rotor surface and short-circuited at each end, as shown in Fig. 11. Only the conductor arrangement is shown in Fig. 11, and the ferromagnetic material has been omitted. The system looks like a cage and for that reason, the arrangement is termed a cage rotor (formerly termed a squirrel-cage rotor).

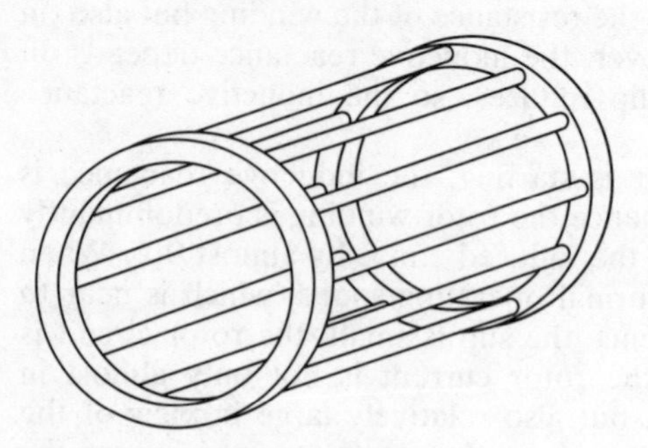

Fig. 11 Conductor system of a cage rotor

The torque/speed characteristic for a three phase cage-rotor induction motor is shown in Fig. 12. Due to the relatively small rotor currents, the torque at starting is not particularly good, but as the rotor currents increase, the torque builds up to a peak, after which the decreasing rotor e.m.f. reduces the rotor currents, and hence the torque falls away comparatively quickly until there is no torque when synchronous speed is reached. This sharp fall in the characteristic ensures that the motor operates at about the same speed over quite a large range of load torques, although we can see that the speed will drop slightly with increase of load.

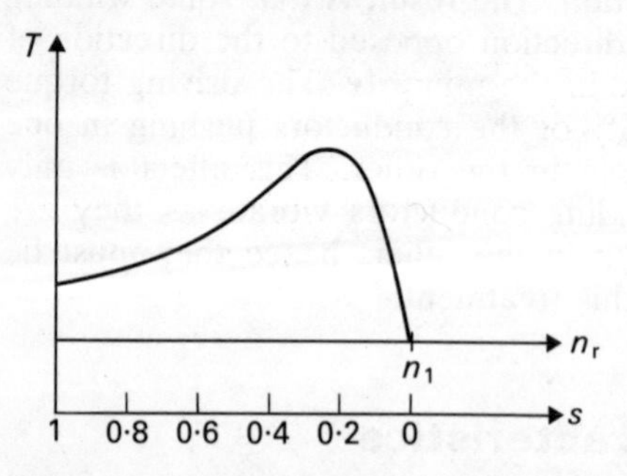

Fig. 12 Torque/speed characteristic of a 3-ph cage-rotor induction motor

The cage rotor is cheap to manufacture and therefore is the most popular form of the motor. However, the starting torque is relatively poor and we can see from Fig. 12 that once the motor has started, much higher values of load torque can be overcome. Thus we may attempt to find applications in which the motor is started and then the load is applied, e.g. driving a lathe, or find applications where the load is proportional to the speed of rotation, e.g. driving a fan. Once the motor is running, however, additional load can be applied up to the point at which the motor develops its maximum torque. The application of further load would cause the motor to stall; hence that value of torque is termed the pull-out torque.

In order to overcome the problem of poor starting, the winding on the rotor can be made in a similar manner to the winding on the stator, the windings being connected as shown in Fig. 13. The ends of the windings are brought out through slip rings and connected to a bank of variable resistors. In this way, it is possible to increase the rotor circuit resistance, which improves the torque developed by the rotor. However, having extra resistance in the rotor circuit leads to I^2R losses; thus when the motor speed has risen to a value tending towards synchronous speed, the resistance bank is short circuited and the motor then performs in a manner similar to that of the cage-rotor machine.

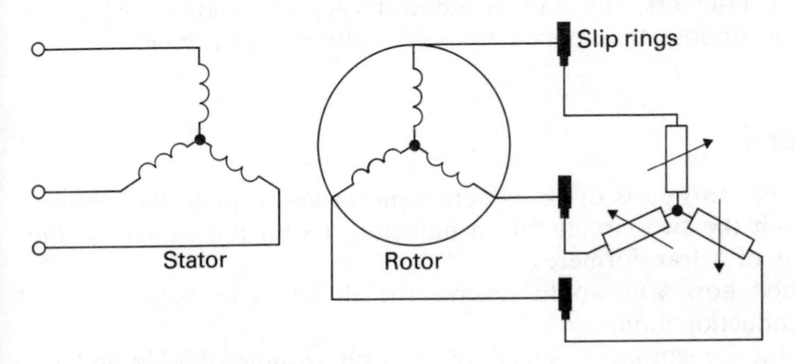

Fig. 13 Wound-rotor induction motor

Because the windings on the rotor have to be wound, the arrangement using slip rings is called a wound rotor, hence the motor is termed a 3-ph wound-rotor induction motor. It is more expensive to manufacture; thus this form of motor is only used in applications where high starting torque is required, or in applications in which speed control is required.

The torque/speed characteristics of a wound-rotor induction motor are shown in Fig. 14 and we can see that effect of increasing the circuit resistance is to slow down the rotor speed. The change of speed is quite small; thus a very fine control of the speed can be effected by the application of the wound-rotor induction motor.

Finally, if we again consider the wound rotor, it has a 3-ph winding into which is induced a system of 3-ph alternating currents.

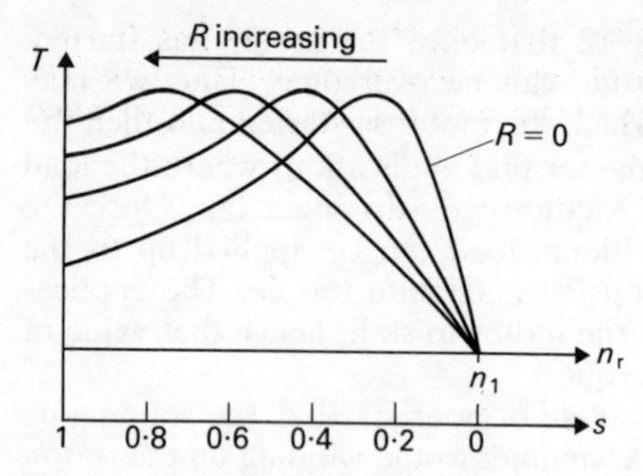

Fig. 14 Torque/speed characteristics of wound-rotor induction motor

It follows that the rotor produces a rotating field. This rotor field rotates in space with the same speed as the synchronous speed of the stator field, and the slip is a measure of the degree to which the rotor is being left behind by its rotating field. This applies to both the wound-rotor and to the cage-rotor machines. Both the stator and the rotor fields are sinusoidally distributed about the air gap so that as they pass the conductors, the e.m.f.s induced vary sinusoidally, which is required to oppose the applied sinusoidal alternating voltages.

Problems

1. Name the two parts of an induction motor and explain the manner in which the two circuits of an induction motor are similar to the circuits of a transformer.
2. Describe how you would reverse the direction of rotation of a 3-ph induction motor.
3. Find the synchronous speed of a 3-ph, 8-pole, 50-Hz induction machine.
4. A cage-rotor induction motor operates at full load with a speed of 1400 rev/min. Given that the motor has four poles and that the slip speed is 100 rev/min, determine the supply frequency.
5. A 3-ph, 6-pole, 60-Hz induction motor has a full-load slip of 0·04. Find the rotor speed under this condition.
6. An 8-pole, 50-Hz cage-rotor induction motor is loaded to operate at 740 rev/min. What is the rotor frequency?
7. Explain, with the aid of suitable sketches, why the rotor of a three-phase induction motor rotates in the same direction as the rotating magnetic field set up by the stator windings. Why is the rotor speed less than the speed of the rotating field?

   A 6-pole, 3-ph, 415-V, 50-Hz induction motor develops 30 kW at 920 rev/min. Calculate the slip of the rotor.
8. Write illustrated notes describing how a 3-ph system of voltages

may produce a rotating field in an electric machine. Give diagrams of the cross-section of such a machine, clearly indicating the direction of rotation of the field relative to the windings. Define 'synchronous speed' and 'slip'.

Answers

3. 750 rev/min
4. 50 Hz
5. 1152 rev/min
6. 0·67 Hz
7. 0·08

Chapter 10

Three-phase synchronous machines

If the importance of an electrical machine were to be judged by the number manufactured, the 3-ph synchronous machine would be so poorly quoted that it could easily be overlooked. Yet if by some magic stroke all 3-ph synchronous machines could be made to disappear, then virtually all supplies of electricity would cease, because most electrical energy is generated by 3-ph synchronous generators.

Although most engineers never have direct experience of such machines, we still require an understanding of their principles of operation. It is this understanding which helps us to appreciate the electricity supply we receive under both normal and fault conditions, and these affect most engineers.

1 Principle of operation

The 3-ph synchronous machine has a stator constructed in a manner similar to that of the induction machine, but instead of having the currents induced into the rotor conductors, the synchronous machine depends on an external supply of direct current to the rotor winding. A simple form of this arrangement is shown in Fig. 1, although the source of direct current has been omitted.

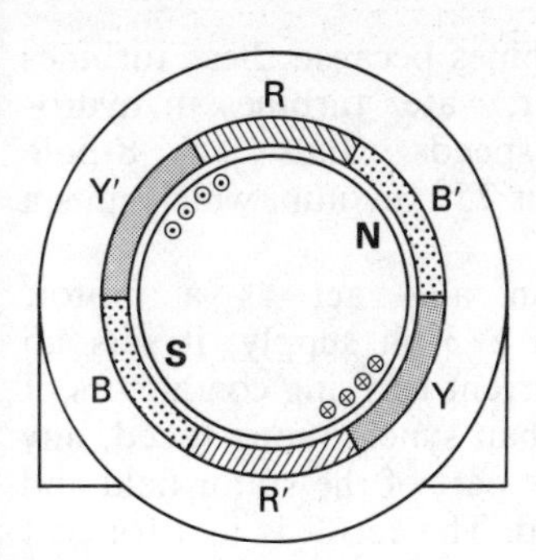

Fig. 1 Simple 3-ph synchronous machine

Let us consider first of all the action of the machine as a generator. The rotor winding is excited by direct current; thus a magnetic field is set up around the rotor. This field is constant in magnitude and its position is fixed relative to the rotor. If we drive the rotor at a steady speed, the rotor field is made to rotate at the same steady speed and it sweeps past the stator windings.

This motion of the field past the stator windings causes e.m.f.s to be induced in them. Since the windings are placed at equal intervals around the stator, this causes a 3-ph system of e.m.f.s to be generated. Provided that the distribution of the rotor field around the surface of the rotor is sinusoidal, the induced e.m.f.s are sinusoidal, which is invariably the case in practical machines.

For the machine shown in Fig. 1, each time the rotor field is rotated, it induces one cycle of alternating voltage in each of the stator windings. The field is therefore synchronised to the induced e.m.f., but since the rotor moves at the same speed as its field, it follows that since the rotor speed is synchronised to the induced e.m.f.s, the rotor speed is also the synchronous speed of the machine. It is because the rotor cannot slip relative to the synchronous speed that the machine is called a synchronous machine.

As with the induction machine, the synchronous machine can have more than two poles. The frequency of the induced e.m.f.s is given by the relation

$$f = pn_1 \tag{1}$$

where again n_1 is the synchronous speed of the machine given in revolutions per second.

When operating as a generator, the frequency of the induced e.m.f.s is therefore determined by the speed of the rotor. In the United Kingdom, almost all systems operate at 50 Hz; thus for a 2-pole generator, the rotor speed is also 50 rev/s or 3000 rev/min. This is the fastest possible operating speed for a 50-Hz system, which is awkward for generators driven by steam or gas turbines because these turbines operate better at higher speeds. However, water turbines in hydro-electric systems operate at lower speeds, thus an 8-pole generator has a rotor speed of 12·5 rev/s or 750 rev/min, which suits a water turbine.

The 3-ph synchronous machine can also act as a motor, because when the stator is excited from a 3-ph supply, it sets up a rotating field which interacts with the current-carrying conductors of the rotor. However, at any speed other than synchronous speed, any rotor conductor spends half its time in one part of the stator field and the other half in the other part of the field. The result is that for half the time it experiences a force in one direction and for the other half it experiences a force in the opposite direction. The overall result is that the average force developed is zero. However, if the rotor rotates at exactly the synchronous speed, then any rotor conductor always remains in the same position relative to the rotating stator field and therefore the force is consistently in the one direction. By rotating in the same direction as the rotating stator field, the rotor can therefore experience a steady torque and hence the machine can operate as a motor.

This principle of operation indicates the importance that the motor action is possible only at the synchronous speed of the machine. The motor is therefore unsuitable for operating with variable speed, but even more important is the difficulty of starting the motor, since it cannot develop any useful torque until synchronous speed is attained. However, it is possible to drive the rotor up to synchronous speed, e.g. by letting it run initially as an induction motor. Once the motor has been brought up to the synchronous speed, it is synchronised to the supply, i.e. the d.c. excitation is applied to the rotor and it then locks in to the stator field, thus ensuring its further operation at the synchronous speed. The load torque can then be applied.

With such a complicated method of starting, clearly the motor is not likely to find applications which require much stopping and starting. However, the advantage of the synchronous motor is that it can operate at a leading power factor, which means that not only can we obtain electrical energy to drive the motor at the most attractive cost from the supply authority, but also it can help to improve the overall power factor of the installation.

The power factor at which the motor operates depends on the rotor excitation. This observation also applies to the generator. Most synchronous generators are connected to the National Grid; thus increase of excitation of one machine is insufficient to increase the operating voltage of the entire Grid. Instead, the variation of the excitation of the machine varies the power factor of the generated power and this can be either leading or lagging. Because most loads operate at a lagging power factor, it is generally necessary that the generators also operate at a lagging power factor.

The stator construction of a synchronous machine is similar to that of the induction machine, although in generators driven by steam or gas turbines the stator is especially long, in order to increase the induced e.m.f., since these machines can develop up to 33 kV.

There are two forms of rotor construction—round rotor and salient pole rotor. A simplified cross-section of a round rotor is shown in Fig. 2. In order to obtain the sinusoidal distribution of the developed flux, the conductors are set into the surface at irregular intervals so that the field distribution is sinusoidal. Most round rotors are made for operation at high speeds, e.g. 3000 rev/min, and the cylindrical formation is particularly suited, since it is structurally very strong and capable of withstanding the high centrifugal forces. Even allowing for this, however, the rotor diameter cannot be too large and therefore the rotor surface is relatively limited, with the result that usually round rotors appear only in the 2-pole form shown in Fig. 2.

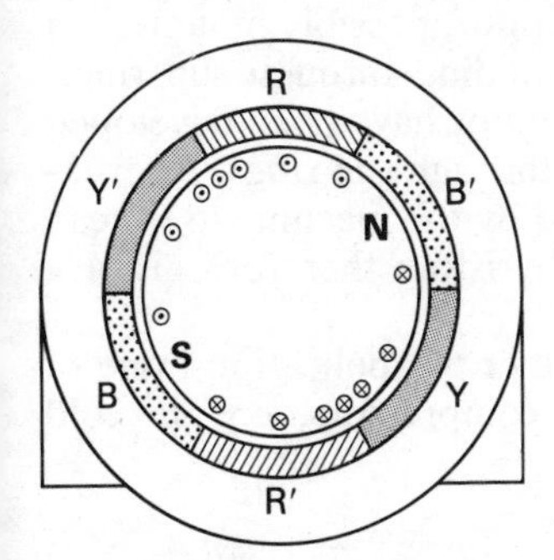

Fig. 2 Round-rotor synchronous machine

In cases where slower operating speeds are encountered, the salient pole arrangement shown in Fig. 3 is the more suitable, especially as it permits higher operating flux densities. The distribution of the rotor field is achieved by shaping the pole shoes. Mechanically, this arrangement is not so robust as the round rotor, and therefore it is associated with multi-pole machines operating at quite low speeds.

The round-rotor machines usually operate as generators driven by high-speed devices such as steam turbines. These machines produce most of the electrical energy available throughout the world, the steam being the result of burning coal or oil or of nuclear reactors. They can be rated as high as 1100 MW.

The salient-pole machines either operate as water driven generators or as motors. As generators, they are part of the many hydro-electric generating systems throughout the world. Unlike the round-rotor generators, they can often be found rotating vertically

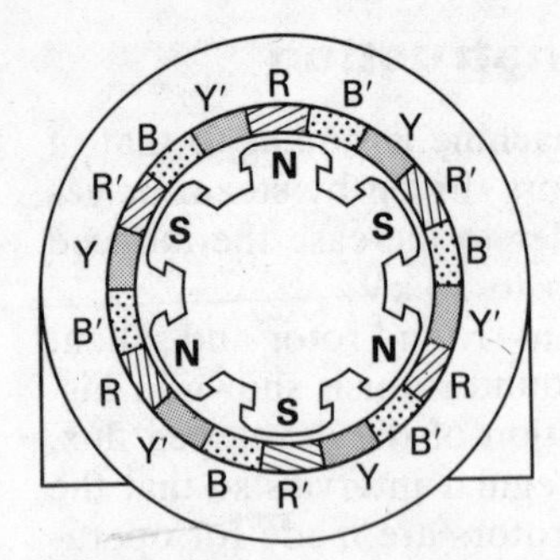

Fig. 3 Salient-pole synchronous machine

instead of horizontally. In some hydroelectric schemes, the machines operate both as generators and as motors, pumping water to a higher level so that it is available to generate electricity at a more suitable subsequent time.

In either form of machine, the rotor winding is supplied from an external d.c. supply. This may be a d.c. generator, possibly mounted on the same shaft, which supplies the rotor winding through slip rings, although some specialised forms of d.c. generator have been developed that do not require the external connection through slip rings. Alternatively, an a.c. supply, possibly that generated by the machine itself, can be rectified and controlled by means of thyristors therefore given a source of variable direct current.

Synchronous motors have a limited number of applications, one of the most common being to drive rotary compressors, e.g. in cold storage plants.

Problems

1. How is the rotor field of a 3-ph synchronous generator excited? Upon what factors do the generated e.m.f.s depend, and in normal operation which of these factors is variable?
2. A synchronous generator operates at 375 rev/min and the frequency of the output supply is 50 Hz. How many poles does the generator have?
3. What is the operating speed of a 2-pole, 50-Hz synchronous generator?
4. Describe, with the aid of sketches, the construction of the magnetic circuit of a typical 3000-V, 50-Hz, 3-ph synchronous generator designed to operate at 3000 rev/min. Comment on the reasons which led to the evolution of such a machine.
5. What is the active power (in kilowatts) rating of a 3-ph, 1-MVA synchronous generator operating at
   (*a*) unity power factor;

(*b*) 0·8 power factor leading;
(*c*) 0·5 power factor lagging?

6. A 3-ph, 415-V, 1-MVA synchronous generator supplies its full rated load at a power factor of 0·75 lagging. The efficiency of the generator is 0·9. What is the input power to the generator?

Answers

2. 16 poles
3. 3000 rev/min
5. 1000 kW, 800 kW, 500 kW
6. 830 kW

Chapter 11

Thyristors

The return to popularity of the d.c. machine was ascribed to the introduction of the thyristor. This important device requires some explanation as to what it is and how it operates, although it can be simply described as a switch which has such high speeds of operation that it can be made to switch a circuit on at any preselected instant throughout a cycle of alternating current. However, it also has the disadvantage that it can only conduct in one direction; thus for a.c. operation two thyristors are required, one for each direction of conduction. In order to evaluate these properties, we must first of all understand the construction and principles of operation of the thyristor, which used to be known as a silicon-controlled rectifier or s.c.r.

1 Construction of a thyristor

The thyristor is somewhat similar to a transistor except that it has four layers of doped semiconductor material instead of three. The device is therefore of the form p-n-p-n as indicated in Fig. 1 and three connections are made as shown, being the anode, the cathode and the gate.

The thyristor appears to be similar to three diodes connected in series so that two are forward biased and the middle diode is reverse

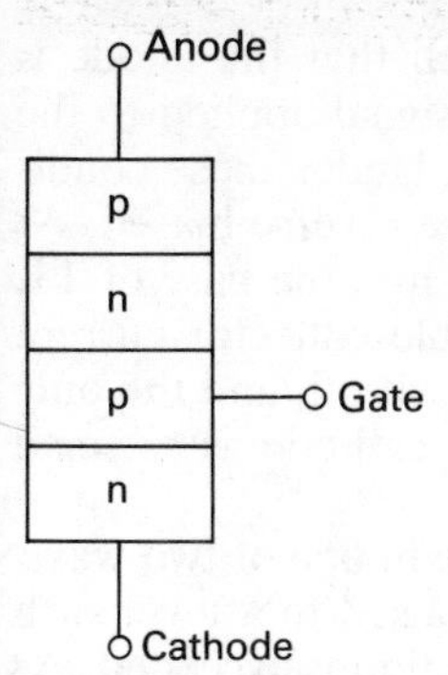

Fig. 1 Construction of a thyristor

biased. As such, the arrangement does not readily pass current in either direction. In order to pass current from the anode to the cathode, it is necessary to overcome the reverse-biased diode, in which case conduction takes place as though the overall device were a forward-biased rectifier.

However, the action of the thyristor is more definite than this simple explanation would make it appear. When a diode is made to conduct when reverse-biased, it still requires a considerable voltage drop to maintain the breakdown current; yet the thyristor requires very little voltage across it in order to maintain the flow of current from the anode to the cathode. Let us therefore look at the thyristor action in greater detail.

2 Principle of thyristor operation

The thyristor can also be thought of as being equivalent to two transistors connected as shown in Fig. 2. In this case, the collector current of the p-n-p transistor T1 passes into the base of the n-p-n transistor T2, and conversely the collector current of transistor T2 is the base current of transistor T1.

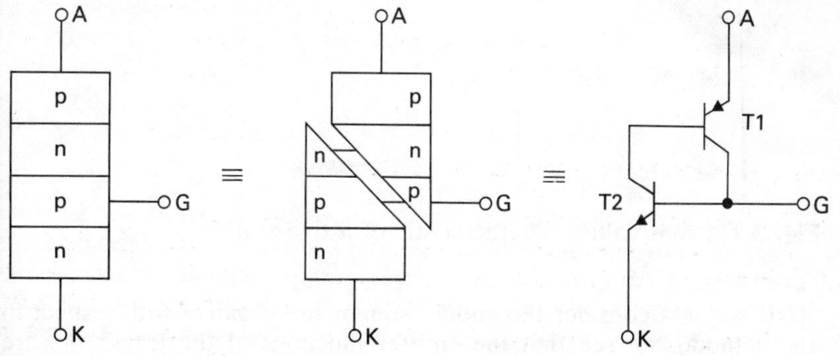

Fig. 2 Transistor equivalent circuit of a thyristor

When a voltage is applied to the thyristor such that the anode is positive with respect to the cathode, and with no signal applied to the gate, the thyristor acts like an open-circuit switch. Under these conditions, the collection junctions of both transistors are reverse biased. As a result, T2 is cut off and passes negligible current into the base of T1, which is therefore also cut off and passes negligible collector current into the base of T2. Both transistors are therefore cut off, and the only current which can flow from the anode to the cathode is a small leakage current.

Current can be made to flow through the device in one of two ways. First of all, the anode to cathode voltage can be raised to a level such that the breakdown voltage of either collector junction is exceeded. As soon as this happens, current passes through that transistor into the base of the other transistor, thereby switching it on. However this in turn causes the other transistor to pass current to the base of the first; thus the action is regenerative, each transistor now holding the other in the on condition. When this action has occurred, the thyristor then continues to conduct so long as the anode remains slightly positive with respect to the cathode. It is this regenerative action which requires little voltage across the device to ensure the continuation of conduction, and the characteristic of the thyristor is shown in Fig. 3.

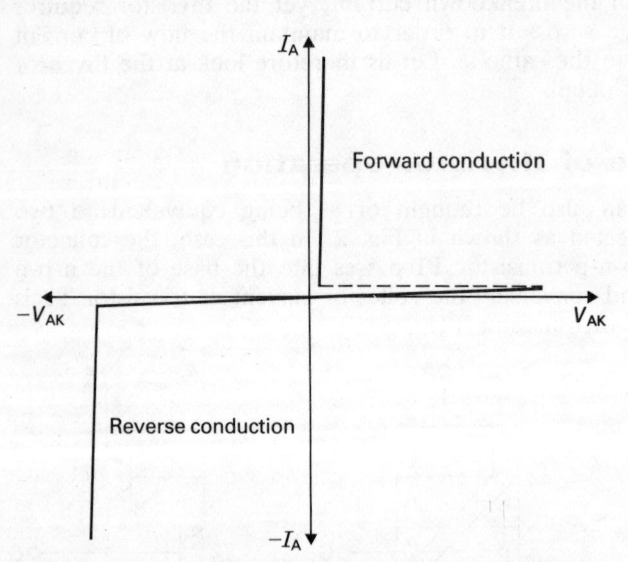

Fig. 3 Current/voltage characteristic of a thyristor

If we next consider the anode being made negative with respect to the cathode, we see that the emitter junctions of the transistors are reverse biased and again the thyristor blocks the passage of current apart from some leakage current. However, if a sufficient reverse voltage is applied, then the thyristor breaks down and passes a considerable current due entirely to the applied voltage. This quickly causes the thyristor to overheat and readily results in the complete failure of the device.

The basic thyristor is therefore a device that can pass current readily only when a substantial voltage is applied; yet when conduction is initiated in one direction, it will then continue quite readily with only a small voltage drop, whilst if conduction is initiated in the reverse direction, it will quickly lead to the thyristor being burnt out.

However, if a current is supplied to the gate when the thyristor anode is positive with respect to the cathode, then the thyristor can be triggered into the conducting state without requiring the high anode-to-cathode voltage to initiate conduction. When the gate current is supplied to the thyristor, regeneration quickly turns the thyristor on and the maximum circuit current flows between the anode and the cathode. As is the case with an ordinary silicon diode, there is a small voltage drop of one or two volts across the thyristor when the current is flowing.

The gate current arises when the gate is made positive with respect to the cathode. Provided that the gate current is sufficiently large, T2 is biased to the on condition and its collector current flows through the base of T1. This base current is amplified by T1 and is fed back to the base of T2, where it is again amplified and fed to the base of T1. This regenerative action builds up and rapidly both transistors are saturated. The time required for this switch-on action is a few microseconds.

Once the thyristor has reached this regenerative condition and current is freely passing from anode to cathode, the gate current no longer serves any purpose and can be discontinued. More important, it follows that only a brief pulse of positive current need be injected into the gate in order to switch on the thyristor.

In case this self-sustaining condition is not self-evident, you should observe that the T2 base drive required to sustain the operation in the saturated state is provided from the T1 collector; thus the condition remains in the on condition even when the gate drive is removed. The gate connection has a certain resistance in it and this ensures that the base current continues to flow in T2 even if the gate and cathode terminals are short-circuited or even reverse biased. Thus not only does the gate no longer require to supply current, but we now find that the gate has lost all control once the thyristor has switched on, i.e. the gate cannot be used to switch the thyristor off again.

The thyristor can only be switched off by the interruption of the supply to the anode, either by the reduction of the anode current to zero, which is normally the case, or by at least reducing the anode current to a value known as the minimum holding current. This latter value lies on the forward conduction characteristic shown in Fig. 3 and corresponds to the voltage across the thyristor being less than that required for forward conduction under switch-on conditions. Generally it is difficult to recognise the difference between conduction ceasing at the minimum holding value and at zero because the anode current is being interrupted anyway. For example, if the thyristor is connected to an a.c. supply, then the current ceases to flow when the current waveform passes through zero. Essentially the current ceases to flow when the minimum holding value is reached, but this is almost at the same instant that the current would be zero anyway, and hence the thyristor action would be interrupted.

Generally thyristors are either switched off because they operate from an a.c. supply, in which case the current is interrupted every half cycle, or because they are effectively short-circuited by an uncharged capacitor. In the first instance, the thyristor is switched off every half cycle and therefore it is necessary to switch it on again every time that it can conduct in the appropriate direction. However, this cannot be achieved with a d.c. supply; thus the connection of a large uncharged capacitance across the thyristor temporarily reduces the voltage across it and the current falls below the minimum holding value. In practice it is better to use a charged capacitor and to discharge it through the thyristor circuit by applying it in opposition to the thyristor, thus ensuring the complete interruption of charge flow through the thyristor.

3 Some thyristor circuits

A simple switch-on arrangement incorporating a thyristor is shown in Fig. 4. In this case, when the switch S1 is closed, the thyristor immediately blocks the passage of current through the lamp. When switch S2 is temporarily closed, some milliamperes of current flow into the gate and switch on the thyristor. Although in this instance the thyristor current is likely to be only a few amperes at most, this arrangement could be adapted to switch on currents in excess of 1000 A, yet the gate current could even be less than a milliampere. Once the thyristor is switched on, it is no longer necessary to keep switch S2 closed.

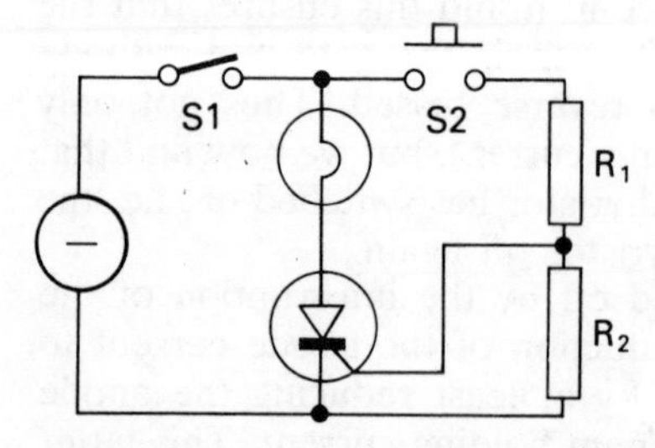

Fig. 4 Simple d.c. thyristor switch-on circuit

In order to switch off the lamp in Fig. 4, we could open switch S1, but this raises the question – is switch S1 not doing the same work as the thyristor and have we not duplicated our switching? Really we should be looking to S1 being replaced by the thyristor, and this can be achieved if we modify the circuit to that shown in Fig. 5. The closure of switch S3 connects the capacitor so that it tries to discharge in opposition to the passage of current through the thyristor with the result that for an instant there is no current passing through the thyristor. This short interruption is sufficient to stop the thyristor conducting and it therefore returns to the off condition.

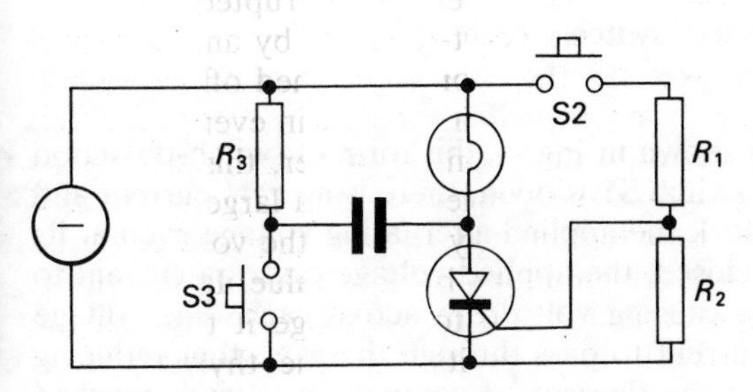

Fig. 5 Simple d.c. thyristor on–off circuit

When operating from an a.c. supply, the thyristor does not always receive sufficient gate current to switch it on. Instead we find that the forward conduction characteristic changes as the gate current is increased; thus if we have insufficient gate current to switch on the thyristor, it is nevertheless noticeable from the characteristic shown in Fig. 6 that less anode–cathode voltage is required to raise the leakage current to a sufficient level that the avalanche breakdown takes place and the thyristor switches on of its own accord.

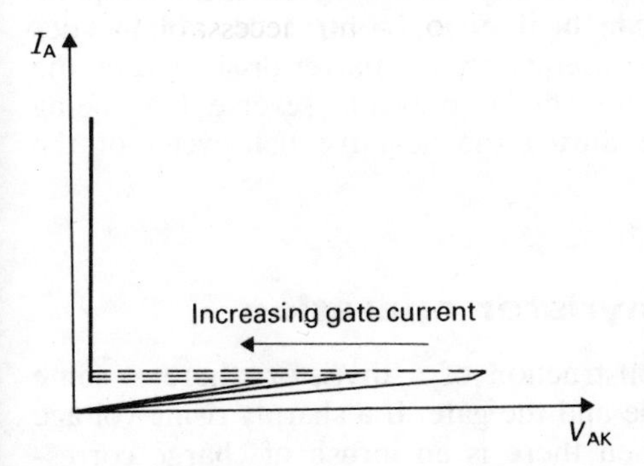

Fig. 6 Forward conduction characteristic of thyristor

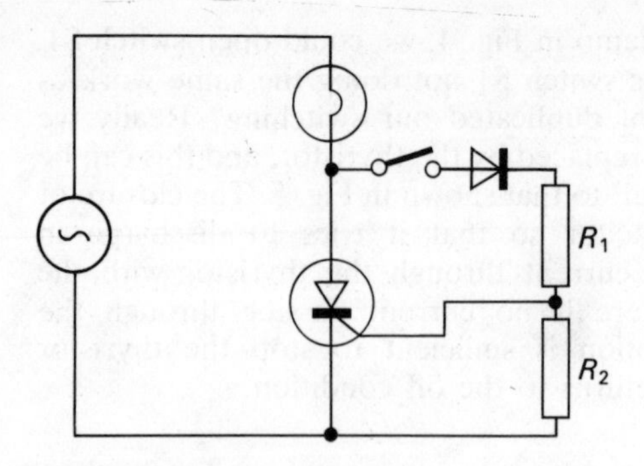

Fig. 7 Simple half-wave a.c. switch circuit

For the simple circuit shown in Fig. 7, this form of switch-on action can be employed. When switch S1 is open, there is no gate current and the thyristor is able to block the applied alternating voltage even at its peak value. When S1 is closed, the applied voltage causes a current to pass through R_1 and R_2, causing volt drops across each. The voltage across R_1 causes some current to pass through the gate, thus reducing the breakdown voltage of the thyristor. Eventually a point is reached at which either there is sufficient gate current to directly switch on the thyristor or the gate current sufficiently reduces the breakdown voltage of the thyristor that it again switches on – usually we experience this latter form of action.

Once the thyristor commences to conduct, the volt drop across it falls almost to zero; thus the gate current almost disappears. When the half cycle of the supply is completed, the anode current falls to zero and the thyristor switches off.

During the subsequent half-cycle, the thyristor cannot conduct; thus the lamp receives approximately a half-wave supply, depending on how soon the thyristor switches on after the commencement of the positive half-wave. By varying R_1, the instant of switch on can be delayed; thus the thyristor can be used to control the voltage developed across the lamp and consequently the power dissipated by the lamp. The diode in the circuit shown prevents reverse bias being applied to the thyristor gate during the negative half-cycles of the supply.

4 Limitations to thyristor operation

Due to the nature of the construction of a thyristor, there is some capacitance between the anode and the gate. If a sharply rising voltage is applied to the thyristor, then there is an inrush of charge corresponding to the relation $i = C(dv/dt)$. This inrush current can switch on the thyristor, and it can arise in practice due to surges in the supply system, e.g. due to switching or due to lightning. Thus thyristors may be inadvertently switched on, and such occurrences can be avoided by providing *C–R* circuits in order to divert the surges from the thyristors.

The leakage current in any p–n junction depends on the temperature of the junction. It follows that if we were to raise the temperature of a thyristor, the leakage current would rise, and it approximately doubles for every 8°C rise in temperature. If the temperature is permitted to rise too much, again the leakage current could inadvertently switch on the thyristor; thus precautions must be taken in order to maintain the operating temperature of a thyristor at a reasonably low level. Alternatively, it is possible to make use of this observation and to use the thyristor as a switch which will complete a circuit should a predetermined temperature be exceeded.

5 The thyristor in practice

The thyristor has therefore been seen to be a switch, and you may wonder why we should use this form instead of the normal mechanical device, especially since we see from the simple circuits that we still must retain mechanical switches. The advantage of the thyristor is that it can operate without involving arcing; thus there are no parts being worn out either as a result of motion or of the burning of the arc. The switches used to turn the thyristor on and also off do not involve the interruption of large currents, so they are less prone to wear and tear.

Apart from the aspects of reliability and safety due to the lack of moving parts, the thyristor as a switch is much more definite in its action; thus we can determine, for instance, the instant at which it will commence to pass current during each cycle of an alternating current. This permits its use as a control device, thus making the thyristor available as a means of regulating the speed of a machine, regulating a voltage supply and regulating a host of other variable quantities necessary to the control engineer.

Problems

1. Explain the construction of a thyristor and hence explain its forward and reverse conduction characteristics when the gate is disconnected.
2. Explain the action of introducing a gate current to a thyristor when its anode is positive with respect to its cathode.
3. Explain how a thyristor can be used to control the current in a resistive load connected to an a.c. supply.

4. Explain how a thyristor can be used to switch on and to switch off the current in a resistive load connected to a d.c. supply.
5. Explain why no gate current is required to maintain the flow of charge through a thyristor once it has commenced to pass a current from anode to cathode.
6. Describe a method whereby you could switch off a thyristor operating

   (*a*) from an a.c. supply;
   (*b*) from a d.c. supply.

Chapter 12

Electronic amplifiers

The purpose of an amplifier is to produce gain whereby a small input signal power controls a larger output signal power. It does not necessarily concern us as to how the amplifier operates, although it may depend on a transistor, a thyristor or even a thermionic valve. The advent and extensive use of the integrated circuit and of the microprocessor, where the components are formed on one small piece of circuitry, precludes interest in the make-up of the amplifier and it is sufficient to consider the amplifier as a block with input and output terminals, leaving little or nothing said about the contents of the block. Although there is no need to know the circuit details of an amplifier to be able to use it in electronic systems. This chapter is therefore an introduction to the systems approach to electronic engineering.

1 Basic amplifier principles

Figure 1 shows an amplifier with a resistive load R_L connected across the output terminals. The basic parameters of the amplifier are

$$\text{Voltage gain } G_V = \frac{\text{Output signal voltage}}{\text{Input signal voltage}}$$

$$G_V = \frac{V_2}{V_1} \tag{1}$$

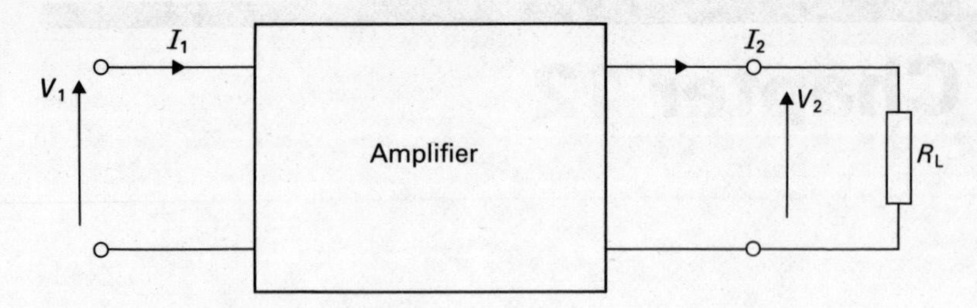

Fig. 1 Amplifier block diagram

$$\text{Current gain } G_I = \frac{\text{Output signal current}}{\text{Input signal current}}$$

$$G_I = \frac{I_2}{I_1} \qquad (2)$$

$$\text{Power gain } G_P = \frac{\text{Output signal power}}{\text{Input signal power}}$$

$$G_P = \frac{P_2}{P_1} \qquad (3)$$

$$= \frac{V_2 \times I_2}{V_1 \times I_1}$$

$$G_P = G_V G_I \qquad (4)$$

Figure 2 shows typical waveforms where the signals are assumed to vary sinusoidally with time. The waveforms met with in practice tend to be more complicated but it can be shown that such waves are formed from a series of pure sine waves, the frequencies of which are exact multiples of the basic or fundamental frequency. These multiple-frequency sine waves are known as harmonics.

Examination of the waveforms in Fig. 2 shows that the output signal

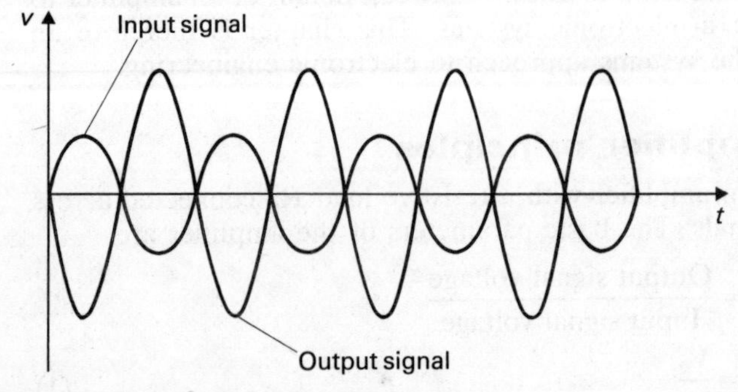

Fig. 2 Amplifier signal waveforms

voltage is 180° out of phase with input signal voltage. Such an amplifier is known as a phase-inverting one and the gain as defined by relation (1) is negative. These amplifiers are very common in practice although non-inverting types, where the output and input signals are in phase, are also to be found. Some amplifiers have alternative inputs for inverting and non-inverting operation.

So far only the input and output signals have been considered in the operation of the amplifier. It is necessary to provide a source of power from which is obtained the output signal power fed to the load. The magnitude of this signal power is controlled by the magnitude of the input signal. The source itself has to be a direct-current one and can be a battery or a rectified supply. The magnitude of this source voltage depends on the type of device used in the amplifier; thus for a transistor circuit of an integrated circuit it is often in the range 3–30 V, whilst for thermionic valves it can be in the order of hundreds of volts.

The voltage range within which the output signal voltage can vary is limited, being very much dependent on the value of the source voltage. This means therefore that there is a maximum value of input signal that will produce an output signal, the waveform of which remains an acceptable replica of the input signal waveform. Increasing the input signal beyond this level will produce clipping of the input signal waveform as illustrated in Fig. 3, although the clipping levels need not be symmetrical about the zero level.

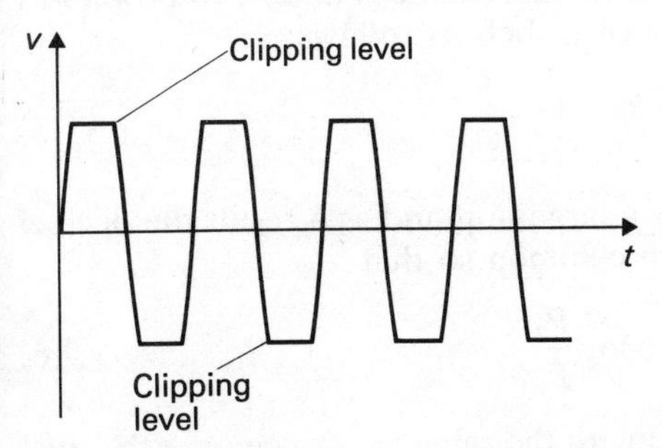

Fig. 3 Clipping of output signal waveform

Examination of the gain of an amplifier shows that it does not remain constant with variation of frequency of the input signal. Some amplifiers exhibit a reduction in gain at both high and low frequencies while others have a reduction at high frequencies only. In both cases there is a considerable frequency range over which the gain remains essentially constant. It is within this frequency range that the amplifier is designed to operate. The effect of the unequal gain is to produce another form of waveform distortion since the harmonics present in a complex input signal waveform may not be amplified by an equal amount. Figure 4 shows typical gain/frequency characteristics in which

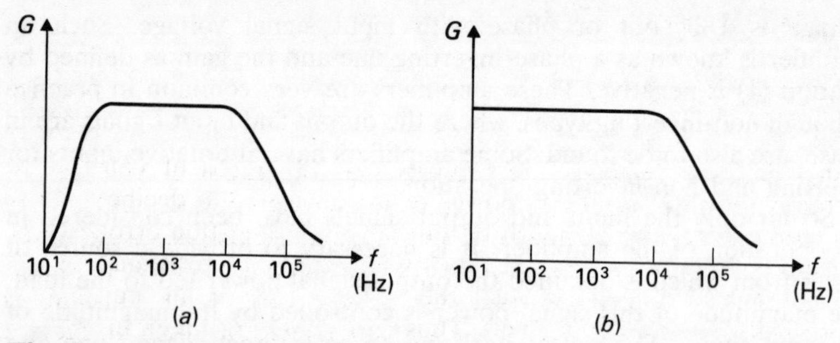

Fig. 4 Gain/frequency characteristic of (a) a capacitance-coupled amplifier, (b) a direct-coupled amplifier.

it should be noted that the frequency scales are logarithmic. The advantage of an amplifier with a characteristic as illustrated in Fig. 4(b) is that it is capable of amplifying signals at the very low frequencies experienced in many industrial applications. It is known as a direct-coupled amplifier and integrated-circuit amplifiers are of this type.

2 Logarithmic units

It is sometimes found convenient to express the ratio of two powers P_1 and P_2 in logarithmic units known as bels as follows:

$$\text{Power ratio in bels} = \log \frac{P_2}{P_1} \tag{5}$$

It is found that the bel is rather a large unit and as a result the decibel (i.e. one-tenth of a bel) is more common so that

$$\text{Power ratio in decibels (dB)} = 10 \log \frac{P_2}{P_1} \tag{6}$$

If the two powers are associated with the same resistance or with equal resistances, then

$$P_1 = \frac{V_1^2}{R} = I_1^2 R \qquad \text{and} \qquad P_2 = \frac{V_2^2}{R} = I_2^2 R$$

where V_1, I_1, V_2 and I_2 are the voltages across and the currents in the resistance(s). Therefore

$$G_P = 10 \log \frac{P_2}{P_1} = 10 \log \frac{V_2^2/R}{V_1^2/R} = 10 \log \frac{V_2^2}{V_1^2}$$

$$= 20 \log \frac{V_2}{V_1} \tag{7}$$

Similarly

$$G_P = 20 \log \frac{I_2}{I_1} \tag{8}$$

Although these relations are expressed as the ratios of voltages or currents, they still represent power ratios measured in decibels. They are used extensively, although by fundamental definition erroneously, to express voltage and current ratios where common resistance values are not involved. For example, the voltage gain of an amplifier is expressed sometimes in decibels. Thus care should be taken in this use of decibels, since the expression of power gain in decibels ought to be in complete agreement with the basic definition.

The use of decibels gives a representation of one power (or voltage) with reference to another. If P_2 is greater than P_1, then P_2 is said to be $10 \log P_2/P_1$ decibels up on P_1. For P_2 less than P_1, $\log P_2/P_1$ is negative. Since $\log P_1/P_2 = -\log P_2/P_1$, it is usual to determine the ratio greater than unity and P_2 is said to be $10 \log P_1/P_2$ down on P_1.

Example 1 The voltage gain of an amplifier when it feeds a resistive load of 1·0 kΩ is 40 dB. Determine the magnitude of the output signal voltage and the signal power in the load when the input signal is 10 mV.

$$G_P = 20 \log \frac{V_2}{V_1} = 40$$

$$\log \frac{V_2}{V_1} = 2{\cdot}0$$

$$\frac{V_2}{V_1} = 100$$

$$V_2 = 100 \times 10 = 1000 \text{ m}V = \underline{1{\cdot}0 \text{ V}}$$

$$P_2 = \frac{V_2^2}{R_L} = \frac{1{\cdot}0^2}{1000} = 0{\cdot}001 \text{ W} = \underline{1 \text{ mW}}$$

Example 2 Express the power dissipated in a 15-Ω resistor in decibels relative to 1 mW when the voltage across the resistor is 1·5 V r.m.s.

$$P_2 = \frac{V^2}{R} = \frac{1{\cdot}5^2}{15} = 0{\cdot}15 \text{ W} = 150 \text{ mW}$$

$$G_P = \frac{P_2}{P_1} = \frac{150}{1} = 150$$

and $G_P = 10 \log 150 = 10 \times 2{\cdot}176 = \underline{21{\cdot}76 \text{ dB}}$

Problems

1. Draw the block equivalent diagram of an amplifier and indicate on it the input and output voltages and currents. Hence produce statements of the voltage gain G_V and the power gain G_P.
2. An amplifier has a voltage gain of 50 dB. Determine the output voltage when the input voltage is 2·0 mV.
3. The output of a signal generator is calibrated in decibels for a resistive load of 600 Ω connected across its output terminals. Determine the terminal voltage to give
   (*a*) 0 dB corresponding to 1 mW dissipation in the load;
   (*b*) +10 dB;
   (*c*) −10 dB.
4. Express in decibels the gain of an amplifier which gives an output of 10 W from an input of 0·1 W.

Answers

2. 0·63 V
3. 0·775 V, 2·45 V, 0·245 V
4. 20 dB

Chapter 13

Measuring instruments and measurements

The question of success or failure of an electrical or electronic circuit to perform the function for which it was constructed may simply be whether it works or not. This is often the simple criterion of success for a lamp circuit or a heating circuit. However just to operate may not be enough; for instance, two apparently identical radios may give quite different levels of output sound when tuned to the same station and adjusted to their maximum outputs. Such differences require detailed investigation of the electronic arrangements to ascertain why the performance of one set is poorer than that of the other and to subsequently rectify this deficiency. To make any investigation into the performance of an electronic network, we require to develop our measurement techniques, and in turn this requires that we develop our understanding of the interpretation of our observations.

1 Cathode-ray oscilloscope

The cathode-ray oscilloscope has a very wide range of applications in electronic engineering and also in telecommunication engineering. It is capable of giving either direct or almost direct measurement of voltage, current, time, frequency and phase. As the display afforded by the cathode-ray oscilloscope is directly driven by the input signal information, it can make a visual impact in a manner not possible with other instruments.

There are other forms of oscilloscope, but the cathode-ray oscilloscope is so universal in its application that you may take it that when the term oscilloscope is used, it is the cathode-ray oscilloscope that is intended.

Most oscilloscopes are general-purpose instruments and the basic form of their operation is illustrated in Fig. 1. For simplicity, we shall restrict our interest to displaying one signal, although most oscilloscopes are capable of displaying two.

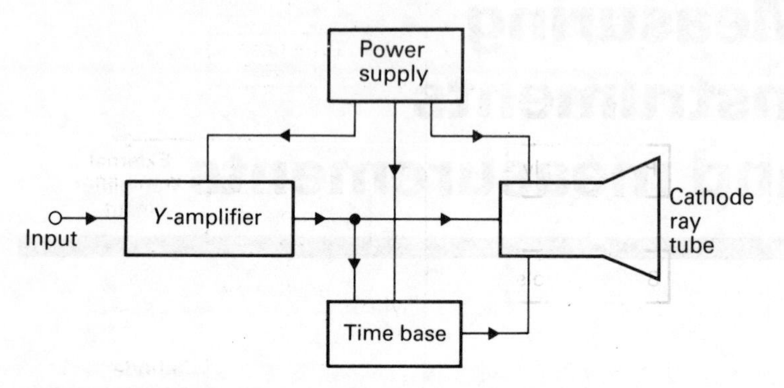

Fig. 1 Basic schematic diagram of cathode-ray oscilloscope

The input signal is amplified by the *Y*-amplifier, so called because it causes the beam to be driven up and down the screen of the cathode-ray tube in the direction described as the *Y*-direction by mathematicians.

The time base serves to move the beam across the screen of the tube. When the beam moves across the screen, it is said to move in the *X*-direction. It would not be appropriate if the movements in the *X*- and *Y*-directions were not coordinated; hence the time base may be controlled by the output of the *Y*-amplifier. This interrelationship is quite complex and therefore requires further explanation.

However, before proceeding, we require another major component which is the power supply. This serves to energise the grid and anode systems of the cathode-ray tube, as well as to energise the brilliance, focus and astigmatism controls of the beam. The power supply also energises the amplifiers for the control of the beam.

Assuming that you are already familiar with the operation of the cathode-ray tube, it remains to consider those parts of the overall instrument which give rise to controls that we must operate in order to use the oscilloscope. A more complete schematic diagram therefore is shown in Fig. 2.

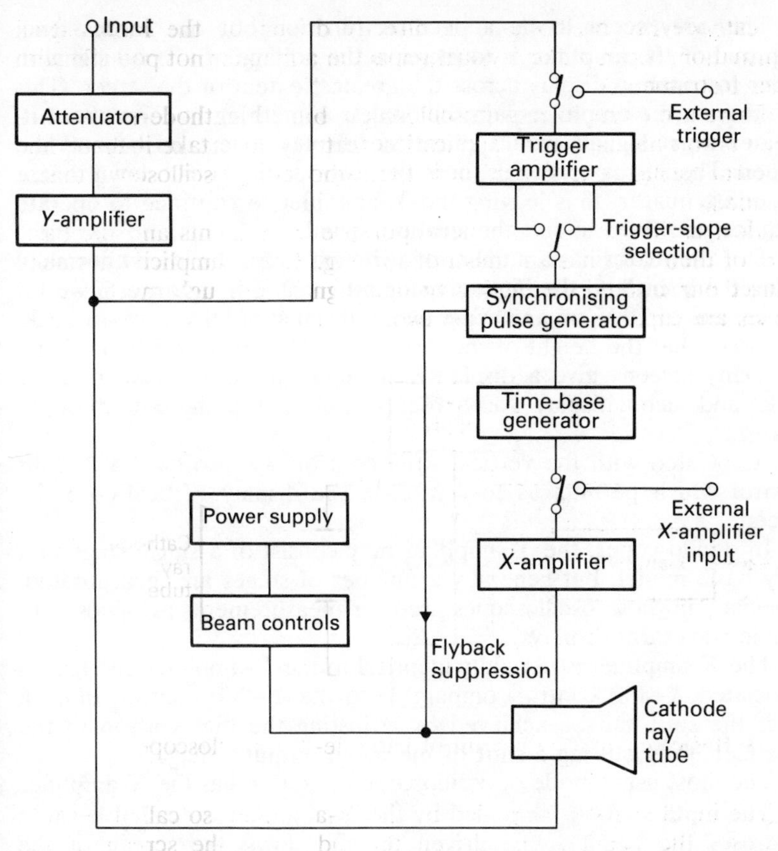

Fig. 2 Schematic diagram of a cathode-ray oscilloscope

The circuitry of an oscilloscope has to be capable of handling a very wide range of input signals varying from a few millivolts to possibly a few hundred volts, whilst the input signal frequency may vary from zero (d.c.) up to possibly 1 GHz, although an upper limit of 10 to 50 MHz is more common in general-purpose instruments.

We have already noted that the input signal drives the display beam in the *Y*-direction. The height of the screen dictates the extent of the possible deflection. The output of the *Y*-amplifier therefore has to be of a sufficient magnitude to drive the beam up and down the screen in order to give as large a display as possible without the display disappearing off the edge of the screen. Nevertheless the *Y*-amplifier operates with a fixed gain and it is necessary to adjust the magnitude of the input signal to the amplifier, this being done with an attenuator. An attenuator is a network of resistors and capacitors, and its function is to reduce the input signal.

This may seem to be a peculiar function but the hardest task required of the amplifier is to increase the voltage of a small signal in order to obtain a display across the greatest extent of the screen. This determines the amplifier gain, but unless something is done about it, greater input signals would cause the display to extend beyond the screen. These greater signals, however, can readily be cut down to size by an attenuator, thus leaving the *Y*-amplifier to continue to operate with its gain fixed to suit the smallest signal.

The attenuator has a number of switched steps, the lowest normally being 0·1 V/cm and the highest being 50 V/cm. For instance, if we set the control to 10 V/cm and apply an input signal of 50 V peak-to-peak, it follows that the height of the display on the screen is $50/10 = 5$ cm. As many screens give a display 8 cm high, this is the best possible scale; and such a display would disappear at the top and bottom of the screen.

Associated with the vertical scale control, we also have a *Y*-shift control which permits us to centralise the display vertically on the screen.

In oscilloscopes, the *Y*-amplifier may consist of a single stage in a very basic model, but generally a number of stages are incorporated, especially in those oscilloscopes used for measurements as opposed to simple waveform displays.

The *X*-amplifier is normally identical to the *Y*-amplifier and has an associated *X*-shift control comparable to the *Y*-shift control. In each case, the shift can be achieved by adjusting the bias voltage to the amplifier, thus causing a shift of the mean output voltage.

The most usual mode of oscilloscope operation has the *X*-amplifier

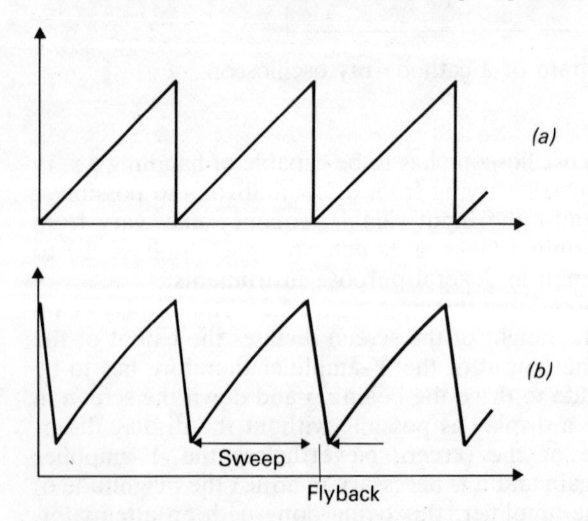

Fig. 3 *X*-amplifier sawtooth waveforms: (*a*) idealised form; (*b*) practical form

fed from a time-base generator, the function of which is to drive the beam at a steady speed across the screen and, when it reaches the right-hand side of the screen, the beam is then made to fly back to the left-hand side and start out again across the screen. To produce such an output, the input signal to the X-amplifier must take the form of a sawtooth waveform, which is illustrated in Fig. 3. As the signal steadily increases, the beam is moved across the screen. When the signal reaches its peak, ideally it should drop to zero, thus instantaneously returning the beam to the beginning of its travel. In practice, there are two (possibly three) differences between the ideal waveform and that actually experienced.

The ramp of the waveform is not linear but is desired from an R–C circuit transient. Provided that the time constant of the R–C circuit is very much greater than that of the time required for the beam to scan across the screen, the ramp is almost linear. The difference has been exaggerated in Fig. 3(b) for clarity.

When the waveform reaches its peak, it is not possible, for reasons which we have observed in our studies of transients, for the signal to suddenly return to zero; thus there is a short period during which the voltage decays. This is called the flyback time, because it is the period during which the beam flys back to the start of its travel.

The flyback time is made as short as possible partly to save the display time lost and partly to reduce the trace of the beam returning across the screen. To help eliminate this unwanted display, the beam current is reduced during the flyback time by means of a flyback suppression pulse.

There may be a short interval between the end of the flyback period and the following scanning period. This is necessary when the time base is controlled from an external trigger source, which is described later. The short delay ensures that the scan starts at the same point in the display waveform, thus causing the display to appear stationary on the screen. In most applications, the time base is controlled by a pulse generator synchronised to the signal from the Y-amplifier, and there is no need to have a delay between the end of the flyback period and the beginning of the scanning period.

The signal to the X-amplifier comes from the time-base generator which may operate in any of the following modes:

(a) self oscillating;
(b) self-oscillating and synchronised;
(c) externally triggered.

In the purely self-oscillating arrangement, the time-base voltage rises to a preset value, at which instant the beam has reached the right-hand extremity of its travel. When the preset value is reached, the flyback is automatically initiated and, as soon as the original value at the beginning of the scan is obtained, the generator starts generating the next sweep across the screen. The problem with the self-oscillating arrangement is that it works independently of the input signal; thus in the first sweep it may start when the input signal is at a positive maximum value yet the next sweep starts at a negative maximum, thus giving a completely different trace. This sort of variation at best gives rise to an apparently moving display and at worst to two or three displays which are superimposed on one another.

To overcome this problem, it is necessary to synchronise the time-base generator to the frequency of the supply. In a self-oscillating and synchronised system, the initiation of the flyback is controlled by a synchronisation signal from the Y-amplifier. Because the flyback is controlled by the synchronising signal, it follows that the synchronising signal also controls the start of the sweep of the beam. In some oscilloscopes, this arrangement is fully automatic, but in many of the cheaper general purpose oscilloscopes, there is a stabilising control which sets the level of signal display at which the flyback is initiated.

It is necessary to appreciate the reason why this arrangement operates from the finish of the display and not from the beginning. The time-base generator causes the beam to sweep across the display at regular intervals. Let us assume that this is taking place with a frequency 50 Hz and also let us assume that the frequency of the signal to be displayed is 150 Hz. During the sweep time, the input signal undergoes three cycles; thus we would hope to see these three cycles being displayed. In practice, a bit of one cycle would be lost because not all of the time is available for display, the remaining time being taken up by the flyback period. However, the main problems are to commence the trace at the same point in the input signal each time. Let us assume that we wish to start when the input signal is positive and rising. Ideally this would coincide with the instant at which the sweep was due to commence; thus three cycles (almost) would be displayed, followed by the flyback, and everything would be ready for the next sweep to commence displaying the following three input signal waves.

However, what happens if we just miss the start of an input signal wave? If we wait for the next instant of the signal being positive and increasing, then we have to wait for almost a complete cycle, which would be lost to the display. And even more awkward, what happens if the frequency of the signal to be displayed is 152 Hz? After all, as the signal frequency increases we expect to see more than three cycles, so that if the frequency is 200 Hz for example, we expect the display of four cycles.

The answer is not to wait for the chosen instant but rather to get on with the display up to the time of the chosen instant. In this way; we do not miss anything by waiting (although we shall miss that short period of display during the flyback) but having reached the chosen instant, the beam is caused to fly back and to recommence the sweep with the minimum delay. It now starts no matter what is happening, and continues again up to the chosen point at which the flyback is again initiated. In this way, we can display any number of cycles or fractions of a cycle in excess of one cycle.

The stabilising control has to be adjusted appropriately to synchronise the flyback of the time-base generator to the output of the Y-amplifier. In most cases, this can be readily achieved, but sometimes the quality of the signal to be displayed is not sufficiently reliable, in which case the time-base generator must be controlled from an external source, which provides a suitable trigger. In this case, the trigger initiates each individual time-base sweep and the flyback then follows automatically when the time-base signal has reached a preset value. In this case, the time-base generator remains inactive until the trigger releases another sweep. This means that possibly a significant part of the display can be omitted. For this reason, some oscilloscopes are provided with gain controls to the X-amplifier whereby the display can be expanded and we can examine the display in greater detail.

If we wish to make time or frequency measurements, the time base control must have a calibration setting at which the display time coincides with the control markings. For instance, if the time-base control is set to 10 μs/cm, then the X-amplifier control is set to the calibration mark and we know that each centimetre of the display in the X-direction represents 10 μs of time. A typical range of time-base control settings is 0·5 s/cm to 1 μs/cm.

There are several applications of the oscilloscope in which we do not require the time-base generation at all but instead we drive the X-amplifier from another signal source in a similar manner to the operation of the Y-amplifier. For this reason, many oscilloscopes afford direct access to the X-amplifier and we shall look at such an application in Section 3.

The input impedance of most general purpose oscilloscopes is 1 MΩ shunted by a capacitance of 20 to 50 pF according to the model used. The effect of the capacitance becomes significantly effective only at high frequencies, i.e. in excess of 1 GHz. Such a high input impedance makes the oscilloscope suitable for many measurement techniques, since the oscilloscope scarcely modifies the network into which it has been introduced.

Mention has already been made of the calibration of oscilloscopes, and many have built-in calibration circuits. Generally these give a square or trapezoidal waveform of known peak-to-peak magnitude and cycle duration. This calibration signal is fed into the oscilloscope and the gain of the Y-amplifier is adjusted to give the appropriate vertical display. Similarly the gain of the X-amplifier is adjusted to give a signal display of appropriate length. These adjustments are usually made by potentiometers with a screw adjustment operated by a screwdriver. In this way, calibration adjustment cannot be confused with the other controls of the oscilloscope.

This brief description of the operation of the principal components has indicated the main controls that we require to use when displaying and measuring waveforms and phase differences by means of the oscilloscope.

2 Use of the cathode-ray oscilloscope in waveform measurement

A discussion of the use of the oscilloscope falls naturally in to two parts; the use of the instrument itself and the methods of connecting the instrument to the circuits in which the measurements are to be made. For ease of introduction, let us assume that the signals applied to the oscilloscope are suitable.

Once the connections between the source of the signal and the oscilloscope have been made, the oscilloscope should be switched on and given time to warm up. Generally a trace will appear on the screen, but should this not occur, some useful points to check are that the vertical and horizontal shift controls are centralised, that the brilliance control is centralised, that the trigger is set to the automatic position (where appropriate) and that the stabilising control is varied to ensure that the display time base is operated. Normally these checks ensure that the display appears, but if these do not work, then you have to check out the full procedure in accordance with the manufacturer's operating manual.

Once the display has been established, adjust the brilliance to obtain an acceptable trace which is not brighter than necessary. Too bright a trace, especially if permitted to remain in the one position for a considerable period of time, can damage the fluorescent material on the screen, hence the reason for minimising the brilliance. It also does not harm to check that the beam is focused and that there is minimum astigmatism. These do not vary much with operation but sometimes the controls are adjusted incorrectly.

The display is next centralised vertically and the scale control adjusted to give the highest possible display that can be contained within the screen.

Having the display clearly in view, it may be that the trace is stationary but it could also be slipping slowly in a horizontal direction. In the latter case, the stabilising control requires to be adjusted until the trace is locked in position and remains stationary.

Unless you have some unusual observations to make on the waveform, the X-gain amplifier should be set to the calibration position and the X-shift control readjusted to centralise the display horizontally. This display may contain only part of the waveform or a great many waveforms; this is changed by adjusting the time-base control until the desired number of waveforms are displayed.

This again is a brief description of the setting-up procedure of the oscilloscope and it only serves to highlight the common form of operation. Different models of oscilloscope vary in detail, but the procedure is essentially that indicated. However, words cannot substitute for the practical experience of operating oscilloscopes and you will readily obtain a better appreciation of the oscilloscope from a few minutes' experimentation with one in a laboratory.

To aid you observation of the display on an oscilloscope, a set of squares is marked on the transparent screen cover. This marking is termed a graticule and is illustrated in Fig. 4.

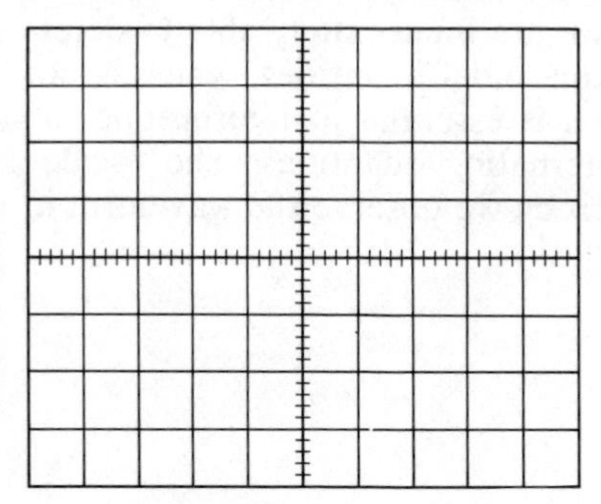

Fig. 4 Cathode-ray oscilloscope graticule

Graticules are marked out with a 1-cm grid and are presently 10 cm across by 8 cm high. Older models had graticules 8 cm by 8 cm or sometimes 10 cm by 10 cm. To avoid parallax error, you should always observe the trace directly through the graticule and not from the side.

Let us now consider the interpretation of the basic forms of display, which are sine waves, square waves and pulses. A typical sine waveform display as seen through a graticule is shown in Fig. 5. To obtain this display, let us assume that the vertical control is set to 2 V/cm and the time-base control to 500 μs/cm.

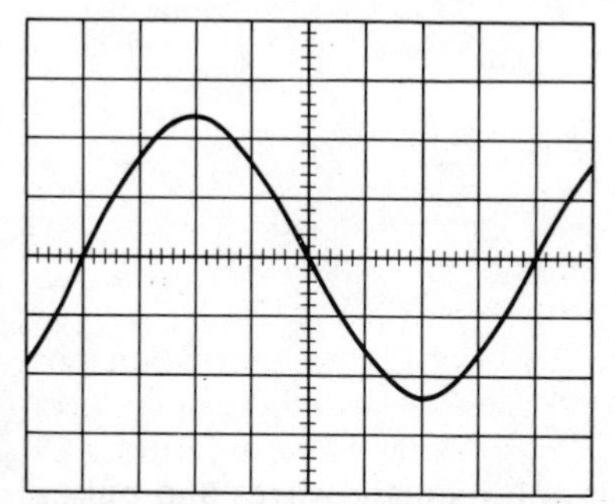

Fig. 5 Sine waveform display on an oscilloscope

The peak-to-peak height of the display is 4·8 cm; hence the peak-to-peak voltage is $4{\cdot}8 \times 2 = 9{\cdot}6$ V. This may be a direct measurement of a voltage or the indirect measurement of, say, a current. In the latter case, if a current is passed through a resistor of known resistance, then the current value is obtained by dividing the voltage by the resistance.

You will note that the voltage measured is the peak-to-peak value. If the signal is sinusoidal, then the r.m.s. value is obtained by dividing the peak-to-peak value by $2\sqrt{2}$, which in this instance gives an r.m.s. value of 3·4 V.

The oscilloscope therefore can be seen to suffer the disadvantage when compared with electronic voltmeters that it is more complex to operate and to interpret. However, we are immediately able to determine whether we are dealing with sinusoidal quantities, which is not possible with other meters and which is essential to interpreting the accuracy of the measurement of alternating quantities. The oscilloscope is therefore an instrument whereby we observe the waveform in

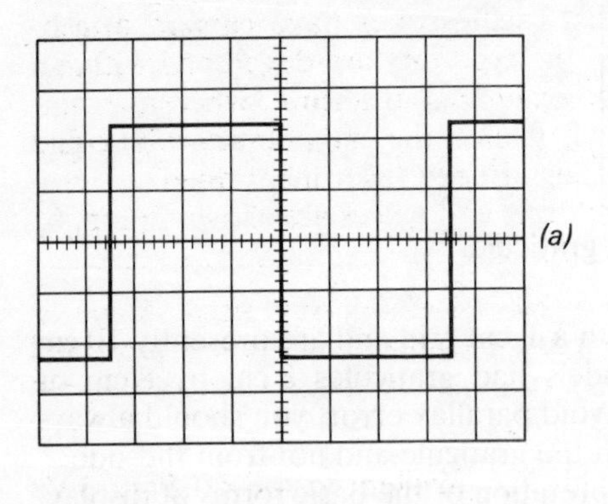

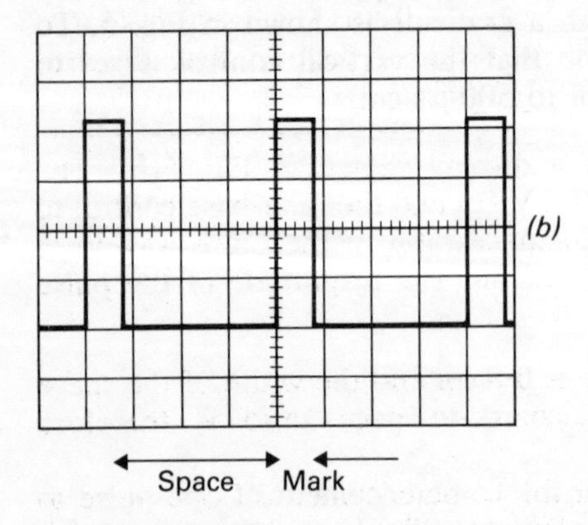

Fig. 6 Cathode-ray oscilloscope displays for square waves and pulses: (*a*) square wave; (*b*) pulse

detail and measurements of magnitude of the signal are essentially those of peak-to-peak values.

Returning to the display shown in Fig. 5, the length of one cycle of the display is 8·0 cm; hence the period of the waveform is

$$8{\cdot}0 \times 500 \times 10^{-6} = 4{\cdot}0 \times 10^{-3}\ \text{s} = 4{\cdot}0\ \text{ms}$$

It follows that the frequency of the signal is

$$1/(4{\cdot}0 \times 10^{-3}) = 250\ \text{Hz}$$

In each case, the accuracy of measurement is not particularly good. At best, we cannot claim an accuracy of measurement on the graticule that is better than to the nearest millimetre; thus the accuracy at best is about 2 per cent.

If we wish to determine the values of a waveform, such as the average and r.m.s. values of non-sinusoidal waveforms or the mark-to-space ratio of a pulse waveform, it is better to take a photograph of the trace. This is easily done as most oscilloscopes have camera attachments which take photographs of the type that are developed within a minute. Such photographs can be examined at leisure, whereas maintaining a trace for such a length of time on the cathode-ray tube could damage the screen, a point that has already been mentioned.

Typical traces for square waveforms and pulses are shown in Fig. 6.

Example 1 The trace displayed by a cathode-ray oscilloscope is shown in Fig. 6(*a*). The signal amplitude control is set to 0·5 V/cm and the time-base control to 100 μs/cm. Determine the peak-to-peak voltage of the signal and its frequency.

Height of display is 4·6 cm. This is equivalent to $4{\cdot}6 \times 0{\cdot}5 = 2{\cdot}3$ V. The peak-to-peak voltage is therefore $\underline{2{\cdot}3\ \text{V}}$.

The width of the display of one cycle is 7·0 cm. This is equivalent to a period of $7{\cdot}0 \times 100 \times 10^{-6} = 700 \times 10^{-6}$ s. It follows that the frequency is given by $1/(700 \times 10^{-6}) = \underline{1430\ \text{Hz}}$.

Example 2 An oscilloscope has a display shown in Fig. 6(*b*). The signal amplitude control is set to 0·2 V/cm and the time-base control to 10 μs/cm. Determine the mark-to-space ratio of the pulse waveform and the pulse frequency. Also determine the magnitude of the pulse voltage.

The width of the pulse display is 0·8 cm and the width of the space between pulses is 3·2 cm. The mark-to-space ratio is therefore $0{\cdot}8/3{\cdot}2 = \underline{0{\cdot}25}$.

The width of the display from the commencement of one pulse to the next is 4·0 cm. This is equivalent to $4{\cdot}0 \times 10 \times 10^{-6} = 40 \times 10^{-6}$ s. being the period of a pulse waveform. The pulse frequency is therefore given by $1/(40 \times 10^{-6}) = 25\,000\ \text{Hz} = \underline{25\ \text{kHz}}$.

The magnitude of the pulse voltage is determined from the pulse height on the display, this being 4·2 cm. The pulse voltage is therefore $4{\cdot}2 \times 0{\cdot}2 = \underline{0{\cdot}84\ \text{V}}$.

---

3 Observation of circuit time constant using an oscilloscope

Most oscilloscopes operate with the body or chassis of the instrument at earth potential. Also most oscilloscopes are connected to the signal source by means of a coaxial cable, the outer conductor of which is connected to the body of the oscilloscope and is therefore at earth potential. It follows that one of the connections from the oscilloscope will connect one terminal of the signal source to earth. The effect of this observation can be illustrated by considering the test arrangement shown in Fig. 7. A resistor and a capacitor are connected in series and supplied from a signal generator. It is usual that one terminal of the signal generator is also at earth potential; thus the connection diagram shown in Fig. 7 is suitable for the circuit and for the oscilloscope. It is suitable because the earth point of the oscilloscope is connected to the earth point of the generator and they are therefore at the same potential.

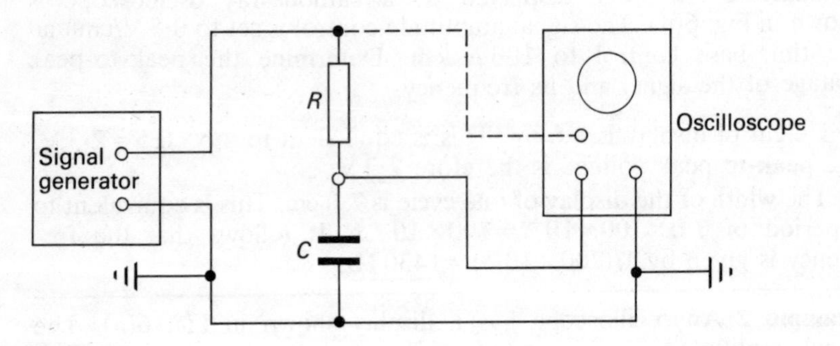

Fig. 7 An experiment involving an oscilloscope and a *C–R* circuit

The oscilloscope displays the waveform of the voltage across the capacitor. Let us assume also that the generator provides a square-wave output signal and that the oscilloscope is synchronised to that wave frequency. This permits us to observe the transient response of the capacitor in a *C–R* circuit. If we are able to involve a double-beam oscilloscope, then it is possible also to display the supply voltage, the extra connection required being indicated by the dotted line in Fig. 7. The resulting display is shown in Fig. 8. If either the capacitor or the

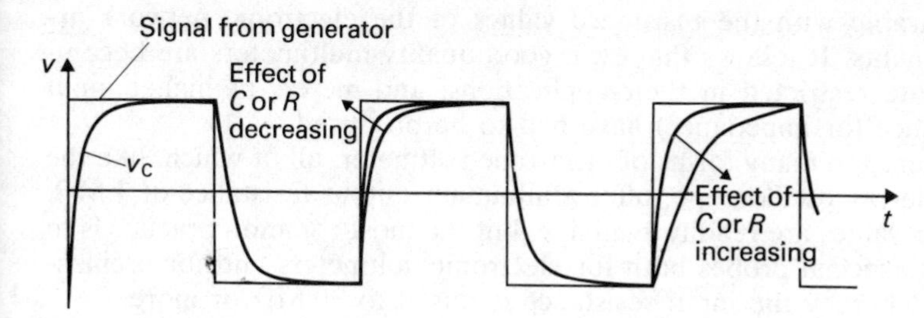

Fig. 8 Oscilloscope display of capacitor voltage transient

resistor can be varied, we may observe the effect of increasing the circuit time constant on the display as is indicated in Fig. 8.

Finally we should note that if we wish to investigate the response of the voltage across the resistor, we can either deduce this as being the difference between the two displays on the oscilloscope or by reconnecting the circuit so that C and R are interchanged (we cannot reconnect the oscilloscope across R in the circuit shown because of the problem of the earth connections). Similar investigations can be carried out with an L–R circuit, bearing in mind that the voltage waveform for the resistor is similar to the current waveform.

4 Loading effects on measurements

One of the most important sources of error in electrical, and especially in electronic, measurement arises from the loading effects of the measuring instruments on the circuits in which measurements are being made. Even though the instruments are perfectly calibrated and observed, i.e. the instrument indicates correctly the signal applied to its terminals and this is correctly observed by the operator, there can still remain an error caused by the instrument itself, since it needs a certain amount of power for its operation, and this power is taken from the circuit being measured, causing the circuit to change.

We have observed in earlier studies that if the power input to the measuring instrument is small, then little error results. If the power to the instrument is even comparable to the circuit power, say to the extent of 1 per cent, then a serious error can arise.

Such disturbance due to the application of meters is most common with the use of voltmeters, although it can also occur with other instruments, especially ammeters and wattmeters. Voltmeters are given a figure of merit which, in the case of general-purpose multimeters, is given in kilohms per volt of full-scale deflection. In low-voltage circuits, especially those associated with miniaturised electronic equipment, the small value of full-scale deflection reduces the effective resistance of the voltmeter to a few thousand ohms, which can be comparable with the resistance values of the electronic network arrangements. It follows that even good quality multimeters are becoming quite restricted in their applications, and meters of higher input resistance (or impedance) have had to be produced.

There are many forms of electronic voltmeter, all of which, like the cathode-ray oscilloscope, offer a minimum output resistance of 1 MΩ. Higher values are readily available, but the most common practice is to provide special probes both for electronic voltmeters and for oscilloscopes whereby the input resistance is raised to 30 MΩ or more.

In order to minimise observation error, many electronic measurement devices provide a digital display, i.e. the indication is displayed in numbers rather than by the position of a needle moving across a scale.

When measuring small currents in circuits with small voltages, the power required to operate an ammeter can be too demanding of the circuit. The alternative method of measurement is to determine the voltage drop across a pure resistive component of the circuit provided that its resistance is known or can be determined. Some electronic ammeters use this principle, since they can measure very small voltage drops across very low values of resistance. This has the advantage of introducing a minimum loading effect into the circuit.

Because electronic instruments have such small loading effects, they provide a better, although more costly, method of measuring voltages and currents in small-power circuits. They are better when compared with the more common and cheaper types of meter such as the multimeter. This advantage may be described rather loosely as being the higher accuracy of the digital instrument when compared with the analog instrument for measuring small voltages and currents.

5 Effects of the source on circuit measurements

In practical applications, most electronic devices depend on some form of transducer to provide the input signal, e.g. a microphone, an aerial, or a sensor. However, for testing purposes we use a signal generator as a source of variable voltage and frequency. Usually signal generators can supply at least sinusoidal and square waveform signals.

From out earlier studies relating to Thévenin's theorem, we have seen the importance of the internal resistance or impedance of the source, and this plays a significant role in the application of signal generators as sources of alternating signals. Most signal generators can have their internal resistances varied to be either 5 Ω or 600 Ω. If the load corresponds to either of these values, i.e. the load is matched to the source, then the signals will be as indicated on the control dials of the generator. However, if the load does not match to either of these values, then the intended power will not be developed in the load and the dials will not give signals in accordance with their calibration.

Generally it is not good practice to rely on the calibration of the controls of signal generators, because we rarely manage to match the loads to the source; but an approximate matching is required, otherwise insufficient power is developed in the load.

The output resistance of the generator can load an amplifier and in many cases it is good practice to insert a high capacitance capacitor in series with the input to the amplifier. If this is not done, then some of the direct current from the power supply to the amplifier can be diverted from some of the bias components and through the generator.

Problems

1. Describe how the Y-plates of a cathode-ray oscilloscope would be connected to give a trace of
   (*a*) the alternating voltage applied to a circuit component (such as a coil);
   (*b*) the current through the component.

   What voltage waveform would normally be applied to the X-plates of the oscilloscope for this purpose?
2. Draw a block diagram showing the principal parts of a cathode-ray oscilloscope amplifier arrangement. What is the purpose of synchronisation in an oscilloscope and why is it essential to the process of waveform display in an oscilloscope?
3. Explain with the aid of a circuit diagram showing the connections made to the cathode-ray oscilloscope, how it may be used to determine the r.m.s. value of an a.c. signal which is
   (*a*) sinusoidal;
   (*b*) non-sinusoidal.

   An oscilloscope is used to display a sinusoidal alternating voltage. The display shows a sine wave of amplitude 3·3 cm and the Y-amplifier sensitivity is set to 5 V/cm. Determine the r.m.s. value of the voltage.
4. Draw a block diagram showing schematically a cathode-ray oscilloscope. Include in your diagram a switch for the selection of internal/external time base, and describe an application of the oscilloscope in which the external time base would be used. Explain the limitation of the Y-amplifier with regard to the accurate observation of rectangular pulses.
5. The display given by a cathode-ray oscilloscope is shown in Fig. 9. Given that the sensitivity of the Y-amplifier is set to 10 V/cm and the time base control to 5 ms/cm, determine
   (*a*) the peak-to-peak voltage;
   (*b*) the period of the signal;
   (*c*) the frequency.

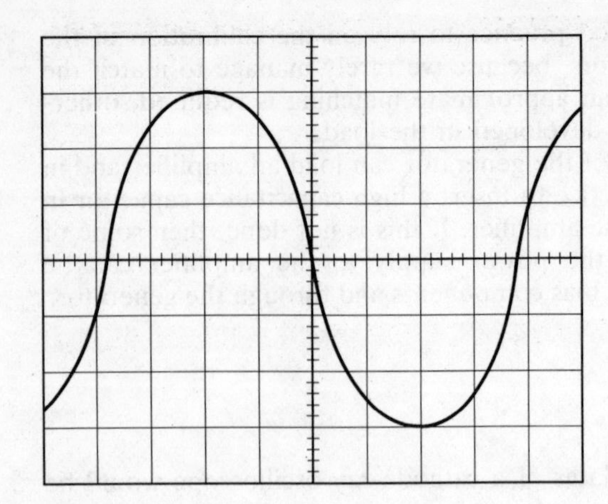

Fig. 9

6. An industrial grade ammeter (0–50 mA) was compared with a precision grade ammeter, and the following test data, after the indications of the precision grade ammeter had been corrected, was obtained

Industrial grade ammeter (mA)	0	10·0	20·0	30·0	40·0	50·0
Precision grade ammeter (MA)	0	9·6	18·4	28·9	39·2	49·5

Determine whether or not this ammeter is within the required ±1·0 per cent tolerance of full scale range.

Answers

3. 11·7 V
5. 60 V, 38 ms, 26·3 Hz
6. Outwith tolerance

Index